NATO ASI Series

Advanced Science Institutes Series

A series presenting the results of activities sponsored by the NATO Science Committee, which aims at the dissemination of advanced scientific and technological knowledge, with a view to strengthening links between scientific communities.

The Series is published by an international board of publishers in conjunction with the NATO Scientific Affairs Division

A Life Sciences B Physics	Plenum Publishing Corporation London and New York
C Mathematical and Physical Sciences D Behavioural and Social Sciences E Applied Sciences	Kluwer Academic Publishers Dordrecht, Boston and London
F Computer and Systems Sciences G Ecological Sciences H Cell Biology I Global Environmental Change	Springer-Verlag Berlin Heidelberg New York London Paris Tokyo Hong Kong Barcelona Budapest

PARTNERSHIP SUB-SERIES

1. Disarmament Technologies	Kluwer Academic Publishers
2. Environment	Springer-Verlag
3. High Technology	Kluwer Academic Publishers
4. Science and Technology Policy	Kluwer Academic Publishers
5. Computer Networking	Kluwer Academic Publishers

The Partnership Sub-Series incorporates activities undertaken in collaboration with NATO's Cooperation Partners, the countries of the CIS and Central and Eastern Europe, in Priority Areas of concern to those countries.

NATO-PCO DATABASE

The electronic index to the NATO ASI Series provides full bibliographical references (with keywords and/or abstracts) to about 50000 contributions from international scientists published in all sections of the NATO ASI Series. Access to the NATO-PCO DATABASE compiled by the NATO Publication Coordination Office is possible in two ways:

- via online FILE 128 (NATO-PCO DATABASE) hosted by ESRIN, Via Galileo Galilei, I-00044 Frascati, Italy.

- via CD-ROM "NATO Science & Technology Disk" with user-friendly retrieval software in English, French and German (© WTV GmbH and DATAWARE Technologies Inc. 1992).

The CD-ROM can be ordered through any member of the Board of Publisher through NATO-PCO, Overijse, Belgium.

Series H: Cell Biology, Vol. 93

Springer
Berlin
Heidelberg
New York
Barcelona
Budapest
Hong Kong
London
Milan
Paris
Tokyo

Modulation of Cellular Responses in Toxicity

Edited by

Corrado L. Galli
Marina Marinovich

Istituto di Scienze Farmacologiche
Via G. Balzaretti 9
I-20133 Milano, Italy

Alan M. Goldberg

Center for Alternatives to Animal Testing
Johns Hopkins School of Public Health
111 Market Place, Suite 840
Baltimore, MD 21202-6709, USA

Springer

Published in cooperation with NATO Scientific Affairs Division

Proceedings of the NATO Advanced Study Institute on the Modulation of Cellular Responses in Toxicity, held in Ponte di Legno, Italy, January 24 – February 3, 1994

Library of Congress Cataloging-in-Publication Data

Modulation of cellular responses in Toxicity / edited by Corrado L. Galli, Marina Marinovich, Alan M. Goldberg.
p. cm. – (NATO ASI series. Sereis H, Cell biology ; vol. 93)
"Published in cooperation with NATO Scientific Affairs Division."
"Proceedings of the NATO Advanced Study Institute on the Modulation of Cellular Responses in Understanding the Toxic Response, held in Ponte di Legno, Italy, January 24–February 3, 1994" – T.p. verso.
Includes bibliographical references.
ISBN-13:978-3-642-79874-0 e-ISBN-13:978-3-642-79872-6
DOI: 10.1007/978-3-642-79872-6
1. Toxology – Congresses. 2. Molecular toxicology – Congresses.
I. Galli, C. L. (Corrado L.) II. Marinovich, Marina. III. Goldberg, Alan M. IV. North Atlantic Treaty Organization. Scientific Affairs Division. V. NATO Advanced Study Institute on the Modulation of Cellular Responses in Understanding the Toxic Response (1994 : Ponte di Legno, Italy) VI. Series.
RA1191.M625 1995
615.9–dc20 95-20208 CIP

ISBN-13:978-3-642-79874-0

Softcover reprint of the hardcover 1st edition 1995

Typesetting: Camera ready by authors/editors
Printed on acid-free paper
SPIN 10101719 31/3136 - 5 4 3 2 1 0

PREFACE

These proceedings are the result of a NATO Advanced Study Institute held in Ponte di Legno, Italy, and was designed to focus on the modulation of cellular responses in understanding the toxic response. To that end, the meeting (and this volume) are focused around three areas. The first of these was to look at methodologies and methodological approaches that are or were expected to be major contributors to the field of toxicology in the immediate future; the second group of chapters deals with major milestones in cellular toxicology, and the last section provides for an examination of target organ toxicity.

We, as editors of this volume, did not edit the manuscripts but structured the program, convene the meeting, and collect the manuscripts. The first section starts with a overview of in vitro toxicology and the need for mechanistically based studies. This approach requires both an understanding of the toxicological process and developing assay methods to assess toxicity. The section then focuses on the molecular biology of cellular toxicity, an examination of receptor mediated mechanisms, structure activity relationships, the necessity for pharmacokinetic modeling, and closes with an example of the use of biosensors in toxicological studies. Clearly, these are not the only approaches that need to be examined. Others are included in later chapters but with focuses on other aspects of the toxicological process.

In the section on Milestones in Cellular Toxicology, we focused on intercellular communication and signal processing, the cytoskeleton, apoptosis and free radical mediated toxicity. Again, this section was not meant to examine all of the milestones in the toxicological process of the last few years, but to highlight those very important aspects that allow for the dissection of mechanisms of toxicity.

In the last section of the book, we examined in vitro methodological approaches through the study of the major organ systems. To this end there are chapters on hepatotoxicity, dermal toxicity, immunotoxicity, nephrotoxicity, neurotoxicity, and respiratory toxicity. Each of these in depth reviews can provide the basis for better understanding of mechanisms of cellular toxicology and provides an opportunity to devise strategies to study these processes.

As participants in the meeting, we were treated to outstanding teachers that provide in depth coverage of their area and raised important questions for us to discuss and to be answered as we more clearly define the mechanisms of the toxicological process.

We hope this volume will prove useful to you. A symposium's proceedings can rarely capture the intensity and excitement of the discussion and the learning experience shared by all. However, this meeting was so structured as to allow maximum conversations, questions, and discussions so that we as students were remarkably influenced. We hope the chapters in this book provide the same stimulation to you.

The Editors

Table of Contents

Modern Methods in Toxicology

Milestones in Cell Toxicology

Target Organ Toxicity

MODULATION OF CELLULAR RESPONSES IN TOXICITY

Alan M. Goldberg
Associate Dean for Corporate Affairs
Professor and Director, Center for Alternatives to Animal Testing
Johns Hopkins School of Public Health and Hygiene
615 North Wolfe Street
Baltimore, Maryland 21205

As a science, toxicology defines the interaction between chemical and/or physical agents and the consequences of the alterations to living tissues, cells, and intact organisms. Another purpose for the study of these biological consequences is to delineate and understand the risk to humans and animals from exposure, either intentionally or accidentally, and to prevent the consequences of that biological-agent interaction.

Correlation Versus Mechanism

The early stages of in vitro toxicology have been focused on descriptive and correlative approaches to methodology development. However, it is becoming increasingly clear that for the science to truly advance we must develop mechanistic understandings of the relationships between xenobiotics and biological systems.

Unfortunately, correlative tests lead to trivial understandings and do not provide true predictive knowledge. This is not to state that correlative tests are without merit. They have merit and are useful as screens, adjuncts and in attempting to make decisions about chemicals within a class. At this point in the development of the science, correlative tests are being used and appropriately so.

Mechanistically based tests, however, will provide us with true understandings and, thus, the ability to predict the consequences of exposure to chemical agents. When we truly have mechanistically based

NATO ASI Series, Vol. H 93
Modulation of Cellular Responses in Toxicity
Edited by C. L. Galli, A. M. Goldberg, M. Marinovich

tests, we will be able to establish rigid criteria for replacement methodology. Further, mechanistically based tests provide the understanding that a single test will rarely replace an in vivo test but that a battery of mechanistically based tests will be able to act in the replacement situation (Frazier, 1990).

The Problem Of Numbers

The National Academy of Sciences in its publication on toxicity testing (1984) described the lack of information available to the general public on safety information regarding a wide range of compounds. As can be seen in figure 1, it is clear that somewhere in the neighborhood of 80% of compounds that are in commerce on a daily basis do not have information available, in the public domain, to make a risk assessment. Clearly, we will not be able to provide this information using two year feeding studies; we must look for new ways to provide quality information. One way to obtain this will be the use of in vitro methodologies.

Issues In In Vitro Toxicology

In a paper by Goldberg and Silber (1992) they identified six issues that are necessary for establishing an in vitro test system (see Table 1).

Table 1

Consideration for Establishing an In Vitro Test System

- Biological System
- Exposure Condition
- Measurement of Effect
- Validation
- Extrapolation
- Risk Assessment

The items they identified included the biological system, exposure conditions, measurement of a specific effect or endpoint, extrapolation, acceptance through peer review and validation. Although most anticipate that the use of human cells will allow direct extrapolation to humans, there is no data to support this yet but it is a testable hypothesis.

Potential Mechanisms of Toxicity

In Goldberg and Silber (1992), they identified a list of potential mechanisms that must be understood, and methods must be developed to measure these mechanisms if we are to be able to utilize in vitro toxicology in risk assessment. The list is not intended to be comprehensive but it is a starting point for a broader discussion.

Table 2

Selected Mechanisms Associated With Toxicity

- Autophagy-Protein Degradation
- Calcium-Mediated
- Cell-Cell Communication
- Cellular Pathways
- Cytoskeleton
- DNA Repair
- Free Radicals
- Membrane Effects
- Programmed Cell Deaths
- Receptor-Mediated Mechanisms

Necessary Components

In Goldberg et al, (1993), and Koeter (1993), several items are identified that will be necessary for incorporation of in vitro methodologies. This requirement, Different End Point, In Vitro Systems to Study Them, Understand Mechanism and Publications in Quality Peer Journals (Table 3). A fifth point that should be included and identified earlier is that in vitro methods will not, except in exceptional circumstances, replace in vivo methods on a one to one basis but that a battery of tests will be necessary.

Table 3

Four Steps to Regulatory Acceptance

- Different End Points
- In Vitro Systems to Study Them
- Understand Mechanism
- Publications in Quality Peer Reviewed Journals

Mechanisms (Some Examples)

Although one can examine several individual or batteries of in vitro tests that have replaced parts of some in vivo tests, the list is not yet that extensive. However included among these are the tests for bacterial endotoxin (Limulus Assay), pregnancy testing, metastatic potential for cancer cells (Albini, 1987) and screening of anti-metastatic agents (Penno, 1992).

However, the work of Grossman and his colleagues (1993 & 1994) to clearly demonstrate the role of biological mechanism, namely a DNA repair defect and thus offers an approach to evaluation of susceptible populations and ways to prevent the toxicity. Grossman, et al demonstrated that individuals with basal cell carcinoma have a decreased ability to repair their DNA from UV damage than those that do not have basal cell carcinoma. They developed a method to measure and identify those populations. Most importantly, and what is evidently clear from this elegant work, is that once you know the mechanism of biological chemical interaction which results in a detrimental effect one is able then to develop strategies and approaches to prevent the toxicity.

Validation

In 1989, Frazier provided the seminal definition of validation and established basic underlying principle for all future validation studies. The definition Frazier proposed is. "validation is the process by which the credibility of a candidate test is established for a specific purpose." The simplicity and clarity of this definition truly allowed significant advances to follow. This was then supplemented by the publication of the Validation Committee of the Johns Hopkins Center for Alternatives to Animal Testing publication by Goldberg, et al. (1993) which provided the framework for the validation and implementation of in vitro toxicity tests. A copy of that framework can be seen in figure 2 .

In continuing the development of the necessary components and implementing a methodology for validation studies, I proposed a modular approach (fig. 3) to the development of in vitro methodologies. There are several features of the validation effort that should be noted:

1. It is a modular study, with each module comprised of one chemical class paired with a single assay (and one endpoint).
2. A unique validation standard (for example, human concurrent studies) is identified for each module, i.e., each module has only one validation standard.

To address the concerns raised in previous studies, this modular approach was developed. It limits each module to a single class of chemicals (or a subclass within a class) which is linked to a predetermined validation standard. The third part of this module will be the Assay (System) and its endpoint measurement. Every module will have only one endpoint (additional endpoints can be used but they will define a new module. Further, a specific endpoint measurement could be the duration time, i.e., time to event of the assay). This approach is under development and I will appreciate any comments or suggestions for its implementation. As you recognize, this is a work in progress and requires much discussion and development

Closing

In closing, I would anticipate that this opening lecture has raised a series of questions that each speaker will address over the next several days. Some of these are:

1) What mechanisms offer us the greatest possibilities for developing good risk assessment approaches?

2) How many mechanisms do we have to know to characterize the toxicity of a compound?

3) Is an understanding of a mechanism necessary to validate methodologies?

4) What strategies do we know or can we develop that will be able to prevent the consequences of toxic exposures?

I look forward to hearing your answers to these questions and to your presentations on mechanisms necessary to understand toxicological responses and the modulation of cellular responses in toxicity.

<u>References</u>

Albini, A., Iwamoto, Y., Kleinman, H.K., Martin, G.R., Aaronson, S.A., Kozlowski, JM, McEwen, RN,(1987) A rapid in vitro assay for quanitating the invasive potential of tumor cells. Cancer Res., 47:3239-3245

Frazier, J.F.,(1989) *Scientific Criteria for Validation of In Vitro Toxicity Tests.* , OECD Publications, Paris.

Goldberg, A.,Silber, P., (1992) Status of In Vitro Ocular Irritation Testing, *Lens and Eye Toxicity Research,* 9(3&4), 151-192

Goldberg, A., Frazier, J., Bruick, D., Dickens, M., Flint, O., Gettings, S., Hill, R., Lipnick, R., Renskers, K., Bradlaw, J., Scala, R., Veronesi, B., Green, S., Wilcox, N. and Curren, R. (1993), Framework for Validation and Implementation of In Vitro Toxicity Tests: Report of the Validation and Technology Transfer Committee of the Johns Hopkins Center for Alternatives to Animal Testing, *Journal of the American College of Toxicology,* Vol. 12, No. 1

Koeter, H.,(1993) International Acceptance of In Vitro Testing, Proceedings of the Symposium: Current Trends: In Vitro Toxicology and Eye Irritancy Testing, J.F. Morgan Research Foundation.

National Academy of Science (1984)- Toxicity Testing, Strategies to Determine Needs and Priorities,National Academy Press, Washington, DC,

Penno, M.B., De Maio, A.,(1992) A murine model for evaluating metastatic potential: Characterization of a 90-110 kDa metastasis binding protein. Exper. Geront. 27:493-501.

Wei Q., Matanoski, G.M., Farmer, E.F., Hedayati, M.A. and Grossman, L.,(1993) DNA Repair and Aging in Basal Cell Carcinoma: A molecular Epidemiology Study. Proc. Nat'l. Academy Science, USA, 90:1614-1618.

Wei, Q, Matanoski, G.M., Farmer, E.R., Hedayati, M.A. and Grossman, L.(1994) DNA Repair: a potential market for cancer susceptibility. Cancer Bull. 46:233-237 .

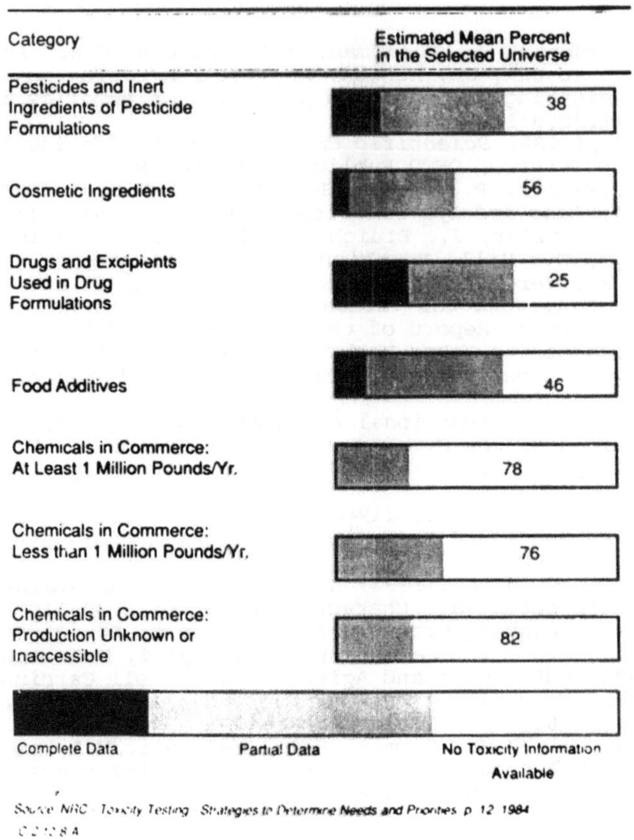

FIGURE 1 Ability to conduct health-hazard assessment of substances in seven categories of select universe

(ref. Toxicity Testing: Strategies to Determine Needs and Priorities. Board on Toxicology and Environmental Health Hazards, National Academy Press, Washington, DC 1984)

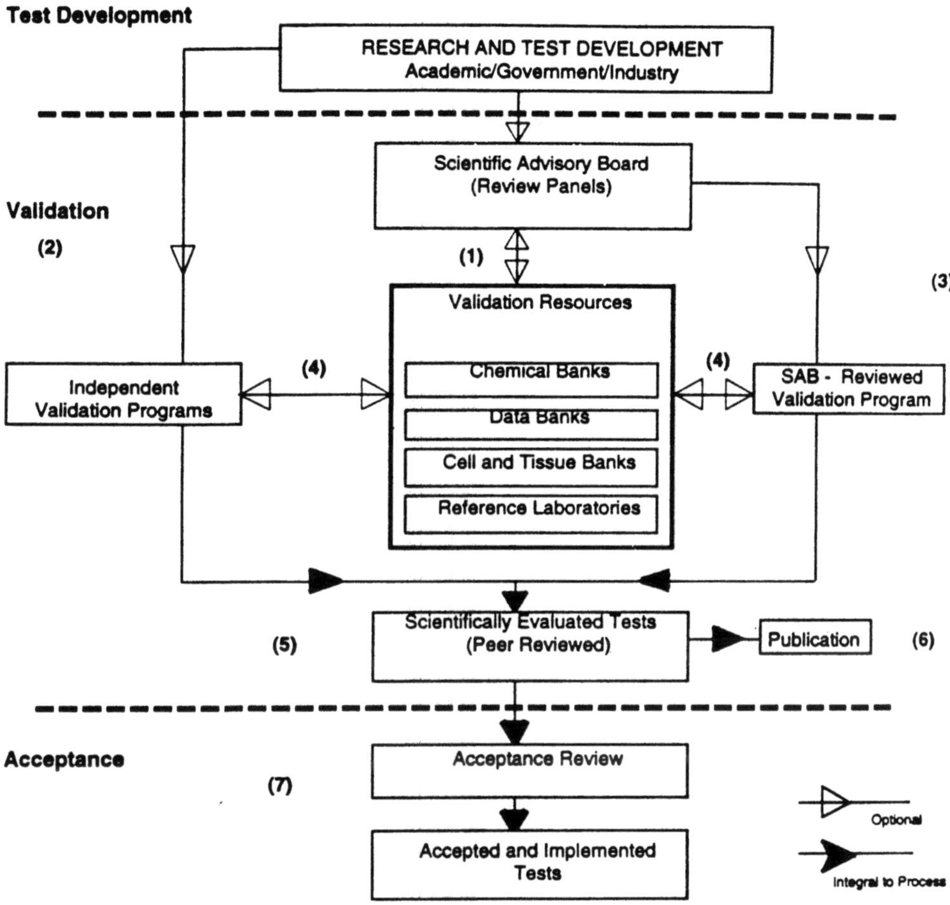

SAB provides oversight of validation resources, expertise and (optional) review of design/conduct of validation programs (1). Developers of tests or other parties may opt to conduct independent (2) or SAB-reviewed validation programs (3), utilization of Validation Resources is optional (4). Peer Review (5) and publication (6) of scientifically evaluated tests is integral to the process. (7) Acceptance of a test is a separate process.

FIGURE 2: Framework for Validation & Implementation of new In Vitro Toxicity Tests
(ref. In Vitro Cell Dev. Biol. 29A:688-692, Sept 1993)

FIGURE 2 Framework for validation and implementation of new in vitro toxicity tests

(ref. In Vitro Cell Dev. Biol. 29A:688-692, Sept. 1993)

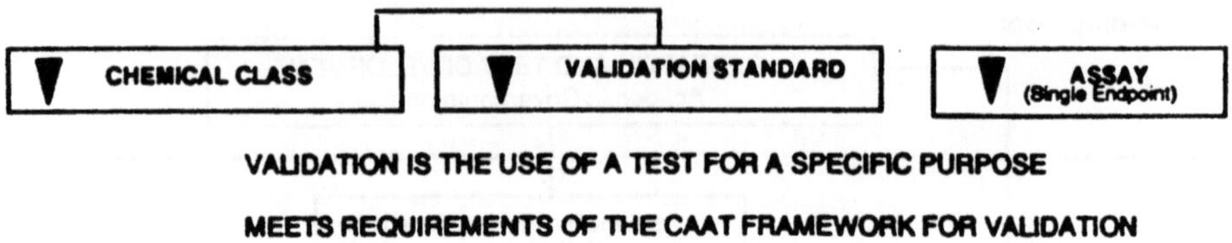

FIGURE 3 The Three components of a module

MOLECULAR BIOLOGY APPROACHES TO CELL TOXICOLOGY

G.Gordon Gibson
Molecular Toxicology Group
School of Biological Sciences
University of Surrey
Guildford, Surrey GU2 5XH
England, U.K.

INTRODUCTION

Toxicology is a hybrid science, drawing on diverse disciplines including physiology, chemistry, biochemistry and pharmacology as the core subjects, and developments in toxicology are directly related to advances made in these component sciences. For example, early toxicology was essentially grossly observational and generally descriptive in nature, but with the emergence of chemical and biochemical approaches, toxicologists were able to rationalise toxic responses to drugs and chemicals at a mechanistic level, thus placing hazard identification and risk assessment on a more rational and scientific basis.The explosion of molecular biology and its associated techniques over the past decade has been stunning and toxicologists have been well aware of the power of molecular biology approaches in dissecting out mechanisms of xenobiotic toxicity. A precise definition of molecular biology is the science that deals with molecular genetics as it relates to DNA and RNA. However a more general and toxicologically-relevant description would be the study of biological processes at the molecular level.

I cannot emphasise enough how important it is to understand molecular mechanisms of toxicity in that, in the absence of this type of information, the science of toxicology would only be phenomenological and of little value in risk assessment. Although chemical and biochemical approaches have been used in mechanistic toxicology studies, they can be substantially limited when studying toxicity in whole organs, tissue homogenates and sub-cellular fractions, particularly in those tissues such as the lung or the kidney that consist of many different cell types. If chemically-induced toxicity arises exclusively in one particular cell type, then the toxicological end-point may well be masked by "dilution" with other

NATO ASI Series, Vol. H 93
Modulation of Cellular Responses in Toxicity
Edited by C. L. Galli, A. M. Goldberg, M. Marinovich

non-affected cell types in whole tissue homogenates when studied by classical biochemical approaches. Accordingly, it is my objective to highlight several aspects of molecular approaches to understanding toxic responses to xenobiotics.

XENOBIOTIC BIOTRANSFORMATION AND ENZYME INDUCTION

As the toxicity of many xenobiotics is either increased or decreased by biotransformation (Gibson and Skett, 1994), it follows that the identification and the toxicological assessment of both the parent compound and its metabolites is an important concept in susceptibility to xenobiotic toxicity. Furthermore, most xenobiotic-metabolising enzymes are inducible and hence enzyme induction and the resultant balance of toxifying and detoxifying enzymes are critical determinants of toxicity (Figure 1).

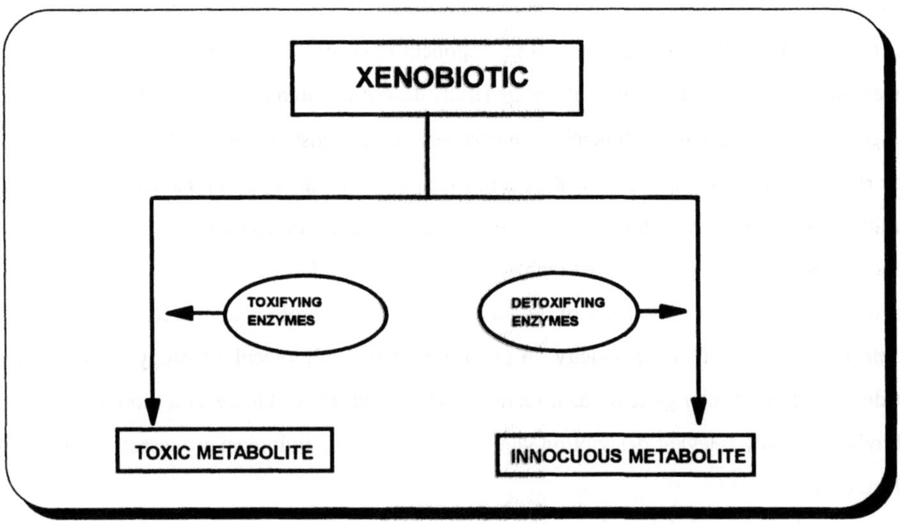

Figure 1. Importance of Biotransformation and Enzyme Induction in Xenobiotic Toxicity.

How then can we assess the relative balance of enzymes ? In addition, if we know that an enzyme in the rat is responsible for the toxicological activation of xenobiotics, can we be assured that a similar enzyme is expressed in man, and if so, at what level ? These problems are exacerbated by the existence of multiple forms (or isoenzymes) of both the activating and deactivating enzymes, which incidentally have been identified, characterised and classified by molecular approaches in the first place. For example, the cytochrome P450 gene superfamily is responsible for the metabolic activation of many toxic xenobiotics (Guengerich, 1991) and to date, approximately 250 - 300 distinct genes encoding distinct enzymes with overlapping substrate specificities have been identified (Porter and Coon, 1991 ; Nelson et al, 1993). How then does one identify if one particular, toxicologically-relevant cytochrome P450 is induced ? Fortunately, molecular probes are available to address this important issue, including monoclonal and polyclonal antibodies and specific oligonucleotide and cDNAs to assess both enzyme protein and their cognate mRNAs respectively (Waxman and Azaroff, 1992 ; Gibson, 1994).

Molecular biology has been relatively successful in characterising human drug metabolising enzymes, an area that has been notoriously difficult to assess. For example, how could one assess the drug metabolising capacity of human liver and the inducibility of the various enzymes ? In vivo, this is problematical and is ethically constrained and is equally difficult to undertake in vitro because of the scarcity of sufficiently large and viable human tissue samples. However, the application of molecular biology techniques in screening human gene and cDNA libraries has identified approximately 40 - 50 human drug metabolising enzyme genes, including the majority of the main human liver cytochrome P450s such as CYP1A2, CYP2E1, CYP2D6, CYP3A4 and CYP4A11. In addition, recent studies have emphasised the importance of isolating the 5' regulatory elements of these human genes either from gene libraries probed with the cognate animal cDNAs or with polymerase chain reaction (PCR) technology using the relevant oligonucleotide primers. This latter approach has led to the current development of "test-tube" methods to assess human enzyme induction wherein the regultory element is coupled to a reporter gene and the resultant recombinant chimera is transfected into a recipient cell line. Subsequent exposure of the transfected cell line to a potential inducer will result in activation of the reporter gene, which then yields an index of inducibilty of the human gene. This type of technology also allows the expression of full length cDNAs encoding human drug metabolising enzymes in heterologous cell systems, whereby fully functional human enzymes are expressed in their native form and hence offers the ability to examine the complete substrate specificity of the enzyme (Fig. 2)

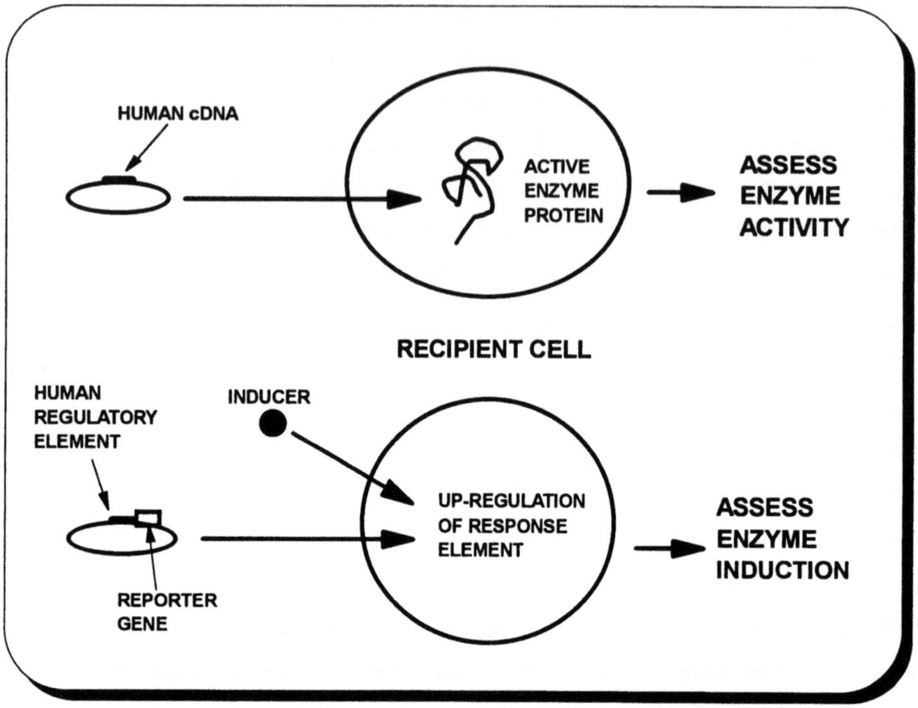

Figure 2. Molecular Strategies for the Assessment of Substrate Specificity and Inducibility of Human Drug Metabolising Enzymes Important in Xenobiotic Toxicity.

HUMAN PHARMACOGENETIC POLYMORPHISMS

It has been known for many years that there is substantial inter-individual variation in drug responses in man, particularly in the ability to metabolise drugs and toxins (Price Evans, 1993 ; Kalow, 1992). This has led to the development of the concept of poor and extensive metabolisers, and depending on the pharmacological / toxicological properties of a drug and its metabolites, may predispose the patient to drug toxicity. For example, the hypotensive drug debrisoquine is mainly cleared from the body by cytochrome P450-dependent metabolism, and in particular the CYP2D6 isoform. In clinical usage, maintenance doses of debrisoquine can vary dramatically between 20 to 400 mg per day, mirrored by the urinary excretion of unchanged drug varying from 8 to 70 % of the dose. Extensive population studies on this phenomenon have revealed that there are at least two sub-populations of debrisoquine metabolisers, designated poor and extensive metabolisers, with the former group being predisposed to toxicity of the drug (effectively showing symptoms of overdosage), as would be expected for a drug that is predominantly cleared by metabolism.

Direct characterisation of the genetic defect associated with debrisoquine hydroxylation status was made possible by cloning, sequencing and comparison of RNA transcipts from defective and normal livers and subsequent expression of the enzyme activity encoded by the cognate cDNAs in heterologous cell systems. The molecular mechanisms of this toxicity were subsequently demonstrated to arise as a result of differential mRNA splicing derived from the human CYP2D6 gene. This differential splicing of the CYP2D6 pre-mRNA results in expression of the "wild type" mRNA which codes for the fully functional enzyme (i.e. an extensive metaboliser) or mutant, truncated mRNA transcripts which are catalytically inactive (i.e. a poor metaboliser). Furthermore, poor metabolisers can be further sub-classified according to the particular nature of the genetic defect.

This precise molecular information can be used in a predictive mode to ascertain a patient's metaboliser status for debrisoquine and other drugs that are metabolised by the same CYP2D6 isoenzyme. To validate this approach, individuals have been phenotyped by the in vivo urinary metabolic ratio of parent drug : 4-hydroxy metabolite. In addition blood samples were taken and lymphocyte DNA digested with several restriction endonulease enzymes and analysed by Southern blotting techniques with a cDNA to CYP2D6 or with specific oligonucleotides that recognise specific mutations in the CYP2D6 gene. In this genetic analysis (known as restriction fragment length polymorphism - RFLP), specific DNA fragments are associated with poor or extensive metabolisers and this genotyping approach correlates well with the above phenotyping methodology (Meyer et al, 1992). Thus this molecular genetic approach has provided not only a coherent explanation for the polymorphic phenotypes, but also the possibility to more readily genotype individuals, and thus predict their probable toxicological responses to these drugs. Again, it must be emphasised that the toxicological relevance of the particular enzyme system under investigation must be established first before meaningful safety assessment issues can addressed and therfore managed.

RECEPTORS

Several xenobiotics produce toxicity via receptor interaction and notable advances in this area have been substantially helped by molecular biology approaches. The peroxisome proliferators (PP) are a sub-class of non-genotoxic carcinogens, which by definition, are negative in all genotoxicity assays investigated to date (Budroe and Williams, 1993), thereby creating a substantial problem in their hazard identification and hence safety assessment. The PPs are toxic in several rodent tissues including the thyroid, testes and the liver, the last organ having been particularly intensively studied for mechanisms of non-genotoxic hepatocarcinogenesis (Gibson and Lake, 1993). Characteristic liver responses to PPs in rodents include hepatomegaly (usually a combination of hypertrophy and hyperplasia), morphometric proliferation of peroxisomes and induction of the endoplasmic reticulum. Proliferation of the latter two organelles is accompanied by substantial enzyme induction of peroxisomal fatty acid beta-oxidation and microsomal fatty acid omega-hydroxylase (a CYP4A-dependent enzyme) activities respectively. Although the molecular mechanisms of PP-dependent rodent liver carcinogenesis still remains the topic of active debate and investigation, it would appear that a combination of the hyperplastic response (essentially stimulated S-phase DNA synthesis) and enzyme induction predispose the liver to malignancy (Gibson and Lake, 1993).

It must be emphasised that the above liver changes appear to be lower rodent-specific phenomena and occur in the rat and mouse and do not readily occur in higher primates such as the monkey and marmoset (Makowska et al, 1991 ; Bentley et al, 1993). Quite clearly, this species difference in response would seem to indicate that peroxisome PPs would not represent a hazard and hence risk to man, and indeed, this is the currently prevailing view (Bentley et al, 1993). However, because of the potential exposure of PPs to man, which includes hypolipidaemic drugs, phthalate ester plasticisers and pesticides amongst others, governmental regulatory agencies were initially sceptical about the well-documented species differences in response and the importance of this to extrapolation to human health. This scepticism was essentially based on the lack of a coherent molecular mechanism underpinning the liver responses, and hence if we did not understand why PPs are toxic in rodents, how could we be assured that PPs were safe to man when we had no hypothysis or mechanism to test in man ?

Substantial progress in dissecting out the molecular mechanisms of PP-dependent pleiotropic responses was made with the discovery of a receptor system in the mouse which mediates the regulation of PP-activated genes. This receptor belongs to the steroid hormone nuclear receptor superfamily of ligand-dependent transcriptional activation factors and was termed the peroxisome proliferator activated receptor (PPAR, Issemann and Green, 1990). The PPAR has a molecular weight of approximately 54 kDa and has been characterised in several species including Xenopus, mouse, rat and man (Motojima, 1993). The PPAR consists of two domains, namely a ligand (PP)-binding domain and a DNA-biding domain that recognises a PP response element (PPRE) in responsive genes and hence switches on gene expression (Figure 3).

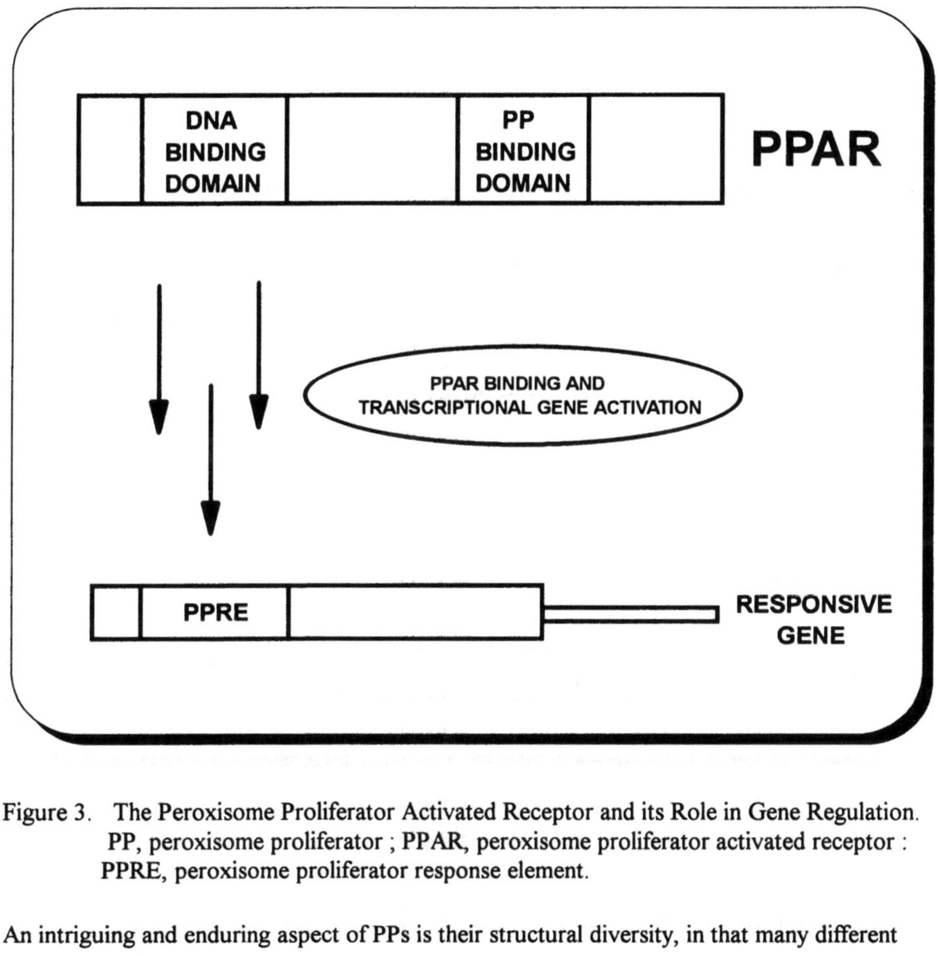

Figure 3. The Peroxisome Proliferator Activated Receptor and its Role in Gene Regulation. PP, peroxisome proliferator ; PPAR, peroxisome proliferator activated receptor : PPRE, peroxisome proliferator response element.

An intriguing and enduring aspect of PPs is their structural diversity, in that many different chemicals all produce essentially the same biological responses (Gibson and Lake, 1993). This broad range of chemical structures is intriguing, and several different hypotheses have been forwarded to sustain this molecular diversity. As shown in Figure 4, the PPAR-dependent regulation of responsive genes may operate through several mechanisms including direct xenobiotic-dependent PPAR activation, or in view of the fact that PPs perturb lipid homeostasis and lipids themselves can activate the PPAR (Gottlicher et al, 1992), via a lipid-mediated mechanism. In addition, emerging evidence indicates that the PP-dependent responses are blocked by calcium channel antagonists and that PPs themselves perturb intracellular calcium homeostasis, would seem to indicate that calcium and its important role as a secocond messenger in intracellular communication may well play a critical role in mediating the diverse liver responses to PPs

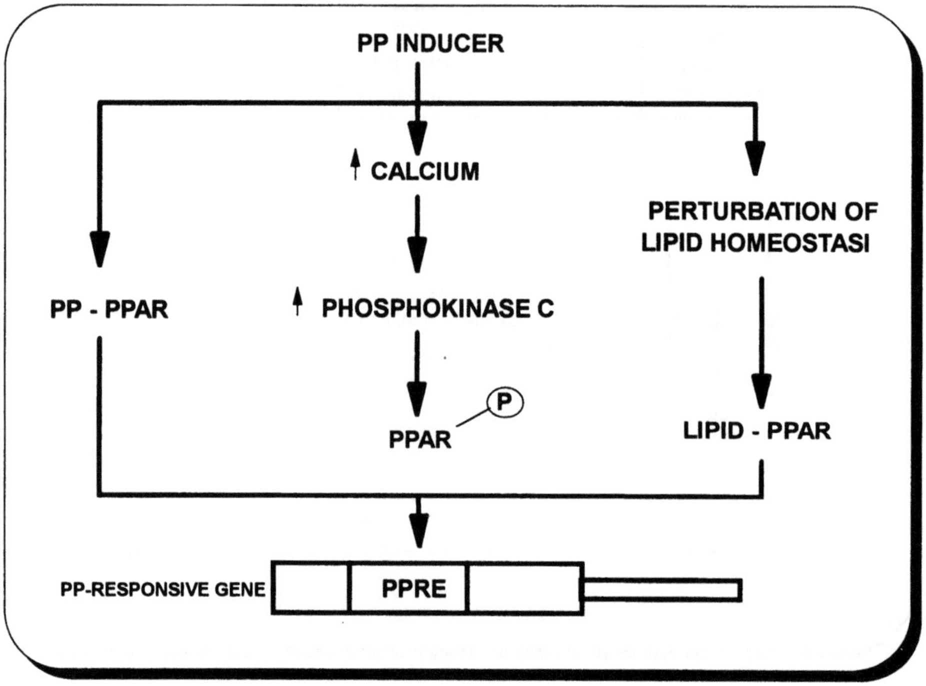

Figure 4. Molecular Mechanisms of PPAR Activation.

With this wealth of molecular information, how does this help us to answer the crucial question of human susceptibility to PP toxicity ? This question is being actively investigated and current emphasis is being placed on the following

- the structural and functional properties of the human PPAR(s)
- role of auxiliary receptor systems (e.g. the retinoic acid receptors) in man

- do human genes contain the necessary PPREs in their regulatory regions ?
- does perturbation of lipid and calcium homeostasis occur in man ?
- what is the role of the PPAR in controlling differentiation, growth and cell division ?
- how do the above relate to the susceptibility to carcinogenesis ?

PERSPECTIVES

Although molecular biology approaches to understanding mechanisms of toxicity is of very substantial value, they must not be considered in isolation and must be integrated with other cellular or whole animal approaches in order to facilitate a comprehensive risk assessment for the safety evaluation of drugs and chemicals.

REFERENCES

Bentley, P., Calder, I., Elcombe, C., Grasso, P., Stringer, D. and Wiegand, H.J.(1993). Hepatic peroxisome proliferation in rodents and its significance for humans. Fd.Chem.Toxicol., 31 : 857 - 907.

Budroe, J.D. and Williams, G.M. (1993). Genotoxicity studies of peroxisome proliferators. In, Peroxisomes : Biology and Importance in Toxicology and Medicine, G.G.Gibson and B.G.Lake (eds.), Taylor and Francis Ltd., London, pps. 525 - 568.

Gibson, G.G. (1993). Molecular Biological Aspects of Toxicology. In, Introduction to Biochemical Toxicology, Second Edition (E.Hodgson and P.E.Levi, eds.), Appleton and Lange, Connecticut, pps. 241 - 263.

Gibson, G.G. and Lake, B. (1993). Peroxisomes : Biology and Importance in Toxicology and Medicine, Taylor and Francis, London.

Gibson, G.G. and Skett (1994). Introduction to Drug Metabolism, Blackie, Glasgow.

Guengerich, F.P. (1991). Reactions and significance of cytochrome P450 enzymes. J.Biol.Chem., 266 : 10019 - 10023.

Issemann, I. and Green, S.(1990). Activation of a member of the steroid hormone receptor superfamily by peroxisome proliferators. Nature : 347, 645 - 650.

Kalow, W. (1992). Pharmacogenetics of Drug Metabolism, Pergamon Press, New York.

Makowska, J.M., Bonner, F.W. and Gibson, G.G. (1991). Comparative induction of cytochrome P4504A1 and peroxisome proliferation by ciprofibrate in the rat and marmoset. Arch.Toxicol. 65 : 106 - 113.

Meyer, U.A., Skoda, R.C., Zanger, U.M., Heim, M. and Broly, F. (1992). The genetic polymorphism of debrisoquine / sparteine metabolism - molecular mechanisms. In, Pharmacogenetics of Drug Metabolism (W.Kalow, ed.), Pergamon Press, New York, pps. 609 - 624.

Motojima, K. (1993). Peroxisome proliferator activated receptor (PPAR) : structure, mechanisms of activation and diverse functions. Cell Struct. Funct. 18 : 267 - 277.

Nelson., D.R., Kamataki, T., Waxman, D.J., Guengerich, F.P., Feyereisen, R., Gonzalez, F.J., Coon, M.J., Gunsalus, I.C., Gotoh, O., Okuda, K. and Nebert, D.W. (1993). The P450 superfamily : Update on new sequences, gene mapping, accession numbers, early trivial names of enzymes and nomenclature. DNA Cell Biol. 12 : 1 - 51.

Porter, T.D. and Coon, M.J. (1991). Cytochrome P450 multiplicity : isoforms, substrates and catalytic and regulatory mechanisms. J.Biol.Chem. 266 : 13469 - 13472.

Price Evans, D.A. (1993). Genetic Factors in Drug Therapy. Clinical and Molecular Pharmacogenetics, Cambridge University Press, Cambridge.

Waxman, D.J. and Azaroff, L. (1992). Review : phenobarbital induction of cytochrome P450 gene expression. Biochem.J. 281 : 577 - 592.

Receptor Mediated Toxicity: The Dioxin Receptor as an Example of Biological Complexity and Experimental Approaches

Thomas R. Sutter, Chris W. Cody, Jonathan A. Gastel, Carrie L. Hayes, Ying Li, Nigel J. Walker and Hong Yin
Department of Environmental Health Sciences
Division of Toxicological Sciences
Johns Hopkins University
School of Hygiene and Public Health
615 North Wolfe Street
Baltimore, Maryland
U.S.A.

During the past decade tremendous advances have been made in understanding the role of receptors as signal transducers, linking cells and their environment in a carefully orchestrated symphony of multicellular life. Through changes in gene expression, receptors facilitate a multitude of complex biological responses including differentiation, mitosis, apoptosis, and quiescence. While the mechanics of receptor activation and subsequent enhancement of transcription have yielded to modern methods of pharmacology and molecular biology, less progress has been made towards understanding causal relationships between changes in gene expression and cellular and tissue responses. Like other critical cellular targets, receptor proteins provide specific, high affinity sites for chemical interaction. In some cases, such chemical interactions have defined the action of receptors in the absence of any knowledge of their endogenous ligand or function. The dioxin receptor is one such example that illustrates both the biological complexity of, and experimental approaches to understanding receptor mediated toxicity.

NATO ASI Series, Vol. H 93
Modulation of Cellular Responses in Toxicity
Edited by C. L. Galli, A. M. Goldberg, M. Marinovich

RECEPTOR-CHEMICAL INTERACTIONS

The coordinate regulation of cell growth and differentiation, of developmental fate and morphogenesis, of local and systemic response to injury and infection, are mediated through a complex circuitry of intra and extracellular signals that must be integrated at the cellular, tissue, and organismal levels. Soluble polypeptide growth factors and chemical hormones represent broad classes of the signal components of this communication network. Through their recognition by, and activation of specific receptors, these signals modulate the sets of coordinately expressed and repressed products of the genome that define both cellular phenotype and function. Within the context of the cell on which they act, such signals are not merely gene switches of on and off, but dictate both intensity and duration (concentration and time), and act in combination with other signals (codes). In addition, these protein and chemical factors, and their corresponding receptors, may form gradients within a tissue, and the gene products whose levels they affect may function within the specific spatial structure of a cell or tissue (space). When healthy, this network supports the cellular homeostasis that maintains multicellular life. When compromised by aging, disease, or environmental insult, these same cellular signal transduction pathways can mediate cancer, birth defects, immune suppression and deficits in neurologic function (Cross and Dexter, 1991; Edwards, 1994; Gudas, 1992; Hunt and Krumlauf, 1992; Walters, 1992).

Since the discovery of enzyme induction by foreign compounds in the 1950's (Conney, 1967) much energy has been expended in deciphering the mechanics of this phenomenon. These efforts have led to numerous important discoveries, including cytochromes P450 which can detoxify or alternatively activate some of these chemical inducers, and novel cellular receptors that mediate these inductions (Conney, 1967; Okey, 1990; Poland and Knutson, 1982; Green, 1992). From these studies of chemical carcinogenesis, and from related studies of tumor promoting agents, has come the understanding that for several classes of xenobiotics, the action of these chemicals is to activate cellular receptors (Poland and Knutson, 1982; Green, 1992). In the case of the phorbol esters, these chemicals mimic the activity of the endogenous second messenger diacylglycerol (Weinstein, 1991). In the cases of the dioxins and peroxisome proliferators, it is not yet clear whether parallel endogenous signals exist, but the specificity of action and structure activity relationships of these classes of compounds suggest that this is true.

Table 1. Examples of Toxin–"Receptor" Interactions[1]

Target	Ligand	Toxin	Action
Intracellular			
Receptor			
Ah	?	Dioxins	LATF
ER	Estrogens	Estrogens	LATF
2nd Messenger			
PKC	DAG	Phorbol esters	Kinase activity increased
G-protein	G–protein coupled receptors	Bacterial exotoxins	Gs or Gi activity increased
Integral Membrane			
Receptor			
Muscarinic	Acetycholine	Belladona alkaloids	Cholinergic antagonist
NMDA	Glutamate	Metals	Allosteric effector
Channel			
Na+	Na+	Tetrodotoxin	Blocks channels
Cationic	Ca++	Maitotoxin	Activates cation channels
Transporters			
Dopamine	Dopamine	Cocaine	Dopamine
Na+/K+ ATPase	Na+/K+	Digitalis	Intracellular Ca^{++} increased
Metabolizing Enzymes			
Acetyl-cholinesterase	Acetylcholine	Organophosphate pesticides	Acetylcholine increased
Myo-inositol mono-phosphatase	Myo-D inositol phosphate	Lithium	Inositol 3-phosphate decreases

[1]Examples compounds and references: Bacterial exotoxins-cholera toxin (Serventi et al., 1992; Madshus and Stenmark, 1992), pertusis toxin (Sekura et al., 985); Belladona alkaloids (Snyder et al., 1975); Cocaine (Kuhar et al., 1988); Digitalis (Hoffman and Bigger, 1990); Dioxins– 2,3,7,8-tetrachlorodibenzo-p-dioxin (Poland and Knutson, 1982; Swanson and Bradfield, 1993); Estrogens–diethylstilbesterol (Greco et al., 1993; Katzenellenbogen et al., 1993); Lithium (Pollack et al., 994); Maitotoxin (Gusousky et al.,1990); Metals–lead (Guilarte et al., 1993; Uteshev et al., 1993); Phorbol esters–12-O-tetradecanoylphorbol-13-acetate (Castagna, 1987; Olson et al., 1993); Tetrodotoxin (Kao and Levinson, 1986)

Abbreviations: Ah, aryl hydrocarbon receptor; DAG, diacylglycerol; ER, estrogen receptor; LATF, ligand activated transcription factor; NMDA, N-methyl-D-aspartate receptor; PKC, protein kinase C.

As a clearer picture develops of how signals are transduced from the cytoplasmic membrane to the nucleus, it is important not to limit our concept of receptor-mediated toxicity to just receptor ligands. As shown in Table 1, this is only one type of toxin-receptor interaction. While this list is not exhaustive, it does exemplify multiple mechanisms of chemical actions, each of which could be considered to be receptor-mediated. For each of these examples, the net effect is to alter cellular receptor-mediated response, and the likely first common effect would be transient changes in gene expression.

In addition to the specific characteristics of the chemical-receptor interaction, multiple response modifiers can influence the net biological response (Table 2). Several isoforms of the receptor may exist within a cell or tissue, and the affinity of the chemical for specific isoforms may vary by several orders of magnitude. Multiple proteins may interact with the receptor, both before and after chemical activation. Such protein:protein interactions may enhance or inhibit the actions of the chemical-receptor complex. Processing enzymes such as kinases or phosphatases may modify the receptor protein or proteins with which it interacts. Distribution or metabolism of the ligand or other substrates can modify the biological response, and binding proteins or transporters can also influence these processes. At the cell and tissue levels, specific localization and distribution of receptor isoforms can modify the net response (Fuller, 1991; Glass, 1994; Clark and Docherty, 1993; Orti et al., 1992).

Table 2. Receptor Response Modifiers

o	multiple receptor isoforms
o	protein:protein interactions
o	processing enzymes
o	ligand metabolism/distribution
o	binding proteins
o	transporters
o	intracellular localization
o	cell type distribution

LIGAND-DEPENDENT ACTIVATION OF THE Ah RECEPTOR

2,3,7,8-Tetrachlorodibenzo-p-dioxin (TCDD or dioxin) is a prototype for a large class of halogenated aromatic hydrocarbons that are both widespread and persistent chemical pollutants. TCDD elicits a wide spectrum of toxic responses, many of which involve alterations of cell growth and differentiation.

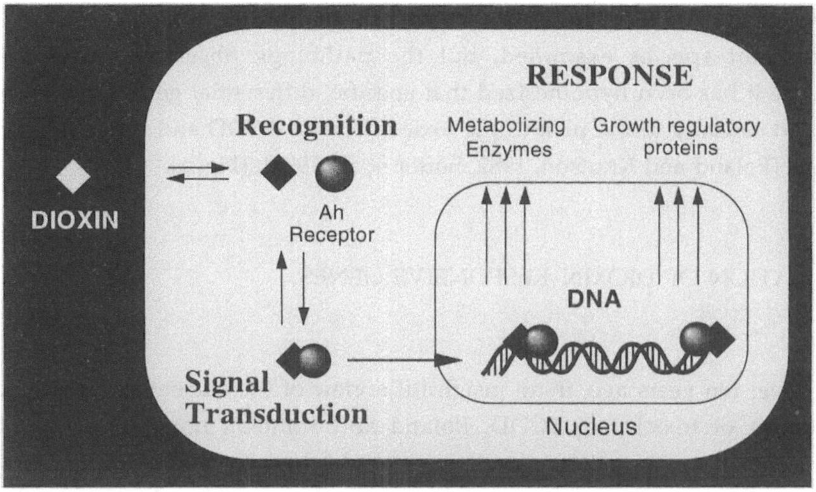

Figure 1. Proposed mechanism of action of TCDD and other Ah receptor agonists. It is known that TCDD enters the cell and binds with high affinity to the Ah receptor (Nebert et al., 1972; Poland and Glover, 1976). Activation and nuclear translocation of the TCDD-Ah receptor complex must occur prior to binding DNA (Ligand Recognition). The Ah receptor, and at least two additional proteins are known to be involved in this process (Burbach et al., 1992; Reyes et al., 1992; Wilhelmsson et al., 1990). Once in the nucleus, this activated complex binds to DNA response elements and begins the process to activate gene transcription (Signal Transduction), (Jones et al., 1985; Shen and Whitlock, 1992). This process results in increased levels of specific gene products that include metabolic enzymes and growth regulatory proteins (Primary Biological Response), (Sutter et al., 1991b). This process may also result in repressed gene expression. The altered expression of specific subsets of genes regulated by the Ah receptor is hypothesized to cause the toxic effects of TCDD (Adapted from Sutter and Greenlee, 1992b).

In rodents, TCDD is a potent carcinogen and tumor promoter. In humans, the primary lesion associated with exposure to TCDD is a skin pathology known as chloracne. The carcinogenic risk to humans however is unclear (Safe, 1990; Huff et al., 1991; Johnson, 1993).

The biological effects of TCDD are mediated through its high affinity and saturable binding to the Ah receptor. This receptor is a member of a distinct class of helix-loop-helix transcription factors (Table 1). At least three different proteins are involved in the ligand binding, nuclear translocation, and transcription enhancer activities of this protein complex (Swanson and Bradfield, 1993). As characterized for CYP1A1 (cytochrome P4501A1, P_1450, P450c), activation of transcription requires accumulation of the TCDD-Ah receptor complex in the nucleus, and interaction of this complex with specific DNA response elements found upstream of the mRNA initiation site (Whitlock, 1993).

In general, pharmacokinetic distribution of TCDD, tissue distribution and number of the Ah receptor, and ligand binding affinities appear to be similar in the different species examined, but the pathology observed varies greatly. Therefore it has been hypothesized that specific, differential gene expression may be linked causally to the pleiotropic toxic effects of TCDD and other Ah receptor agonists (Poland and Knutson, 1982, Sutter et al., 1991), (Fig. 1).

REGULATION OF DIOXIN–RESPONSIVE GENES

Over ten years ago, in an insightful review of the current knowledge of the mechanism of toxicity of TCDD, Poland and Knutson noted that the TCDD-dependent increase in the levels of several known metabolizing enzymes (CYP1A1, UDP-glucuronosyl transferase, menadione oxidoreductase, etc.) could not account for the broad spectrum of tissue- and species-specific responses that followed chemical exposure. They postulated that such enzyme induction represented a limited pleiotropic response that may be common to many species, while specific toxicities represented additional responses that were normally restricted. In this model, the Ah receptor affected the coordinate regulation of specific gene subsets, some of which could mediate toxicity (Poland and Knutson, 1982).

Since the early 1980's, Greenlee and co-workers have been investigating the mechanism of action of TCDD in human keratinocytes. Both the prevalence of chloracne in individuals exposed to TCDD through industrial accidents, and the ability to culture phenotypically normal keratinocyte cell lines and primary epidermal keratinocytes, have made this an excellent model of human responses to TCDD. In addition to enzyme induction, Greenlee and coworkers were also able to investigate the more complex cellular response of differentiation (Greenlee et al., 1987; Greenlee et al., 1987b). This well characterized dioxin-responsive cell model allowed us to test the gene subset hypothesis. Using the method of differential hybridization, five dioxin-responsive cDNA clones were isolated from a keratinocyte cell line (Sutter et al., 1991; Sutter et al., 1991b; Greenlee et al., 1994). One of the clones was the cDNA for CYP1A1, a gene previously shown to be under the direct transcriptional control of the Ah receptor (Whitlock, 1993). Two additional clones were identified as genes encoding proteins involved in inflammatory responses and growth regulation: plasminogen activator inhibitor-2 (PAI-2), a regulator of extracellular matrix proteolysis, and interleukin-1β (IL-1β), a cytokine. The rate of transcription of the PAI-2 gene was shown to be increased within 90 minutes following treatment with TCDD. Both PAI-2 and IL-1β are inducible by treatment with TPA. These results support the concept that actions common to dioxin and other tumor promoting agents, such as the induction of inflammatory responses and cellular differentiation, may occur as the result of the overlapping regulation of a subset of genes that code for multifunctional proteins (Sutter et al., 1991; Sutter et al., 1991b; Greenlee et al., 1994).

During this same period, Nebert and Gonzalez introduced the concept of the Ah gene battery (Nebert and Gonzalez, 1987). They identified six Ah receptor-agonist-responsive genes whose products, RNA or protein, were the result of either increased rates of gene transcription, or whose production segregated with the Ah-responsive mouse genotype. These included two Phase I products, CYP1A1 and CYP1A2, and four Phase II products, NAD(P)H:menadione oxidoreductase, aldehyde dehydrogenase, UDP-glucuronosyl transferase and glutathione transferase. They proposed that this battery represented a subset of genes coordinately regulated by the Ah receptor, and thus implied a mechanism analogous to the Ah receptor-DRE interactions so well characterized for CYP1A1 induction.

In a subsequent review, Nebert noted that several additional gene products, inducible by either polycyclic aromatic hydrocarbons or TCDD, had been identified in the literature. However, he chose not to include them in the gene battery because of insufficient evidence of an Ah receptor mediated mechanism. He reasoned that such protein increases could be the result of responses to secondary signals generated by the increased enzyme levels of the Ah gene battery (Nebert, 1989).

With the discovery of additional human dioxin-responsive genes (Sutter et al., 1991b), we reviewed the literature on such products and identified 26 potential members of the Ah gene battery (Sutter and Greenlee, 1992). From the perspective of establishing a molecular basis of dioxin action at the level of target tissues, it seemed important to broadly define the set of dioxin-responsive genes, and then, to group these into subsets based on the available experimental evidence supporting common mechanism(s) of regulation (Sutter and Greenlee, 1992; Greenlee et al., 1994), (Table 3).

It is possible that the TCDD related response characterized by Poland and Knutson as a "restricted pleiotropic response" represents the net action of a subset of genes that are completely regulated by secondary signals. However, it seems more likely that such restricted responses represent multiple mechanisms and levels of response regulation. The comparison of multiple gene responses may

Table 3. Classification of Dioxin-Responsive Genes[1]

Transcription	**RNA**	**Protein**
Ah Receptor [2]	αhCG[a,b]	Ah Receptor [2]
Cyp1A1	Interleukin-1β [a,b]	ODC
Cyp1A2	TGF–α [a,b]	Ugt-1
Gst-Ya		
Nmo-1	c-erbA related	AhR Agonist
T-ALDH	GST-Yb	AHBP
	GST-Yc	ALAS
	MDR-1	Choline kinase
AhR Agonist	Testosterone	Malic enzyme
CYP1B1	7α-OHase	Phospholipase A2 [b]
PAI-2 [a,b]		Protein Kinase C [b]
		Enzyme pp60$^{c\text{-}src}$
		60kDa esterase

[1]Source: Adapted from Sutter and Greenlee (1992). The groupings shown in this table are based on experimental data describing increases in the rate of gene transcription, RNA levels or protein levels.
[2]The classification is based on experimental data showing that the transcriptional regulation by TCDD is mediated via the aromatic hydrocarbon receptor (AhR), as determined either biochemically or by mouse genetics.
Abbreviations: AHBP, aryl hydrocarbon binding protein; ALDH, aldehyde dehydrogenase; ALAS, δ-aminolevulinic acid synthetase; CYP, cytochrome P450; GST, glutathione S–transferase; hCG, human chorionic gonadotrophin; IL, interleukin; MDR, multidrug-resistance; Nmo, menadione oxidoreductase; ODC, ornithine decarboxylase; PAI-2, plasminogen activator inhibitor; PKC, protein kinase C; TGF, transforming growth factor; Ugt, UDP-glucuronyl transferase.
[a]Secreted protein
[b] Growth regulatory protein

facilitate the identification of such multiple regulatory mechanisms. Currently, more mechanisms can be proposed, than can be substantiated by experimental evidence. These include: 1) Ah receptor dependent and DRE mediated, 2) Ah receptor dependent and DRE independent, 3) Ah receptor dependent and DRE mediated, but requires additional cell or tissue specific transcription factors, 4) Ah receptor dependent, post-transcriptional regulation, 5) Ah gene battery dependent, secondary responses, 6) Ah receptor independent modifiers of Ah receptor dependent responses (either at the level of receptor regulation or enzyme activity). Concerning Table 3, several updates are noteworthy. For CYP1A2 (Quattrochi et al., 1994) and class 3 aldehyde dehydrogenase (T-ALDH), (Asman et al., 1993; Takimoto et al., 1992), additional experimental evidence supports the Ah receptor dependent and DRE mediated mechanism of gene regulation (Class 1A in the previous classification scheme; Sutter and Greenlee, 1992). Two human dioxin-responsive cDNA clones previously identified as clones 1 and 141 (Sutter et al., 1991) are now known to be nonoverlapping cDNA clones corresponding to CYP1B1, a new subfamily of cytochromes P450 (Sutter et al., 1994). Further studies of dioxin-responsive TGF-α (Choi et al., 1991) have provided evidence for regulatory mechanisms other than direct transcriptional activation. Gaido and co-worker (Gaido et al., 1992) have shown that the dioxin-dependent increases of TGF-α mRNA involves mRNA stabilization, and not increased rates of gene transcription. Two examples of repressed gene expression are now documented. In rats, mRNA levels of hepatic phosphoenolpyruvate carboxykinase were shown to be significantly decreased following treatment with TCDD (Stahl et al., 1993) In a human keratinocyte cell line, both levels of mRNA and rate of transcription of TGF-β2 were decreased following the treatment of cells with 10 nM TCDD (Gaido et al., 1992).

FUNCTION OF DIOXIN-RESPONSIVE GENES: FROM CELLULAR TO TISSUE RESPONSE

In Table 3, several secreted proteins are identified including: PAI-2, a proteinase inhibitor; hCG, a common subunit of several glycoprotein hormones; IL-1β, a cytokine, and TGF-α, a growth factor. The Ah receptor-mediated production of secreted proteins indicates that the actions of TCDD on one cell can affect additional cells by paracrine or endocrine pathways. Such cells need not have an Ah receptor, only receptors or sites of action for these secretory proteins. In addition, these secreted proteins may act via autocrine pathways to alter the behavior of the cell in which they are produced. Such factors could act either as stimulatory or inhibitory modifiers of the Ah receptor-mediated cellular actions of TCDD.

Three family 1 cytochromes P450, CYP1A1, CYP1A2 and CYP1B1, are inducible by TCDD. While the role of these enzymes in the detoxication and activation of xenobiotics is well characterized (Okey, 1990; Savas et al., 1994), their normal physiologic activity remains unknown. While it is possible that the sole purpose of the inducible CYP1 family of cytochromes P450 is to metabolize foreign compounds (Nebert and Gonzalez, 1987), current knowledge of the diversity of signal transduction pathways suggest that these enzymes may also metabolize substrates of physiologic relevance like steroid hormones or eicosanoids (Nebert, 1991; Nebert, 1994). Recent studies of TCDD-inducible CYP1B1 in the MCF-7 breast cancer cell line shows a strong correlation between the induction of CYP1B1 and the low K_m 17β-estradiol 4-hydroxylase activity (Spink et al., 1994). These results indicate possible endocrine regulatory roles for

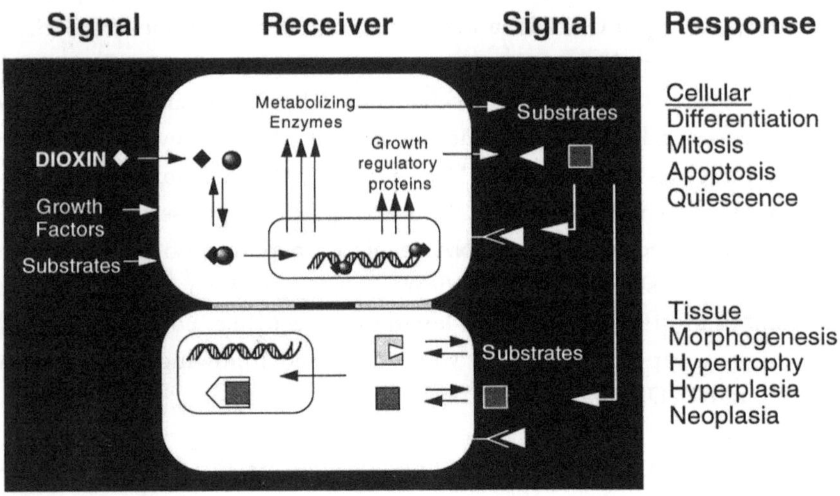

Figure 2. An integrated signal approach to the receptor-mediated actions of dioxin. Dioxin, through its high-affinity interaction with the Ah receptor, results in enhanced and repressed expression of multiple proteins. Within this dioxin-responsive cell, specific biochemical changes occur. These changes may be the direct effect of proteins that have growth regulatory potential, or may be affected through changes in cellular metabolism of signal molecules or other critical cellular substrates. The primary signal is not simply dioxin, but includes the net signal being received by the cell. Components of this signal include dioxin, other growth factors and receptor ligands, substrate pools of cellular enzymes, and intracellular communication networks. The primary dioxin-receiver cell must possess Ah receptors. However, the context in which the dioxin signal is received will be dependent upon the net cellular content of receptors, enzymes, and intracellular communication machinery. The primary signal, once received, will be integrated into a net cellular response. This initial response will be characterized by changes in gene, and subsequent protein expression. In addition to specific intracellular changes, the receiver cell may change how it interfaces with its environment by altering its expression of receptors and secretory proteins, through the metabolism of substrate pools, and by changes in intracellular communication, for example, gap junction. This secondary signal may extend the action of dioxin to cells that do not have an Ah receptor, and may change the cellular and tissue context of this iterative signalling process. It is proposed that through such integrated signal mechanisms, dioxin elicits such complex biological responses as chloracne and cancer.

the recently discovered CYP1B subfamily (Sutter et al., 1994; Savas et al., 1994) and perhaps of the CYP1 family in general.

Collectively, such molecular responses and determinants of responsiveness to TCDD are integrated at the level of the cell, the tissue, and the organism. This continuum of biological action and reaction, of imbalance and homeostatic regulation, results in either compensation or pathology. To experimentally approach this continuum we are currently forced to segment such processes into discrete mechanistic steps. In the case of dioxin, significant advances have been made in understanding the receptor recognition of ligand, transduction of this signal to the nucleus, and enhanced transcription (Fig. 1). Increased knowledge of the number and kinds of genes that may be regulated by this pathway has come from numerous laboratories (Table 3), and it is believed that these studies will provide new insight into the integrated biological actions of TCDD at the cellular and tissue levels (Fig. 2).

One of the greatest challenges currently faced by toxicologists is to understand the causal relationships between changes in gene expression and complex biological response. Such knowledge is necessary if we are to understand the mechanisms of receptor-mediated toxicity, and more importantly, the determinants of human susceptibility to such compounds. As seen for the example of dioxin, we must understand both the action of the chemical and the biology of the cell.

ACKNOWLEDGEMENTS

We acknowledge our collaboration with William F. Greenlee (Purdue University) and thank Bill for his continued encouragement and support of this work. Research is this laboratory is supported by grants from the NIH (ES06071), the American Forest and Paper Association, and the Johns Hopkins Center for Alternatives to Animal Testing.

Asman DC, Koichi T, Pitot HC, Dunn TJ, Lindahl R (1993) Organization and characterization of the rat class 3 aldehyde dehydrogenase gene. J Biol Chem 268: 12530-12536

Burbach KM, Poland A and Bradfield CA (1992) Cloning of the Ah-receptor cDNA reveals a distinctive ligand-activated transcription factor. Proc Natl Acad Sci 89:8185-8189

Castagna M (1987) Phorbol esters as signal transducers and tumor promoters. Biol of the Cell 59:3-14

Choi EJ, Toscano DG, Ryan JA, Riedel N, Toscano Jr WA (1991) Dioxin induces transforming growth factor-α in human keratinocytes. J Biol Chem 266: 9591-9597

Clark AR and Docherty K (1993) Negative regulation of transcription in eukaryotes. Biochem J 296:521-541

Conney AH (1967) Pharmacological implications of microsomal enzyme induction. Pharmacological Reviews 19:317-366

Cross M and Dexter TM (1991) Growth factors in development, transformation, and tumorigenesis. Cell 64:271-280

Edwards DR (1994) Cell signalling and the control of gene transcription. Trends in Pharmacol Sci 15:239-244

Fuller P (1991) The steriod receptor superfamily: mechanisms of diversity. FASEB J 5:3092-3099

Gaido KW, Maness SC, Leonard LS and Greenlee WF (1992) 2,3,7,8-Tetrachlorodibenzo-p-dioxin-dependent regulation of transforming growth factors-α and $-\beta_2$ expression in a human keratinocyte cell line involves both transcriptional and post-transcriptional control. J Biol Chem 267: 24591-24595

Glass CK (1994) Differential recognition of target genes by nuclear receptor monomers, dimers, and heterodimers. Endocrine Reviews 15:391-407

Greco TL, Duello TM, and Gorski J (1993) Estrogen receptors, estradiol, and diethylstilbestrol in early development: the mouse as a model for the study of estrogen receptors and estrogen sensitivity in embryonic development of male and female reproductive tracts. Endocrine Reviews 14:59-71

Green S (1992) Nuclear receptors and chemical carcinogenesis. Trends in Pharmacol Sci 13:251-255

Greenlee WF, Osborne R, Dold KM, Ross L and Cook JC (1987) TCDD: mechanisms of altered growth regulation in human epidermal keratinocytes. Banbury Report 25: Nongenotoxic mechanisms in carcinogenesis 25:247-255

Greenlee WF, Osborne R and Dold KM (1987b) Altered regulation of epidermal cell proliferation and differentiation by 2,3,7,8-tetrachlorodibenzo-p-dioxin (TCDD). Rev Biochem Toxicol 8: 1-35

Greenlee WF, Sutter, TR and Marcus C (1994) Molecular basis of dioxin action on rodent and human target tissues. In Receptor-Mediated Biological Processes: Implications for Evaluating Carcinogenesis, Spitzer HL, Slaga, TJ, Greenlee WF, McClain M (eds) Recent Prog Clin and Biol Res, Wiley-Liss, Vol 387, pp. 47-57

Gudas LJ (1992) Retinoids, retinoid-responsive genes, cell differentiation, and cancer. Cell Growth and Differentiation 3:655-662

Guilarte TR, Miceli RC, Altmann L, Weinsberg F, Winneke G and Wiegand H (1993) Chronic prenatal and postnatal Pb^{2+} exposure increases $[^3H]$MK801 binding sites in adult rat forebrain. Eur J Pharmacol-Environ Toxicol and Pharmacol Section 248:273-275

Gusousky F, Daly JW, (1990) Maitotoxin: A Unique Pharmacological Tool for Research in Calcium Dependent Mechanisms, Biochem. Pharm. 39(11):

1633-9
Hoffman BF, Bigger JT (1990) Digitalis and Allied Cardiac Glycosides. In The Pharmacological Basis of Therapeutics, Goodman Gilman A, Rall T, Nies A, and Taylor P (eds) , Pergamon Press, New York, pp. 814-840
Huff JE, Salmon AG, Hooper NK and Zeise L (1991) Long-term carcinogenesis studies on 2,3,7,8-tetrachlorodibenzo-p-dioxin and hexachlorodibenzo-p-dioxins. Cell Biol and Toxicol 7:67-94
Hunt P and Krumlauf R (1992) HOX codes and positional specification in vertebrate embryonic axes. Annu Rev Cell Biol 8:227-256
Johnson, ES (1993) Important aspects of the evidence for TCDD carcinogenicity in man. Environ Health Perspect 99:383-390
Jones PBC, Galeazzi DR, Fisher JM and Whitlock JP, Jr (1985) Control of cytochrome P_1-450 gene expression by dioxin. Science 227:1499-1502
Kao CY and Levinson SR (eds) (1986) Tetrodotoxin, Saxitoxin, and the molecular biology of sodium channel. Annals of New York Acad. Sci. Vol 479
Katzenellenbogen BS, Bhardwaj B, Fang H, Ince BA, Pakdel F, Reese JC, Schodin D and Wrenn CK (1993) Hormone binding and transcription activation by estrogen receptors: analyses using mammalian and yeast systems. J Steriod Biochem Molec Biol 47:39-48
Kuhar MJ, Ritz MC, Sharkey J, (1988) Cocaine Receptors on Dopamine Transporters Mediate Cocaine-Reinforced Behavior, in Mechanisms of Cocaine Abuse and Toxicity, Clouet D, Khurseed A, Brown R (eds) NIDA Res. Monograph, vol. 88
Madshus IH and Stenmark H (1992) Entry of ADP-ribosylating toxins into cells. Current Topics in Microbiol and Immunol 175:1-26
Nebert DW (1989) The Ah locus: genetic differences in toxicity, cancer, mutation, and birth defects. Crit Rev Toxicol 20:153-174
Nebert DW (1991) Proposed role of drug-metabolizing enzymes: Regulation of steady state levels of the ligands that effect growth, homeostasis, differentiation, and neuroendocrine functions. Mol Endocrinol 5: 1203-1214
Nebert DW (1994) Drug-metabolizing enzymes in ligand-modulated transcription. Biochem Pharmacol 47: 25-37
Nebert DW and Gonzalez FJ (1987) P450 genes: structure, evolution and regulation. Annu Rev Biochem 56:945-993
Nebert DW, Goujan FM, Gielen, JE (1972) Aryl hydrocarbon hydroxyalse induction by polycyclic hydrocarbons: simple autosomal dominant trait in the mouse. Nature 236:107-110
Okey AB (1990) Enzyme induction in the cytochrome P-450 system. Pharmac Ther 45:241-298
Olson EN, Burgess R and Staudinger J (1993) Protein kinase C as a transducer of nuclear signals. Cell Growth and Differentiation 4:699-705
Orti E, Bodwell JE and Munck A (1992) Phosphorylation of steroid hormone receptors. Endocrine Reviews 13:105-128
Poland A and Knutson JC (1982) 2,3,7,8-Tetrachlorodibenzo-p-dioxin and related halogenated aromatic hydrocarbons: examination of the mechanism of toxicity. Ann Rev Pharmacol Toxicol 22:517-554
Poland A, Glover E (1976) Stereospecific, high affinity binding of 2,3,7,8-tetrachlorodibenzo-p-dioxin by hepatic cytosol. J Biol Chem 251:4936-4946

Pollack SJ , Atack JR, Knowles MR, McAllister G, Ragan CI, Baker R, Fletcher SR, Iversen LL, and Broughton HB (1994) Mechanism of inositol monophosphatase, the putative target of lithium therapy. Proc Natl Acad Sci USA 91: 5766-5770

Quattrochi LC, Vu T and Tukey RH (1994) The human CYP1A2 gene and induction by 3-methylcholanthrene. J Biol Chem 269: 6949-6954

Reyes H, Reisz-Porszasz S and Hankinson O (1992) Identification of the Ah receptor nuclear translocator protein (Arnt) as a component of the DNA binding form of the Ah receptor. Science 256:1193-1195

Safe S(1990) Polychlorinated biphenyls, dibenzo-p-dioxins, dibenzofurans, and related compounds: environmental and mechanistic considerations which support the development of toxic equivalency factors. Crit Rev Toxicol 21:51-88

Savas U, Bhattacharyya KK, Christou M, Alexander DL and Jefcoate CR (1994) Mouse cytochrome P-450EF, representative of a new 1B subfamily of cytochrome P-450s. J Biol Chem 269: 14905-14911

Sekura R, Moss J, Vaughan M (eds) (1985) Pertusis Toxin, Academic Press, Orlando, FL, pp. 1-251

Serventi IM, Moss J and Vaughan M (1992) Enhancement of cholera toxin-catalyzed ADP-ribosylation by guanine nucleotide-binding proteins. Current Topics in Microbiol and Immunol 175:43-67

Shen ES and Whitlock JP, Jr (1992) Protein-DNA interactions at a dioxin-responsive enhancer. J Biol Chem 267:6815-6819

Snyder SH, Chang KJ, Kuhar MJ, Yamamura HI (1975) Biochemical Identification of the Mammalian Muscarinic Cholinergic Receptor, Fed. Proc. 34(10): 1915-21

Spink DC, Hayes CL, Young NR, Christou M, Sutter TR, Jefcoate CR and Gierthy JF (1994) The effects of 2,3,7,8–tetrachlorodibenzo-p-dioxin on estrogen metabolism in MCF-7 breast cancer cells: Evidence for induction of a novel 17β-estradiol 4-hydroxylase. J Steroid Biochem Molec Biol, in press.

Stahl BU, Beer DG, Weber LWD and Rozman K (1993) Reduction of hepatic phosphoenolpyruvate carboxykinase (PEPCK) activity by 2,3,7,8-tetrachlorodibenzo-p-dioxin (TCDD) is due to decreased mRNA levels. Toxicology 79: 81-95

Sutter TR and Greenlee WF (1992) Classification of members of the Ah gene battery. Chemosphere 25: 223-226

Sutter TR and Greenlee WF (1992b) Identification of a human dioxin-responsive cDNA (clone 1) as a new member of the cytochrome P450 superfamily. Organohalogen Compounds 10: 221-224

Sutter TR, Andersen ME, Corton JC, Gaido K, Guzman K and Greenlee WF (1991) Development of a molecular basis for dioxin risk assessment in humans. Banbury Report 35: Biological basis for risk assessment of dioxins and related compounds 35:379-388

Sutter TR, Guzman K, Dold KM and Greenlee WF (1991b) Targets for dioxin:genes for plasminogen activator inhibitor-2 and interleukin-1β. Science 254:415-418

Sutter TR, Tang YM, Hayes CL, Wo Y-YP, Jabs EW, Li X, Yin H, Cody CW and Greenlee WF (1994) Complete cDNA sequence of a human dioxin-inducible mRNA identifies a new gene subfamily of cytochrome P450 that maps to chromosome 2. J Biol Chem 269: 13092-13099

Swanson HI and Bradfield CA (1993) The Ah-receptor: genetics, structure and function. Pharmacogenetics 3:213-230

Takimoto K, Lindahl R and Pitot HC (1992) Regulation of 2,3,7,8-tetrachlorodibenzo-p-dioxin-inducible expression of aldehyde dehydrogenase in hepatoma cells. Arch Biochem Biophys 298: 492-497

Uteshev V, Busselberg D and Haas HL (1993) Pb^{2+} modulates the NMDA-receptor-channel complex. Naunyn Schmiedebergs Arch Pharmacol 347:209-213

Walters MR (1992) Newly identified actions of the vitamin D endocrine system. Endocrine Reviews 13:719-764

Weinstein IB (1991) Nonmutagenic mechanisms in carcinogenesis: role of protein kinase C in signal transduction and growth control. Environ Health Perspect 93:175-179

Whitlock JP, Jr (1993) Mechanistic aspects of dioxin action. Chem Res Toxicol 6:754-763

Wilhelmsson A, Cuthill S, Denis M, Wikstrom, A-C, Gustafsson J-A, and Poellinger L (1990) The specific DNA binding activity of the dioxin receptor is modulated by the 90kd heat shock protein. EMBO J 9:69-76

PERSPECTIVE ON THE USE OF STRUCTURE-ACTIVITY EXPERT SYSTEMS IN TOXICOLOGY

Herbert S. Rosenkranz[1] and Gilles Klopman[2]

[1]Department of Environmental and Occupational Health
University of Pittsburgh
Pittsburgh, PA 15238

and

[2]Department of Chemistry
Case Western Reserve University
Cleveland, OH 44106

NATO ASI Series, Vol. H 93
Modulation of Cellular Responses in Toxicity
Edited by C. L. Galli, A. M. Goldberg, M. Marinovich

SUMMARY

MULTICASE and META, two computer–based expert systems, are described and applications illustrated.

MULTICASE, in conjunction with specific data bases, can be used to predict the biological activity (*e.g.* toxicity) of yet untested molecules as well as to gain mechanistic insight. Heretofore, in excess of 70 toxicity data bases have been successfully analyzed by MULTICASE.

META is an expert system that has assimilated 665 enzymic and 286 spontaneous reactions. It can be used to identify biotransformation pathways and putative metabolites.

When operated in tandem, META and MULTICASE allow the prediction of the toxicity of parent molecules as well as of their metabolites.

INTRODUCTION

Ultimately the application of the knowledge derived from the study of toxicological phenomena should be to reduce the uncertainty in the estimation of the risk to humans of exposure to physical, chemical and even biological agents. To achieve this aim we usually approach our discipline by investigating dose–response relationships and by probing the basis of the mechanism of action of specific toxicants. Indeed, the recent explosion in our understanding of cellular processes at the molecular level as well as the development of molecular probes capable of reflecting dose to the target tissue has permitted great progress in our discipline. However, societal expectations of practitioners of our discipline have been extended further to include the early detection of diseases of environmental and occupational etiology so as to lead to early preventative measures. In this aspect, this may lead to the development and application of early markers of disease such as possibly the detection of the activation of oncogenes, the inactivation of suppressor genes or difference in the levels or ratios of cytokines. Additionally, we are being asked to participate in the development of environmentally benign and human friendly new products and technologies as well as in the remediation of contaminated industrial and military toxic waste sites.

We are assigned these tasks amidst an uncertain global economy and at a time when past expenditures and the societal benefits derived therefrom are undergoing close scrutiny. Thus, recently I was questioned by members of the legislative branch of the U.S. government about the recent costs to the United States taxpayer of the various federally supported toxicological assay programs. My assessment of approximately U.S. $1.4 billion was considered by some of my colleagues to be a low estimate. I was then asked what the scientific community had learned as a result of these expenditures and whether I foresaw a need for the continuation of these expenditures at the current level if the purpose is to protect public health.

Actually, looking at the data resulting from these systematic bioassay efforts from the perspective of an experimentalist interested in structure–activity relationships (SAR), my answer was that the availability of reliable data had made possible the application of recently developed computer–based structure–activity relational expert systems to the understanding of toxicological phenomena and had even permitted the prediction of the properties of yet untested chemicals. I also emphasized that such computer–based expert systems can only be used as aids in conjunction with human experts and that they could not replace altogether toxicological testing. On the other hand, they could be used to help deploy our resources more efficiently by identifying classes of chemicals which have as yet not been tested as well as to prioritize chemicals and technologies for safety containment, regulatory action and/or remediation.

At this time, there is a consensus that present computer–based predictive approaches have reached a limit in their development. The rate–determining factors derive primarily from a lack of sufficient reliable data and from oversimplification in the interpretation of the results of bioassays. The latter point is exemplified by the fact that we assign the same attributes of "carcinogenicity" to a chemical that causes renal neoplasias in male rats, or leukemia to female mice, or cancers in multiple organs of female and male mice and rats. This consolidation of possibly different phenomena reflects lack of understanding on our part of the fundamental processes responsible for carcinogenicity. Accordingly, we are limited in identifying the attributes that are suitable for analysis by expert systems.

Still, even with these qualifications, there are a number of applications in which the newer computer–based structure–activity relational expert systems have shown great potential. Some of these are discussed further herein.

The availability of validated SAR models predictive of toxicological endpoints would capitalize on the expenditures already made in the development, validation and systematic testing programs such as the extensive efforts of the U.S. National Toxicology Program. The demonstration of such capabilities would be highly effective in justifying to governmental representatives past as well as hopefully future expenditures. Thus, in fact, as mentioned earlier, the current SAR models developed and validated by us which we use routinely in mechanistic and predictive studies are based upon data bases which are worth approximately U.S. $1.4 billion.

The use of such SAR models, once the data bases have been created and the predictive computer algorithms developed, is both economical and rapid and has a number of attractive features. Thus, such SAR models are useful in regulatory contexts. In fact, examination of SAR, albeit it may be at a relatively superficial stage, is already part of the regulatory process in many countries. Additionally, SAR models are able to provide rapidly a potential toxicological profile of chemicals as yet untested (Table 1). Thus, they are able to provide guidance in prioritizing the large backlog of chemicals in commerce and industry that are, as yet, untested (National Academy of Sciences, 1984). The same approach is applicable to the selection of toxic waste sites for remedial action.

SAR has been demonstrated to be an effective tool for selecting candidate chemicals in the early product development phase. Moreover, the expert system available to us (CASE/MULTICASE) has the ability to maximize the beneficial effects (such as therapeutic activity) and minimize unwanted side effects (such as toxicity). Obviously, SAR approaches have the potential for satisfying societal concern by providing an efficient and reliable alternative to extensive animal testing. Finally, for the experimentalist, SAR expert systems provide the means for generating testable hypotheses related to the mechanism of toxicological activity.

The fundamental concept underlying SAR is that there is a structural basis for the biological and toxicological activities of molecules. However, given the number of possible permutations, the moieties responsible for activity are not always readily recognized. Additionally, the unsystematic accumulation of toxicological data that need to be assessed are well beyond human capability for analysis. In order to fulfill the criteria and functions listed above, the SAR paradigm must provide mechanistic insight as well as be predictive of the activity of molecules as yet untested. The SAR approach (CASE/MULTICASE) described herein is a knowledge–based expert system, *i.e.* it derives all of its information from a "learning set" (*i.e.* data base) composed of chemical structures and indices of their biological activity. While the activity index could be a binary one, *i.e.* active or inactive, maximum information and predictivity is achieved when the activity is expressed on a continuous scale using units such as mg/kg/day, millimoles/kg/day, minimal or 50% inhibitory concentrations (expressed in μg/ml). If activity is given in a continuous scale, it is one of the functions of the "human expert" to select the cut–off between active and marginally active, and between marginally active and inactive chemicals.

One of the most important functions in developing a "learning set" is the establishment, *a priori*, of rigid criteria for exclusion and inclusion of data. This requires strict adherence to a predetermined experimental protocol, inclusion of concurrent positive and negative controls, statistical analyses of the significance of the results, etc. Unless such criteria are determined before a data base is assembled, the whole SAR modeling exercise may result in a time–consuming and labor intensive wasteful and futile exercise.

In assembling a date base, it is most useful to determine both the intra– as well as inter–laboratory reproducibility of the toxicological assay under consideration. Obviously, an SAR model cannot be expected to be more predictive of the activity of yet untested chemicals than the reproducibility of the assay upon which it is based. Thus, under standard conditions, using coded chemicals, the interlaboratory reproducibility of the *Salmonella* mutagenicity assay has been determined to be approximately 85% (Piegorsch and Zeiger, 1991). Accordingly, the predictive accuracy of the SAR model based upon that data base cannot be expected to be better than 85%. In fact, it was approximately 82% (Klopman and Rosenkranz, 1992).

The SAR system CASE/MULTICASE described herein is one that was developed in these laboratories (Klopman, 1984, 1992; Klopman and Rosenkranz, 1984). This method possesses a number of features that are uniquely appropriate to the requirements listed above and some of these are discussed herein.

Because it is completely automatic, the CASE/MULTICASE program is independent of operator bias. This is an important consideration in the study of toxicological phenomena, as frequently their mechanistic bases are as yet unknown. Thus, the application of MULTICASE to a toxicological data base may reveal structural features which suggest sites for metabolic activation to ultimate toxicants or which are suggestive of ligand binding requirements. Additionally, a comparison of the structural features identified as associated with apparently different toxicological phenomena may suggest mechanistic commonalities between them (Rosenkranz and Klopman, 1993a; Rosenkranz *et al.*, 1994b).

One of the strengths of the MULTICASE system is its ability to recognize moieties that are larger than the ones commonly "visualized" by the human expert but which may be of significance biologically. We know, for example, that receptors, the active sites of enzymes or immunoglobins, recognize moieties that are much larger than those usually visualized by the chemist. Additionally, once MULTICASE recognizes the structural features (biophores) suggestive of biological activity, it also identifies the structural as well as physical chemical (solubility, octanol:water partition coefficient, HOMO, LUMO and interatomic distances) properties (modulators) which contribute to the likelihood that the activity is expressed and which also determine the extent of that activity, *i.e.* the potency (Klopman and Rosenkranz, 1994a,b).

An added advantage of CASE/MULTICASE is its ability to handle non–congeneric data bases, *i.e.* the learning set can include a wide variety of chemical species as long as they all have been studied in a well–described toxicological assay system and the experimental data used meet the pre–determined criteria for inclusion.

A new feature of the CASE/MULTICASE program is the inclusion of an expert system (META) that has extensive knowledge of a wide spectrum of enzymic as well as of spontaneous chemical reactions (Klopman and Rosenkranz, 1994a,b). Thus, META when coupled to the CASE/MULTICASE system can not only generate putative metabolites but also enable an examination of their potential toxicological effects.

Altogether, MULTICASE has been applied successfully to over 70 data bases (Klopman and Rosenkranz, 1994b), some of which are listed in Table 1. This, in turn, has led to the ability of MULTICASE to generate rapidly and in a cost–efficient fashion the potential toxicological profiles of a variety of molecules (*e.g.* Table 1).

MATERIALS AND METHODS

Data Base:

For illustrative purposes, for the present study, we used a data base of the maximum tolerated doses (MTD) in rats of a series of 284 chemicals (Table 2). The protocol used was that of the U.S. National Toxicology Program (NTP). The chemicals used in this study were those that were also tested by the NTP for carcinogenicity in rodents (Ashby and Tennant, 1991). We are grateful to Dr. Raymond Tennant of the National Institute of Environmental Health Sciences, Research Triangle Park, N.C. (U.S.A.) for making these data available to us. An earlier SAR study of this MTD data base using an earlier version of the CASE program has been reported (Rosenkranz and Klopman, 1993b).

For the present study, the reported MTD values for routes of administration other than by gavage were transformed to gavage equivalent doses using previously determined values (Brown and Ashby, 1990; Gold *et al.*, 1984, 1986). These MTD values, expressed in mg/kg/day were converted (by the CASE program) into mmoles/kg/day. For the purpose of the present exercise, MTD values in excess of 1.0 mmol/kg/day were considered inactive (*i.e.* non–toxic), those in the range of 0.9–1.0 mmol/kg/day as marginally toxic and those molecules with MTD values less than 0.9 mmol/kg/day as toxic. Using the equation

$$\text{CASE Activity} = 218.54 \times \log(1/\text{Activity}) + 20 \qquad \text{(Equation 1)}$$

these MTD values were transformed into arbitrary CASE Activity units wherein inactive molecules were in the range of 10–19, marginally toxic molecules in the range of 20–29 and toxic molecules between 30 and 99 CASE units (Table 2).

Expert System: CASE/MULTICASE:

The CASE methodology has been described on a number of occasions (Klopman, 1984; Klopman *et al.*, 1985a,b, 1987, 1990; Klopman and Rosenkranz, 1984, 1994a,b; Rosenkranz *et al.*, 1984). Basically, CASE selects its own descriptors automatically from a learning set composed of active and inactive molecules. The descriptors are easily recognizable single, continuous structural fragments that are embedded in the complete molecule. The descriptors consist of either activating (biophore) or inactivating (biophobe) fragments. Each of these fragments is associated with a confidence level and a probability of activity which is derived from the distribution of these biophores and biophobes among active and inactive molecules. Once the training set has been assimilated, CASE can be queried regarding the predicted activity of molecules of unknown activity. Thus entry of an unknown chemical will result in the generation of all the possible fragments ranging from 2 to 10 atoms accompanied by their hydrogens and these will be compared to the previously identified biophores and biophobes. On the basis of the presence and/or absence of these descriptors, CASE predicts activity or lack thereof.

The MULTICASE program used in this study is a recent improvement of the CASE methodology (Klopman, 1992; Klopman and Rosenkranz, 1994a,b; Rosenkranz *et al.*, 1993). The initial procedure, that of fragmentation and identification of fragments associated with activity is identical to the CASE methodology. However, upon completion of these analyses,

MULTICASE selects the most important of these fragments as a biophore (see Table 3), *i.e.* the functionality that is responsible for the experimentally observed activity of the molecules that contain it. MULTICASE then, using the molecules containing this biophore (or set of related biophores), will use them as a learning set to identify the chemical properties (*i.e.* structural fragments) or physical chemical properties (log P, solubility , quantum mechanical, interatomic distance) that modulate the activity of the initially identified biophore (see Tables 4 and 5). This will result in a QSAR equation for this subset of molecules. If the data set is a congeneric one, the single biophore and associated modulators may explain the activity of all the molecules in the data base and the program will exit. With non–congeneric data bases, this will usually not occur and there will be a residue of molecules not explained by the single biophore and associated biophobes. When this happens, the program will remove from consideration the molecules already explained by the previous biophore and it will search for the next biophore (and associated modulators). The process is iterated until all of the molecules in the original data base have been explained. In fact, this results in a multiple CASE analysis wherein the molecules in the data base have been reclassified into logical subsets.

The resulting list of biophores (Table 3) is then used to predict the activity of yet untested molecules. Thus, upon submission for evaluation, MULTICASE will determine if the unknown molecule contains a true biophore. If it does not, the molecule will be assumed to be inactive unless it contains a group that resembles chemically one of the biophores, in which case it will be flagged. When the molecule contains a biophore, the presence of modulators for that biophore will be investigated (*e.g.* Figures 1–3 and Tables 4–6). MULTICASE will then make qualitative as well as quantitative predictions of the activity of the unknown molecule (Figures 1–3).

Obviously, while biophores are the determining structures, the modulators may determine whether the biological potential of the chemical is expressed. The modulators may be restricted to physical chemical properties, or they may consist of structural and/or physical chemical properties. Thus, biophore 1 (Table 3) is derived from 15 molecules, 14 of which are toxic (*i.e.* MTD < 0.9 mmol/kg/day). Thus, presence of the biophore in a molecule is associated with a 93% probability of toxicity (Figure 1). The modulator associated with this biophore is related to the ratio of the number of copies of biophore 1 to the molecular weight of the molecule by the equation –21.9 [Ln No. Biophore/mol wt] where Ln is the natural logarithm.

Thus, hexachloroethane (Figure 1) has a 93% probability of toxicity and moreover the presence of the biophore is associated with –25.7 CASE Activity units. However, the modulating activity contributes

$$-21.9 \; [\ln 6/236.79 = -21.9 \; [-3.68] = 80.3 \text{ CASE Activity units}$$

for an overall activity of 55 CASE units, which using Equation 1 corresponds to an MTD value of 0.53 mmol/kg/day or 125 mg/kg/day.

Similarly, biophore 12 (Table 3), derived from molecules Nos. 24, 269, 273, 276, 277, 279 and 283 (Table 3) is associated with an 86% probability of toxicity (Figures 2 and 3). However, the expression of this potential is modulated solely by the factor –2.3 $(\log P)^2$. Thus based upon this physical chemical property, tris(2–chloroethyl)phosphate is predicted to be extremely active (Figure 2) while tris(2–ethylhexyl)phosphate is predicted to have negligible activity (Figure 3).

On the other hand, biophore 4 (Table 3) derived from 15 molecules, 80% of which are active, is modulated by 3 structural modulators which, if present, cause a decrease in potency as well as by the electronegativity of the molecules (HOMO + LUMO/2) (Table 4). Thus, succinic acid–2,2–dimethylhydrazine, which contains biophore 4, has an 80% probability of activity (Figure 4). However, the concomittant presence of modulator 1 (COOH) (Table 4) and the fact that the contribution of (HOMO + LUMO/2) is 32.7 (–0.20) = –6.6 results in an overall prediction of no or only negligible activity (*i.e.* 10 CASE units) (Figure 4).

An overview of the biophores and modulators associated with the ability to induce systemic toxicity at levels below 0.9 mmoles/kg/day suggests a number of possibly different mechanisms of action. Thus as indicated earlier, biophore 1 appears to be related to the degree of halogenation per unit of molecular weight. Similarly, biophore 18 is modulated by a 2–dimensional biophore wherein the distance between halogen atoms is 2.6 Å. As a matter of fact, that descriptor which is present in 28 molecules (molecules 29, 96, 99, 100, 107, 111, 114, 116, 117, 118, 119, 120, 131, 147, 148, 180, 183, 196, 226, 227, 269, 280, 226, 129, 226, 10, 129, 191), 25 of which are toxicants, is the most significant descriptor found (p = 7.6 x 10^{-6}). These findings suggest that an important determinant of systemic toxicity by halogenated molecules involves interatomic distances and branching. This suggests a possible receptor site or possibly a channel for penetration of these molecules. These findings indicate directions for further experimentation.

The modulators associated with biophores 2–7 are identical (Table 4). Examination of the molecules containing these biophores suggests that they are electrophiles or can be metabolized to electrophiles (*i.e.* N–methylarylamines, reactive alkylbromides, etc.). It is not surprising, therefore, that electrophilicity has a modulating influence (Table 4). Similarly, biophore 8 which describes a p–diaminoarene, presumably capable of being metabolized to an electrophilic species, will be "deactivated" by nucleophilicity (*i.e.* HOMO energy) (Table 5).

Biophores 11 and 12 are both modulated by –2.3 (log P)^2 indicating that the greater the lipophilicity, the lower the toxicity; hydrophilicity favors the systemic toxicity of these molecules.

Biophore 13 describes a p–substituted chlorobenzene derivative which is modified by structural and interatomic distances (Table 6). Thus, the 6.9 Å requirement is seen in chlorpropamide, chlorobenzilate and dicofol, again suggesting specific spatial constraints associated with toxicity.

A systematic analysis of the other biophores reveals additional modulating effects by specific structural determinants as well as modulating effects due to solubility in water.

Thus overall, MULTICASE has identified biophores and modulators which account for 84.1% (χ^2 = 129) of the activity displayed by the molecules in the learning set. Many of these can be used to generate experimentally testable hypotheses of the activity of yet untested molecules.

META Metabolism Program:

The CASE/MULTICASE program is adept at identifying the structural basis of toxicological activity and even pinpointing the portion of the molecule that is in need of metabolic activation. Thus, in an earlier study it identified the bay–region of PAH as the site of metabolism (Rosenkranz *et al.*, 1985; Rosenkranz and Klopman, 1988). It is up to the human expert then to identify the actual metabolite. In some instances this may be quite simple. Thus, in a recent preliminary SAR study involving the developmental toxicity of a small group of chemicals, we found (Takihi *et al.*, 1994) in a comparison of the developmental toxicity of organic acids and their esters in mice and humans that while in mice both acids and their esters were developmental toxicants, in humans only the acids were predicted to be developmental toxicants while the esters appeared to lack that activity. Obviously, this suggests the presence of specific esterases in the mouse and their absence in humans. Such a hypothesis is readily testable experimentally. Moreover, the relevance of this observation would need to be confirmed using a greatly expanded data base.

However, in general, the metabolites may be unknown or partially unknown and we may wish not only to identify putative metabolite products but also to be able to predict their toxicological profiles. To accomplish this, a companion expert system META was developed (Klopman and Rosenkranz, 1994a,b). This program, as has been mentioned earlier, has a dictionary of 665 enzyme catalyzed as well as 286 spontaneous reactions. Thus, the combination of MULTICASE with META allows the identification of putative metabolites as well as a determination of their potential for toxicity.

To illustrate the usefulness of META, we applied it to a study of the possible basis of the carcinogenicity of phenylbutazone (Figure 5). In rodent carcinogenicity bioassays, this chemical was shown to cause cancers in both mice and rats. In general, this is a property exhibited by "genotoxic" carcinogens (Ashby and Tennant, 1991). However, phenylbutazone is neither a *Salmonella* mutagen nor does it exhibit "structural alerts" for genotoxicity, two properties commonly associated with "genotoxic" carcinogens (Ashby and Tennant, 1991). Nor are any of the molecules which share a biophore with phenylbutazone mutagens (Table 7). These findings suggest that either phenylbutazone, in spite of its carcinogenic spectrum, is not a "genotoxic" carcinogen or that in rodents (and perhaps humans) it is metabolized to a mutagen but that this biotransformation cannot occur under the experimental procedures used for the *Salmonella* mutagenicity assay.

In order to explore the latter hypothesis, we used META to generate a series of putative metabolites of phenylbutazone (Figures 6 and 7). All of these were analyzed by MULTICASE for potential mutagenicity in *Salmonella* (using a large *Salmonella* mutagenicity data base generated under the aegis of the U.S. National Toxicology Program). None of the 120 metabolites tested showed any potential for mutagenicity in *Salmonella*. This then reinforces the conclusion that the carcinogenicity of phenylbutazone is not due to a "genotoxic" mechanism of action. This may be of relevance to the assessment of risks to humans, as some investigators suggest that "non–genotoxic" carcinogens present less of a risk to humans than "genotoxic" ones (Andersen *et al.*, 1990; Ashby and Morrod, 1991; Netherlands, 1980; Williams, 1987, 1990; Wilson, 1989).

Informational Content:

In earlier studies we had demonstrated that the predictivity of the SAR model was dependent upon the number of chemicals present in the data base as well on the functionalities represented therein (Rosenkranz *et al.*, 1991). To evaluate the "informational content" of a data base, we assembled a collection of approximately 5,000 chemicals representative of the "universe of chemicals". We took advantage of one of the features of the CASE/MULTICASE program which involves recognition of structures which are "unknown" to the SAR model, *i.e.* they are not represented in the "learning set". The presence of such unknown moieties (Figures 8 and 9) decreases the predictivity of the SAR model (Rosenkranz *et al.*, 1991). Thus, enumeration of the number of predictions accompanied by "warnings" of the presence of "unknown" moieties provides a measure of the "informational content" of the data base (Takihi *et al.*, 1993a).

Thus, it has been demonstrated (Rosenkranz *et al.*, 1991; Takihi *et al.*, 1993a,b) that the informational content as well as the predictivity of a data base is a function of both the size of the learning set and the diversity of chemical moieties present therein.

Data bases, however, are rarely the result of a random testing program. They reflect either the bias of the selection process or the fact that they were assembled to test specific hypotheses, such as "carcinogens are mutagens" (Clayson and Arnold, 1991). Data bases containing the same number of chemicals may have different informational contents. Thus, while the data base of sensory irritation in mice and the one of induction of chromosomal aberrations in cultured cells each contain approximately the same number of chemicals (Table 8), the informational content of the chromosomal aberration data base is significantly greater than that of the structural alert data base. The net result is that the sensory irritation SAR model is restricted in its applicability to the spectrum of the universe of chemicals, *i.e.* it is less predictive of chemicals in general.

However, because MULTICASE also identified the structural moieties that are absent in the data base (Figure 8), this can be used to compile the structures most frequently absent in the SAR model (Table 9). Accordingly, if there is a desire to increase the applicability of the data base to a broader spectrum of chemicals (*i.e.* to increase the informational content), this can be accomplished by including in the bioassay chemicals containing the major structural moieties heretofore absent from this model and thereby increasing predictivity (Takihi *et al.*, 1993c).

Approach to Understanding the Mechanistic Basis of Systemic Toxicity as Expressed by the MTD:

One of the useful capabilities of CASE/MULTICASE is its ability to compare the structural determinants associated with different data bases. A number of studies have indicated that the greater the structural overlap between data bases, the greater the mechanistic similarity. Thus, in previous studies we found that the structural determinants associated with carcinogenicity in rodents overlapped partially with those identified with mutagenicity in *Salmonella* and with those identified with cellular toxicity, yet there was no significant overlap between the determinants of mutagenicity and cellular toxicity (Rosenkranz and Klopman, 1994; see also Rosenkranz *et al.*, 1994b). These results were taken to support the hypothesis that carcinogenicity could be the result of DNA–damage (*i.e.* a "genotoxic" event) and/or cell toxicity presumably leading to cell proliferation.

In the present study we analyzed the overlaps between the structural determinants associated with MTD in rats and other toxicological endpoints. Some of the results are summarized in Table 10. These results indicate that there was a significant overlap (29%) between the structural determinants associated with the MTD and those associated with cell toxicity. This clearly indicates that systemic toxicity may result from injury at the cellular level. On the other hand, there was no or only little overlap between structural determinants of MTD and those identified as associated with either $\alpha 2\mu$–induced nephropathy in male rats, sensory irritation in mice and inhibition of human cytochrome P4502D (Table 10).

Obviously, analyses of the type described herein can lead to testable hypotheses related to mechanism of systemic toxicity.

CONCLUSIONS

Modern SAR techniques have been applied to a variety of data sets. Methods have been developed to measure as well as to improve the predictivity of such models. The point must be made, however, that to be most effective, human intelligence and expertise is needed not only to assemble and develop the data base but also to interpret the SAR projections as well as to lead to further refinements in applications. Finally, although the newer SAR techniques can be used to increase mechanistic understanding of toxicological phenomena, the ultimate use of SAR in regulatory contexts is a societal decision. Chemicals that will see widespread human exposure may be deemed to require experimental testing in addition to analysis by expert systems (Lave *et al.*, 1988).

ACKNOWLEDGMENTS

This investigation was supported by the U.S. Environmental Protection Agency (R818275) and Concurrent Technologies Corporation/National Defense Center for Environmental Excellence in support of the U.S. Department of Defense (Contract No. DAAA21-93-C-0046).

REFERENCES

Andersen, M.E., J. Higginson, D. Krewski, I.C. Munro, A.E. Pegg, H.S. Rosenkranz, K.R. Solomon, E. Weisburger, G.M. Williams and G.N. Wogan (1990) Safety assessment procedures for indirect food additives. An Overview. Regulatory Toxicology and Pharmacology, 12: 2–12.

Ashby, J. and R.S. Morrod (1991) Detection of human carcinogens. Nature, 352: 185–186.

Ashby, J. and R.W. Tennant (1991) Definitive relationships among chemical structure, carcinogenicity and mutagenicity for 301 chemicals tested by the U.S. National Toxicology Program. Mutation Research, 257: 229–306.

Brown, L.P. and J. Ashby (1990) Correlations between bioassay dose-level, mutagenicity to *Salmonella*, chemical structure and sites of carcinogenesis among 226 chemicals evaluated for carcinogenicity by the U.S. NTP. Mutation Research, 244: 67-76.

Clayson, D.B. and D.L. Arnold (1991) The classification of carcinogens identified in the rodent bioassay as potential risks to humans: What type of substance should be tested next? Mutation Research, 257: 91-106.

Gold, L.S., C.B. Sawyer, R. Magaw, G.M. Backman, M. deVeciana, R. Levinson, N.K. Hooper, W.R. Havender, L. Bernstein, R. Peto, M.C. Pike and B.N. Ames (1984) A Carcinogenic Potency Data Base of the standardized results of animal bioassays. Environmental Health Perspectives, 58: 9-319.

Gold, L.S., M. deVeciana, G.M. Backman, R. Magaw, P. Lopipero, M. Smith, M. Blumenthal, R. Levinson, L. Bernstein and B.N. Ames (1986) Chronological supplement to the Carcinogenic Potency Data Base: Standardized results of animal bioassays published through December 1982. Environmental Health Perspectives, 67: 161-200.

Houser, J.J. and G. Klopman (1988) A new tool for the rapid estimation of charge distribution. Journal of Computational Chemistry, 9: 893–904.

Klopman, G. (1984) Artificial intelligence approach to structure–activity studies. Computer Automated Structure Evaluation of biological activity of organic molecules. Journal of the American Chemical Society, 106: 7315–7321.

Klopman, G. and H.S. Rosenkranz (1984) Structural requirements for the mutagenicity of environmental nitroarenes. Mutation Research, 126: 227–238.

Klopman, G., R. Contreras, H.S. Rosenkranz and M.D. Waters (1985a) Structure–genotoxic activity relationships of pesticides: Comparison between the results of several short-term assays. Mutation Research, 147: 343–356.

Klopman, G., M.R. Frierson and H.S. Rosenkranz (1985b) Computer analysis of toxicological data bases: Mutagenicity of aromatic amines in *Salmonella* testers strains. Environmental Mutagenesis, 7: 625–644.

Klopman, G., M.R. Frierson and H.S. Rosenkranz (1990) The structural basis of the mutagenicity of chemicals in *Salmonella typhimurium*: The Gene–Tox Data Base. Mutation Research, 228: 1–50.

Klopman, G. and S. Wang (1991) A Computer Automated Structure Evaluation (CASE) approach to calculation of partition coefficient. Journal of Computational Chemistry, 12: 1025–1032.

Klopman, G. (1992) MULTICASE 1. A hierarchical Computer Automated Structure Evaluation program. Quantitative Structure–Activity Relationships, 11: 176–184.

Klopman, G. and H.S. Rosenkranz (1992) Testing by artificial intelligence: Computational alternative to the determination of mutagenicity. Mutation Research, 272: 59–71.

Klopman, G., S. Wang and D.M. Balthasar (1992) Estimation of aqueous solubility of organic molecules by the group contribution approach. Application to the study of biodegradation. Journal of Chemical Information and Computer Sciences, 32: 474–482.

Klopman, G. and H.S. Rosenkranz (1994a) Prediction of carcinogenicity/mutagenicity using MULTICASE. Mutation Research, 305: 33–46.

Klopman, G. and H.S. Rosenkranz (1994b) Toxicity estimation by chemical substructure analysis: The Tox II program. In: Decision Support Methodologies for Human Risk Assessment of Toxic Substances, in press.

Lave, L.B., F.K. Ennever, H.S. Rosenkranz and G.S. Omenn (1988) Information value of the rodent bioassay. Nature, 336: 631–633.

National Academy of Sciences (1984) Toxicity Testing: Strategies to Determine Needs and Priorities. National Academy Press, Washington, D.C.

Netherlands (1980) Health Council of the Netherlands, Report of the Evaluation of the Carcinogenicity of Chemical Substances. Government Printing Office, the Hague.

Piegorsch, W.W. and E. Zeiger (1991) Measuring intra–assay agreement for the Ames *Salmonella* assay. In: Lecture Notes in Medical Information, Statistical Methods in Toxicology (O. Rienhoff and D.A.B. Lindberg, eds.) Springer, Berlin, pp. 35–41.

Rosenkranz, H.S., G. Klopman, V. Chankong, J. Pet–Edwards and Y.Y. Haimes (1984) Prediction of environmental carcinogens: A strategy for the mid 1980's. Environmental Mutagenesis, 6: 231–258.

Rosenkranz, H.S., C.S. Mitchell and G. Klopman (1985) Artificial intelligence and Bayesian decision theory in the prediction of chemical carcinogens. Mutation Research, 150: 1–11.

Rosenkranz, H.S. and G. Klopman (1988) CASE, the Computer Automated Structure Evaluation system, as an alternative to extensive animal testing. Toxicology and Industrial Health, 4: 533–540.

Rosenkranz, H.S., N. Takihi and G. Klopman (1991) Structure activity–based predictive toxicology: An efficient and economical method for generating non–congeneric data bases. Mutagenesis, 6: 391–394.

Rosenkranz, H.S. and G. Klopman (1993a) Structural evidence for a dichotomy in rodent carcinogenesis: Involvement of genetic and cellular toxicity. Mutation Research, 303: 83–89.

Rosenkranz, H.S. and G. Klopman (1993b) Structural relationships between mutagenicity, maximum tolerated dose, and carcinogenicity in rodents. Environmental and Molecular Mutagenesis, 21: 193–206.

Rosenkranz, H.S., J. Pangrekar and G. Klopman (1993) Similarities in the mechanisms of antibacterial activity (Microtox® Assay) and toxicity to vertebrates. ATLA, 21: 489–500.

Rosenkranz, H.S., Y.P. Zhang and G. Klopman (1994b) Evidence that cell toxicity may contribute to the genotoxic response. Regulatory Toxicology and Pharmacology, in press.

Takihi, N., Y.P. Zhang, G. Klopman and H.S. Rosenkranz (1993a) An approach for evaluating and increasing the informational content of mutagenicity and clastogenicity data bases. Mutagenesis, 8: 257–264.

Takihi, N., Y.P. Zhang, G. Klopman and H.S. Rosenkranz (1993b) Development of a method to assess the informational content of structure–activity data bases. Quality Assurance: Good Practice, Regulation, and Law, 2: 255–264.

Takihi, N., H.S. Rosenkranz and G. Klopman (1993c) Identification of chemicals for testing in the rodent cancer bioassay. Quality Assurance: Good Practice, Regulation, and Law, 2: 232–243.

Takihi, N., H.S. Rosenkranz, G. Klopman and D.R. Mattison (1994) Structural determinants of developmental toxicity, in press.

Williams, G.M. (1987) Definition of a human cancer hazard. In: Non–genotoxic Mechanisms in Carcinogenesis, Banbury Report 25 (B.E. Butterworth and T.J. Slaga, eds.) Cold Spring Harbor Laboratory, Cold Spring Harbor, New York, pp. 367–380.

Williams, G.M. (1990) Screening procedures for evaluating the potential carcinogenicity of food–packaging chemicals. Regulatory Toxicology and Pharmacology, 12: 30–40.

Wilson, J.D. (1989) Assessment of low–exposure risk from carcinogenesis: Implications of the Knudson–Moolgavkar Two–Critical Mutation Theory. In: Biologically–Based Methods for Cancer Risk Assessment (C.C. Travis, ed.) Plenum Press, pp. 275–287.

TABLE 1

Summary of MULTICASE Predictions

NAME: Phenylbutazone CAS# 50339

SYSTEM	Expt Result	Prediction Overall	Prediction Percent
Rodent Carcinogenicity (NTP)	A	3	97.
Mouse Carcinogenicity (NTP)	3	3	66.
Rat Carcinogenicity (NTP)	3	3	71.
Rat Maximum Tolerated Dose (NTP)	0.3247	3	83.
Mouse Maximum Tolerated Dose	0.9728	3	67.
Mutagenicity: Salmonella	1	1	18.
Structural Alert	1	3	80.
Sister Chromatid Exchanges/CHO	1	3	70.
Chromosomal Aberrations/CHO	3	2	53.
Balb 3T3 Transformation		1	0.
Balb 3T3 Cytotoxicity		3	92.
Cytochrome P4502D Inhibitor		1	0.
Alpha 2mu Nephrotoxicity/Male Rat		3	67.
Unscheduled DNA synthesis		3	69.
Micronucleus (Bone Marrow *in vivo*)	3	3	96.
Human Reproductive Toxicant		2	57.
Rabbit Reproductive Toxicant		3	71.
Rat Reproductive Toxicant		3	80.
Mouse Reproductive Toxicant		3	99.
Sensory Irritation (Mouse)		3	75.
Eye Irritation		1	0.
Respiratory Allergen		1	0.
Contact Allergen		3	74.
Ecotoxicity/Minnow		3	66.
Somatic mutation & recombin Drosophila		1	0.
Biodegradability		3	93..

(1:negative; 2:marginal; 3:positive; - or -0-:null)

Carcinogenicity:	Male Rat:	-0-
	Female Rat:	Kidney
	Male Mouse:	Liver Carcinoma
	Female Mouse:	-0-

LUMO energy:	-0.8649
HOMO energy:	0.4174
Water Solubility:	-2.7
Log P:	3.0578

The potential activity of phenylbutazone, using some of the data bases available to us, was predicted using MULTICASE.

TABLE 2

Physical Chemical Properties and Activities of the Chemicals Present in Data Base

No.	Chemical	MolWt	Log P	WSolub	CASE Units
1	Benzofuran	118.14	2.81	0.84	18
2	p-Chloroaniline.HCl	127.57	1.89	1.15	99
3	3,3-Dimethylbenzidine,2HCl	212.30	2.94	-0.77	99
4	Pentaerythritol Tetranitrate	316.14	4.30	2.45	10
5	2,4-Dichlorophenol	163.00	2.93	0.97	10
6	l-Epinephrine Hydrochloride	151.21	1.30	2.28	99
7	p-Vinyl Toluenes	118.18	3.14	-0.47	99
8	m-Vinyl Toluenes	118.18	3.14	-0.47	99
9	Hydroquinone	110.11	1.22	3.64	94
10	Bromoform	252.75	3.04	1.06	42
11	N,N-Dimethylaniline	121.18	1.92	1.98	99
12	Diphenylhydramine Hydrochlorid	255.36	3.87	-1.17	99
13	Rhodamine 6G, HCl	442.56	6.31	-4.74	99
14	D-Limonene	136.24	3.46	-1.08	10
15	3,3'-Dimethoxybenzidine	244.30	1.99	1.04	99
16	Allyl glycidyl ether	114.15	0.30	1.90	20
17	o-Chlorobenzalmalononitrile	188.62	2.28	1.32	99
18	Iodinated glycerol	201.99	0.43	3.53	10
19	Benzaldehyde	106.13	1.60	1.36	10
20	Chloroacetophenone	154.60	2.54	1.22	99
21	N-Methylolacrylamide	101.11	-1.04	2.89	99
22	Tetranitromethane	196.03	0.09	1.07	20
23	Ethyl Chloride	64.52	1.51	2.21	99
24	Tris(2-chloroethyl)phosphate	285.49	2.07	-0.04	99
25	Phenylbutazone	308.38	3.06	-2.70	99
26	Furosemide	331.76	1.11	-0.19	99
27	Hydrochlorothiazid	299.76	-1.27	2.25	99
28	D-L-Amphetamine Su	135.21	2.24	0.39	99
29	Hexachloroethane	236.74	4.58	-0.46	57
30	Bromoethane	108.97	1.72	2.00	99
31	Pentachlorophenol	266.34	4.77	-1.26	99
32	Furfural	96.09	0.95	2.62	64
33	alpha-Methylbenzyl	122.17	1.88	1.85	10
34	4-Vinyl-1-cyclohex	140.18	0.94	0.88	99
35	Succinic Anhydride	100.07	0.58	3.36	20
36	Toluene	92.14	2.50	0.61	99
37	Nalidixic Acid	232.24	0.76	-0.26	57
38	Glycidol	74.08	-0.66	4.03	18
39	Dimethoxane	174.20	-0.01	2.37	10
40	Allyl Isovalerate	142.20	1.87	0.66	98
41	2-Aminoanthraquinone	223.23	2.05	1.96	10
42	3-Amino-9-ethylcarbazole Hydro	210.28	2.94	-1.67	99
43	1-Amino-2-methylanthraquinone	237.26	2.37	1.52	99
44	o-Anisidine HCl	123.16	1.12	2.37	10
45	Benzene	78.11	2.18	1.04	10
46	3-Chloromethylpyridine HCl	127.57	1.59	0.18	10
47	4-Chloro-m-phenylenediamine	142.59	0.99	1.76	10
48	C.I. Basic Red 9	287.37	3.49	-2.70	99
49	C.I. Disperse Yellow 3	269.31	3.36	1.06	10
50	Cinnamyl Anthranilate	253.30	3.54	-1.09	10
51	p-Cresidine	137.18	1.44	1.94	10
52	Cupferron	138.13	0.57	1.51	34
53	2,4-Diaminoanisole sulfate	138.17	0.22	2.98	10
54	2,4-Diaminotoluene	122.17	0.70	2.07	99
55	4-Chloro-o-phenylenediamine	142.59	0.99	1.76	10
56	1,3-Dichloropropene	110.97	2.45	1.35	95
57	Di(2-ethylhexyl)phthalate	390.57	6.05	-6.01	10

No.	Chemical	MolWt	Log P	WSolub	CASE Units
58	1,4-Dioxane	88.11	-0.34	2.76	10
59	HC Blue No.1	255.28	-0.17	3.41	92
60	Hexachlorodibenzodioxin-Isomer	390.87	7.75	-4.78	99
61	Hydrazobenzene	184.24	2.30	0.85	99
62	4,4'-Methylenebis(N.N'-dimethy	254.38	3.91	-0.30	99
63	4,4'-Methylenedianiline Dihydr	198.27	2.62	-0.48	99
64	2-Methyl-1-nitroanthraquinone	267.24	3.00	0.40	99
65	Michler's Ketone	268.36	3.01	1.26	99
66	1,5-Naphthalenediamine	158.20	1.66	0.79	99
67	Nithiazide	216.22	0.05	-0.73	99
68	Nitrilotriacetic Acid	191.14	-1.57	6.07	10
69	5-Nitroacenaphthene	199.21	3.83	-2.11	91
70	5-Nitro-o-anisidine	168.15	0.85	1.86	10
71	Nitrofen	284.10	5.04	-2.22	85
72	p-Nitrosodiphenylamine	198.23	2.54	-1.33	21
73	4,4'-Oxydianiline	200.24	2.29	1.00	99
74	Phenazopyridine.HCl	213.24	2.42	2.48	10
75	Phenesterin	616.76	10.00	-11.76	99
76	Polybrominated Biphenyl	627.62	9.03	-7.69	99
77	Reserpine	608.70	4.28	-3.02	99
78	Sulfallate	223.79	3.63	-0.53	99
79	2,3,7,8-Tetrachlorodibenzo-P-D	321.98	6.52	-3.29	20
80	4,4'-Thiodianiline	216.31	2.85	-1.00	10
81	o-Toluidine	107.16	1.60	1.46	10
82	2,4,6-Trichlorophenol	197.45	3.54	0.23	10
83	2,4,5-Trimethylaniline	135.21	2.24	0.60	99
84	11-Aminoundecanoic Acid	201.31	1.93	-0.50	10
85	Aniline.HCl	93.13	1.28	1.90	10
86	Azobenzene	182.23	4.52	1.30	99
87	Bis(2-chloro-1-methylethyl)eth	171.07	2.47	0.41	10
88	5-Chloro-o-toluidine	141.60	2.21	0.72	10
89	D and C Red 9	377.83	3.98	-0.56	99
90	3,3'-Dimethoxybenzidine-4,4'-d	296.29	0.45	0.46	10
91	2,4-Dinitrotoluene	182.14	1.96	-0.17	99
92	1,2-Epoxybutane	72.11	0.60	2.21	99
93	Estradiol Mustard	760.64	10.00	-11.20	99
94	5-Nitro-o-toluidine	152.15	1.33	0.95	99
95	4,4'-Sulfonyldianiline	248.31	0.97	0.09	99
96	1,1,2-Trichloroethane	133.41	2.74	1.13	64
97	Trifuralin	335.29	3.71	-2.33	26
98	Zearalenone	318.37	3.22	1.06	99
99	Captan	300.59	2.24	-1.26	42
100	Chlordane	409.78	8.37	-4.47	99
101	Chlorobenzilate	325.19	4.72	-1.84	99
102	Chlorothalonil	265.91	3.47	-0.41	10
103	4-Chloro-o-toluidine.HCl	141.60	2.21	0.72	10
104	CI Disperse blue 1	268.28	-0.65	3.78	38
105	m-Cresidine	137.18	1.44	1.94	10
106	Decabromophenyl Oxide	959.22	10.00	-11.15	10
107	Dichlorodiphenyldichloroethyle	318.03	7.20	-4.55	99
108	2,6-Dichlorophenylenediamine	177.03	1.61	1.02	10
109	Di(2-ethylhexyl)adipate	370.58	5.41	-5.86	10
110	N,N'-Diethylthiourea	132.23	0.60	3.32	99
111	Heptachlor	373.32	7.76	-3.92	99
112	5(6)-Nitrobenzimidazole	163.14	0.74	0.65	10
113	N-Nitrosodiphenylamine	198.23	3.44	-0.76	42
114	Pentachloroethane	202.30	3.97	0.11	48
115	Pivalolactone	100.12	0.91	2.40	10
116	1,1,1,2-Tetrachloroethane	167.85	3.35	0.56	10
117	1,1,2,2-Tetrachloroethane	167.85	3.35	0.68	61

Table 2, continued

No.	Chemical	MolWt	Log P	WSolub	CASE Units
118	Toxaphene	411.80	8.37	-3.65	99
119	Trichloroethylene	131.39	2.74	0.98	10
120	Aldrin	364.92	7.79	-4.48	99
121	Allyl isothiocyanate	99.16	1.18	2.37	99
122	3-Amino-4-ethoxyacetanilide	194.24	0.44	0.83	10
123	4-Amino-2-nitrophenol	154.13	0.53	2.56	62
124	2-Amino-5-nitrothiazole	145.14	-0.31	1.61	99
125	p-Benzoquinone Dioxime	138.13	-1.80	3.42	99
126	2-Biphenylamine Hydrochloride	169.23	3.20	-0.52	54
127	Butyl Benzyl Phthalate	312.37	4.45	-2.72	10
128	Chloramben	206.03	1.88	-0.20	10
129	Chlorodibromomethane	208.29	2.84	1.27	99
130	C.I. Vat Yellow 4	334.38	6.16	-3.02	38
131	Dicofol	370.49	6.87	-3.48	99
132	Isophorone	138.21	2.25	1.79	10
133	Melamine	126.12	0.71	2.99	10
134	3-Nitro-p-acetophenetide	224.22	1.07	-0.30	64
135	2-Nitro-p-phenylenediamine	153.14	0.11	1.99	74
136	Piperonyl Sulfoxide	324.49	2.93	-4.23	50
137	Succinic Acid-2,2-dimethylhydr	160.17	-1.31	3.55	10
138	Trimethylthiourea	118.20	0.28	4.59	99
139	Ziram	121.22	1.41	3.37	99
140	p-Anisidine.HCl	123.16	1.12	2.37	10
141	1,2,3-Benzotriazole	119.13	1.05	1.12	10
142	Bisphenol A	228.29	4.11	-0.23	99
143	p-Chloroaniline	127.57	1.89	1.15	99
144	Chlorobenzene	112.56	2.80	0.54	13
145	Diallyl Phthalate	246.27	2.85	-0.91	99
146	2,7-Dichlorodibenzo-p-dioxin	253.09	5.30	-1.80	10
147	Dieldrin	382.93	6.85	-3.60	99
148	Di(p-ethylphenyl)dichloroethan	307.27	7.25	-5.15	19
149	Dimethyl terephthalate	194.19	1.56	1.52	19
150	2,5-Dithiobiurea	150.22	-1.61	6.16	10
151	Eugenol	180.21	2.03	2.20	10
152	Fluometuron	232.21	2.69	-0.07	99
153	3-Nitropropionic Acid	119.08	0.15	2.90	99
154	Picloram	241.46	1.12	-0.99	10
155	Proflavin HCl	209.25	1.42	-0.96	99
156	Propyl Gallate	212.20	1.08	3.80	10
157	Tetrachlorodiphenylethane	285.60	6.59	-4.19	95
158	Acetohexamide	324.40	1.82	-1.69	10
159	Aldicarb	190.27	0.09	1.61	99
160	Anilazine	275.53	3.01	-2.50	99
161	o-Anthranilic Acid	137.14	0.65	1.29	10
162	L-Ascorbic Acid	176.13	-1.61	6.97	10
163	Benzoin	212.25	2.91	0.86	99
164	Butylated Hydroxytoluene	220.36	4.59	-1.73	10
165	n-Butyl Chloride	92.57	2.15	1.08	10
166	Calcium Cvanamide	42.04	-1.22	4.95	99
167	Caprolactam	113.16	0.54	0.53	10
168	Carbromal	237.10	1.07	-0.84	99
169	4-(Chloroacetyl)acetanilide	211.65	1.53	0.24	99
170	2-Chloroethanol	80.51	0.57	3.46	10
171	2-Chloroethytrimethylammonium	122.62	1.57	1.08	23
172	2-Chloromethylpyridine.HCl	127.57	1.59	0.18	10
173	2-Chloro-p-phenylenediamine Su	142.59	0.99	1.76	37
174	3-Chloro-p-toluidine	141.60	2.21	0.72	29
175	Chlorpropamide	276.74	1.73	-1.75	35
176	C.I. Acid Red 14	460.49	2.06	-1.24	10
177	Diarylanilide Yellow	629.51	7.79	-3.59	10

Table 2, continued

No.	Chemical	MolWt	Log P	WSolub	CASE Units
178	Dibenzo-p-dioxin	184.20	4.07	-0.32	10
179	1,2-Dichlorobenzene	147.00	3.41	-0.20	39
180	DDT	354.49	7.81	-5.21	99
181	N,N'-Dicyclohexylthiourea	240.41	3.17	-0.31	10
182	2,4-Dimethoxyaniline HCl	153.18	0.96	2.85	44
183	Endrin	380.91	6.85	-4.15	99
184	Ephedrine Sulfate	165.24	1.62	1.89	99
185	Ethionamide	166.25	0.03	0.49	52
186	EDTA (tri-Na salt)	278.22	-2.51	5.51	14
187	FD and C Yellow 6	410.43	0.77	0.47	10
188	Fenaminsulf	229.26	-1.05	4.27	99
189	HC Blue 2	329.36	-1.09	4.93	10
190	8-Hydroxyquinoline	145.16	1.46	0.71	39
191	Iodoform	393.73	3.84	-1.09	99
192	Lindane	290.83	5.86	-1.51	99
193	Lithocholic Acid	376.58	6.06	-4.20	99
194	Mannitol	182.17	-3.46	9.33	10
195	DL-Menthol	156.27	2.52	0.83	10
196	Methoxychlor	345.66	6.27	-2.77	99
197	Mexacarbate	222.29	1.76	1.06	99
198	N-(1-Naphthyl)ethylenediamine.	186.26	2.30	0.05	99
199	4-Nitroanthranilic Acid	182.14	0.38	0.78	10
200	1-Nitronaphthalene	173.17	3.19	-0.94	99
201	4-Nitro-o-phenylenediamine	153.14	0.11	1.99	99
202	Pentachloronitrobenzene	295.34	4.98	-2.94	10
203	Phenformin	205.26	-1.04	2.00	99
204	Phenol	94.11	1.70	2.46	10
205	p-Phenylenediamine.2HCl	108.14	0.38	2.51	95
206	1-Phenyl-3-methyl-5-pyrazolone	174.20	4.11	-0.04	10
207	N-Phenyl-p-phenylenediamine	184.24	2.30	0.47	99
208	1-Phenyl-2-thiourea	152.22	1.25	2.51	99
209	Phthalamide	164.17	-0.47	0.25	10
210	Phthalic Anhydride	148.12	1.86	1.82	10
211	Piperonyl Butoxide	338.45	2.57	-2.12	10
212	Sodium Diethyldithiocarbamate	149.28	2.05	2.24	59
213	Sulfisoxazole	267.31	-0.39	0.83	10
214	3-Sulfolene	118.16	0.21	1.82	10
215	2,3,5,6-Tetrachloro-4-nitroani	290.92	4.20	-1.72	99
216	Tetraethylthiuram disulfide	296.54	3.84	-0.09	99
217	Tolazamide	311.41	1.50	-2.66	10
218	Tolbutamide	270.35	1.76	-2.00	10
219	2,6-Toluenediamine Dihydrochlo	122.17	0.70	2.07	99
220	2,5-Toluenediamine sulfate	122.17	0.70	2.07	62
221	L-Tryptophan	204.23	-1.15	1.57	10
222	1,2-Dibromoethane	187.87	2.54	1.15	99
223	1,2-Dichloroethane	98.96	2.13	1.58	23
224	Hexachlorodibenzodioxin - Isom	390.87	7.75	-4.78	99
225	1,2-Dibromo-3-chloropropane	236.34	3.48	0.14	99
226	Bromodichloromethane	163.83	2.63	1.48	74
227	Chlorendic Acid	319.96	3.70	0.18	99
228	3-Chloro-2-methylpropene	90.55	2.15	1.18	10
229	1,4-Dichlorobenzene	147.00	3.41	-0.20	10
230	Diglycidyl Resorcinol Ether	222.24	1.27	0.97	99
231	Dimethylvinyl Chloride	90.55	2.15	1.60	10
232	Ethyl Acrylate	100.12	0.91	2.17	10
233	Nitrofurantoin	238.16	2.26	1.97	99
234	Nitrofurazone	198.14	2.37	2.28	99
235	p-Xylene	106.17	2.82	0.18	10
236	Benzyl Acetate	150.18	2.19	0.84	10
237	Monuron	198.65	2.25	0.22	99

Table 2, continued

No.	Chemical	MolWt	Log P	WSolub	CASE Units
238	2-Mercaptobenzothiazole	167.25	2.54	-0.86	10
239	C.I. Solvent Yellow 14	248.29	5.33	0.77	99
240	1,2-Dichloropropane	112.99	2.45	1.20	10
241	Malonaldehyde	72.06	-0.58	3.28	10
242	Methyl Carbamate	75.07	-1.00	3.67	10
243	2-Amino-4-nitrophenol	154.13	0.53	2.56	10
244	2-Amino-5-nitrophenol	154.13	0.53	2.56	10
245	C.I. Acid Orange 3	431.41	1.40	-2.74	10
246	Ampicillin Trihydrate	363.44	1.11	-1.30	10
247	H.C. Red 3	197.20	-0.19	2.63	10
248	4-Hexylresorcinol	194.28	3.15	0.38	61
249	Methyldopa	211.22	-1.52	3.50	10
250	Oxytetracycline	444.45	-0.35	5.73	10
251	N-Phenyl-2-naphthylamine	219.29	4.49	-1.85	40
252	Rotenone	394.43	4.34	-1.32	47
253	Benzyl Alcohol	108.14	1.56	2.11	10
254	C.I. Acid Orange 10	410.43	0.77	0.47	99
255	Erythromycin	733.95	-0.04	3.72	79
256	Geranyl Acetate (A)	198.31	3.16	-0.74	10
257	Geranyl Acetate (B)	182.26	3.30	0.40	10
258	Penicillin VK	350.40	1.21	-0.49	10
259	Phenylephrine Hydrochloride	167.21	0.82	3.45	99
260	Chlorpheniramine Maleate	274.80	3.90	-2.86	99
261	Tetracylcline.HCl	444.45	-0.35	5.22	10
262	m-Xylene	106.17	2.82	0.18	10
263	o-Xylene	106.17	2.82	0.18	10
264	Ethylbenzene	106.17	2.82	0.05	10
265	Fenthion	278.33	2.19	-0.97	99
266	Parathion	291.26	1.38	-0.65	99
267	Coumaphos	362.77	3.06	-1.90	99
268	Dimethoate	229.26	1.94	-0.14	99
269	Dichlorvos	220.98	1.28	1.99	99
270	Dioxathion	456.54	5.66	-3.86	99
271	Azinphosmethyl	317.33	3.31	-2.08	99
272	Malathion	330.36	3.28	-0.57	10
273	Tris(2,3-dibromopropyl)phosphate	697.65	6.12	-4.36	99
274	Methyl Parathion	263.21	0.74	0.48	99
275	Diazinon	304.35	0.77	-2.03	99
276	Trimethylphosphate	140.08	-0.73	3.55	51
277	Tetrachlorvinphos	365.97	4.43	-1.91	99
278	Malaoxon	314.30	1.80	0.01	99
279	Phosphamidon	299.69	0.94	1.09	99
280	Photodieldrin	380.91	6.85	-3.58	99
281	Dimethyl hydrogen phosphite	109.04	10.00	2.92	10
282	Dimethyl Morpholinophosphoram	195.16	-2.48	1.72	10
283	Tris(2-ethylhexyl) phosphate	434.65	6.00	-7.76	10
284	Dimethyl methylphosphonate	124.08	-1.60	2.64	10

Log P, the octanol:water partition coefficient, and solubility in water (WSolub) were calculated as described previously (Klopman and Wang, 1991; Klopman et al., 1992).

Solubility in water is expressed as log (moles solute/10^3 kg).

The activity in CASE units was derived from MTD values (transformed into mmoles/kg/day) using equation 1 (see text).

Table 2, continued

TABLE 3

Major Biophores Associated with the Ability to Induce Systemic Toxicity (MTD) in Rats at Doses Below 0.9 mmoles/kg/day

Fragment 1---2---3---4---5---6---7---8---9---10-------		N	Inactive	Marginal	Active	Av.Act.	Chemicals*
ɪt of BIOPHORE Attributes							
Cl -C -Cl		15	1	0	14	83.0	29,99,100,111,114,116,118,120,131,147,180,183,196,227,280
NH -CH3		8	0	0	8	98.0	6,59,138,159,184,197,259,268
N -CH2-		20	5	1	14	68.9	12,37,42,59,68,75,77,78,93,97,186,189,212,216,217,233,260,271,279,282
N -CH3		15	3	0	12	79.9	11,12,62,65,137,138,139,152,188,197,237,250,255,260,261
NH -CH2-		14	3	0	11	75.4	6,13,21,26,27,67,110,167,175,198,203,218,247,259
Br -CH -		5	0	0	5	82.6	10,129,225,226,273
Br -CH2-		4	0	0	4	99.0	30,222,225,273
NH2-C =CH -CH =C -NH2		5	0	0	5	61.0	104,135,173,205,220
CH =C -C =	<2-CH=>	9	2	0	7	74.0	3,15,17,48,90,98,107,126,177
CH =CH -C. =CH -		10	2	0	8	70.0	1,89,155,187,190,198,200,239,251,254
O -PS -O -	<2-O >	5	0	0	5	99.0	265,266,267,274,275
O -PO -O -	<2-O >	7	1	0	6	79.4	24,269,273,276,277,279,283
CH =CH -C =CH -CH =C -	<3-Cl >	11	1	0	10	84.0	2,101,107,131,143,157,175,180,229,237,260
CH =CH -C =CH -CH =	<3-N >	10	1	0	9	77.0	11,25,52,62,65,75,93,113,188,206
CH =C. -CO -C. =		4	1	0	3	61.0	41,43,64,130
O -C. =CH -C =C -		6	1	1	4	62.0	13,60,79,211,224,252
CH3-C =C -CH =C -		7	1	1	5	76.0	54,83,88,89,91,94,174
Cl -CH -CH -Cl		4	0	0	4	89.0	100,117,147,192
CH2-CH -CH -CH2-		4	0	0	4	99.0	34,75,93,193
CH =CH -C =C. -CH =		5	1	0	4	79.0	66,69,176,198,200
NH -C =CH -CH =CH -		10	2	1	7	67.0	61,72,152,160,177,198,207,208,245,251

Refers to chemicals listed in Table 2.

ɔphore 1 is illustrated embedded in hexachloroethane in Figure 1; biophore 12 is shown embedded in tris(2-chloroethyl)phosphate and tris(2-ethylhexyl)phosphate) in Figures ɪnd 3, respectively. Biophore 4 is shown embedded in succinic acid-2,2-dimethylhydrazine in Figure 4. Biophore 3 is shown embedded in metronidazole in Figure 9.

ː modulators associated with biophores 4, 8 and 13 are shown in Tables 4, 5 and 6, respectively.

TABLE 4

List of MODULATORS related to BIOPHORE 4: CASE Units

N -CH3 Constant= **19.1**

Fragment 1---2---3---4---5---6---7---8---9---10		Nr.
CO -OH	-41.3	1
SO2-NH -	-80.4	2
OH -CH2-CH2-	-22.5	3
(HOMO+LUMO)/2	32.7	

The derivation of biophore 4 is given in Table 3. These modulators are related to biophores 2–7.

HOMO and LUMO energies were calculated as previously described (Houser and Klopman, 1988).

Modulator 1 is shown embedded in succinic acid-2,2-dimethylhydrazine in Figure 4. Modulator 3 is shown embedded in metronidazole in Figure 9.

TABLE 5

List of MODULATORS related to BIOPHORE 8: CASE Units

NH2-C =CH -CH =C -NH2 Constant= **99.1**

Fragment 1---2---3---4---5---6---7---8---9---10		Nr.
NH2-C =C. -	-14.1	1
Cl -C =C -NH2	-48.4	2
HOMO coef.on HA2	-676.7	

The derivation of biophore 8 is given in Table 3.

TABLE 6

List of MODULATORS related to BIOPHORE 13: CASE Units

CH =CH -C =CH -CH =C - <3-Cl > Constant= **98.7**

Fragment 1---2---3---4---5---6---7---8---9---10			Nr.
2D [Cl -] <-- 6.9A --> [CO -]	generic	-31.9	1
2D [Cl -] <-- 6.9A --> [NH -]	generic	-31.9	2
CH =CH -C =CH -C =C -	<3-OH >	-37.7	3
Cl -C =C -Cl		-59.7	4
Cl -C =CH -CH =C -CH =CH -	<5-Cl >	-22.2	5

The derivation of biophore 13 is shown in Table 3.

TABLE 7

Derivation and Mutagenicity of Molecules Contributing to the Prediction of the Carcinogenicity of Phenylbutazone

```
CH2 -CH2
        \
         CH2 -CH
```

No. of Copies		Carcinogenicity	Mutagenicity
2	Di(2-ethylhexyl)phthalate	A	Negative
2	Phenesterin	A	Negative
1	Zearalenone	B	Negative
2	Di(2-ethylhexyl)adipate	C	Negative
1	Phenylbutazone	A	Negative
3	Tris(2-ethylhexyl)phosphate	D	Negative

A: Agents found to be carcinogenic to both mice and rats at one or more sites.

B: Agents found to be carcinogenic to only a single species at two or more sites in one or both sexes.

C: Agents found to be carcinogenic at only a single site in both sexes of a single species.

D: Agents found to be carcinogenic at only a single site in a single sex of a single species.

The biophore contributing to the carcinogenicity of phenylbutazone is derived from the molecules listed. The biophore is shown embedded in phenylbutazone in Figure 5.

TABLE 8

Informational Content of Various SAR Data Bases

Data Base	N*	Informational Content** (%)
Rodent Carcinogenicity (NTP)	283	37.9
Mouse Carcinogenicity (CPDB)	639	19.0
Rat Carcinogenicity (CPDB)	744	17.8
Salmonella Mutagenicity	916	14.1
Micronuclei (*in vivo*)	237	31.5
Sensory Irritation in Mice	226	68.4
Chromosomal Aberrations	223	38.2
Unscheduled DNA Synthesis	270	31.4
Drosophila Somatic Mutations	294	34.1

* Number of chemicals present in data base.

** Percentage of predictions accompanied by a "warning" about the presence of "unknown" moieties.

The informational content was ascertained by testing a panel of approximately 5400 chemicals in the various SAR models.

TABLE 9

Most Abundant Fragments Unknown to the Rat MTD Data Base

N	Fragment
61	*** CO -CH -CH3
57	*** N ≡C -C =
54	*** F -C =CH -
54	*** NH -C" -CH3
51	*** OH -CH -C -
44	*** O -C -CH -
43	*** COH-C =CH -
25	*** OH -CH2-C -
25	*** OH -CH -CH3
25	*** COH-C =C -
23	*** O^ -C -CH -
23	*** CO -CH -O -
23	*** N -CH2-CH =
23	*** NH2-CH -CH -
20	*** OH -CH -O -
20	*** N ≡C -CH2-
19	*** CO -C" -O -
19	*** CO -C" -CH2-
18	*** S -CH =CH -
18	*** NH -C -CH3
17	*** NH2-CH -CH3
15	*** NH -SO2-C =
15	*** N" -N =N -
15	*** NH -C" -CH2-

These fragments were identified by testing a panel of approximately 5400 chemicals with the MTD data base. Unknown fragments (see Figures 8 and 9) were enumerated.

TABLE 10

Structural Overlap Between MTD and Other Toxicological Activities

NUM	Fragment		Activity	S.I.	Cytox	α2μ	P450
1	CH3-N -		act				
2	CH3-NH -		act				
3	O^ -CH -		act				
6	Br -CH -		act				
7	Br -CH2-		act				
8	CO -N -		act		X		
9	CS -N -		act		X		
10	SO2-NH -		inact				
11	N -CH2-CH3		act				X
17	Cl -C -Cl		act				
18	CO -NH -CH -		inact		X		
19	CO -NH -CH3		act	X			
20	NO2-C =C. -		act				
21	CH =C -C =	<2-CH3>	act		X		
24	C =CH -CH =C. -		act				
27	CH =CH -C. =CH -		act		X		
29	CH =CH -C" -O -		act		X		
30	OH -C =C -CH =		inact				
32	Cl -C =C -Cl		act		X		
34	C =C -CH =C -C =		act		X		
37	C =CH -C =C -OH		inact				
40	C =CH -C =C -CH =C -		act	X	X		
41	CH =C -CH =CH -C =CH -		act		X		
42	CH =CH -C =CH -CH =C -		act		X		
43	CH =CH -CH =C -C. =CH -		act				X
44	NH2-C =CH -CH =C -NH2		act		X		
45	CH =CH -CH =C -CH =CH -	<4-N >	act		X		
Overlap				4.4%	28.9%	0%	4.4%

O^ indicates an epoxide; C. indicates a carbon atom shared by two rings. <2-CH3> indicates a methyl substituent on the second (non-hydrogen) atom from the left.

Overlaps between data bases are indicated by X. Overlaps are defined as complete identity or when one fragment is embedded in another one (e.g. NH2-C=CH vs. NH2-C=CH-CH).

S.I. indicates sensory irritation in mice; Cytox, cellular toxicity; α2μ, induced nephropathy in male rats; P450, inhibition of human cytochrome P4502D.

```
********************************************************************************
                              HEXACHLOROETHANE
********************************************************************************

The molecule contains the Biophore   (nr.occ.= 6):

            Cl  -C
                  \
                   Cl

     *** 14 out of the known  15 molecules ( 93%) containing such Biophore
         have an average MTD value in rats of  83. (conf.level=100%)
                                                   Constant is    -25.7
         Ln Nr.Bi/Mol.Wt.    = -3.68 ;          Nr.Bioph/MW contrib.is   80.3

** The probability that this molecule will induce systemic toxicity is 88.2% **

     ** The compound is predicted to be VERY active **
     ** The projected MTD value is   55.0     CASE units **
```

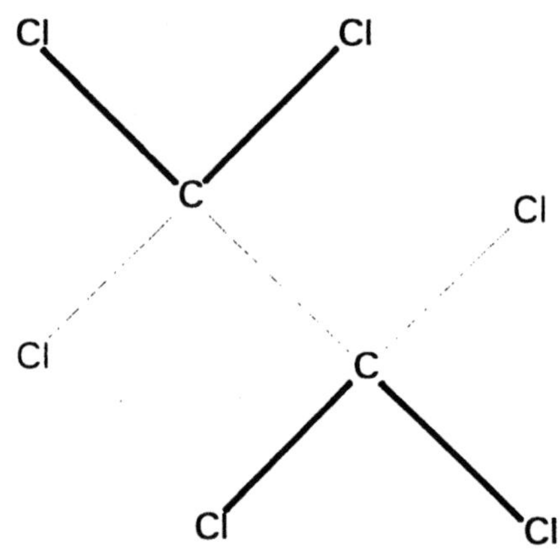

FIGURE 1: Prediction of the systemic toxicity in rats (i.e. MTD < 0.9 mmol/kg/day) of hexachloroethane. The probability of activity is due to the presence of the biophore Cl–C–Cl, which is present 6 times. Two of three are shown in bold. The potency is largely due to the modulator Ln Nr. Biophores/mol. wt. (see text).

```
********************************************************************************
                          TRIS(2-CHLOROETHYL)PHOSPHATE
********************************************************************************

The molecule contains the Biophore   (nr.occ.= 3):
                                  O
                                 /
              O    -PO
                                 \
                                  O

     ***  6 out of the known   7 molecules ( 86%) containing such Biophore
          have an average MTD value in rats of  79. (conf.level= 96%)
                                                     Constant is      99.5
          Log partition coeff.=  2.07 ;          LogP contribution is     -9.7

** The probability that this molecule will induce systemic toxicity is 77.8% **

     ** The compound is predicted to be EXTREMELY active **
     ** The projected MTD value is   90.0     CASE units **
```

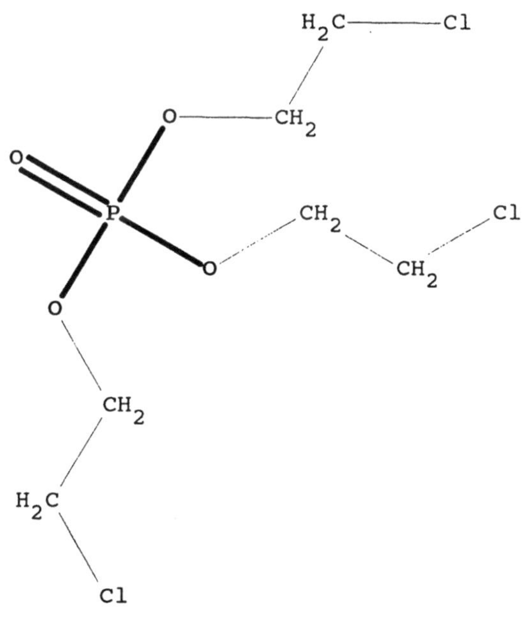

FIGURE 2: Prediction of toxicity of tris(2-chloroethyl)phosphate.

```
********************************************************************************
                          TRIS(2-ETHYLHEXYL)PHOSPHATE
********************************************************************************
```

The molecule contains the Biophore (nr.occ.= 3):

```
               O
              /
  O   -PO
              \
               O
```

*** 6 out of the known 7 molecules (86%) containing such Biophore have an average MTD value in rats of 79. (conf.level= 96%)

Constant is 99.5

Log partition coeff.= 6.00 ; LogP contribution is -81.4

** The probability that this molecule is a systemic toxicant is 77.8% **

** However, the MTD value is expected to be negligible (18) **

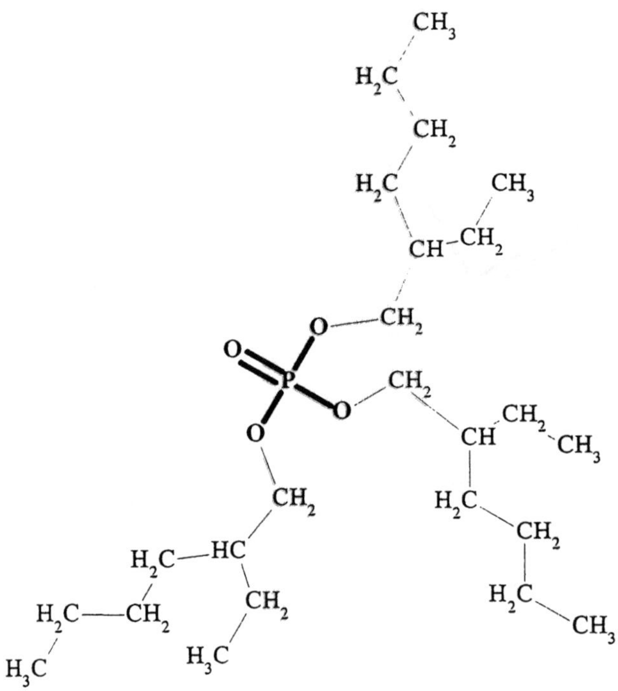

FIGURE 3: Prediction of the negligible toxicity of tris(2-ethylhexyl)phosphate. Even though this molecule contains the same biophore as tris(2-chloroethyl)phosphate (Figure 2) which is predicted to be a potent toxicant, the log P of tris(2-ethylhexyl)phosphate decreases the potency (see text).

```
******************************************************************************
                        SUCCINIC ACID-2,2-DIMETHYLHYDRAZINE
******************************************************************************

The molecule contains the expanded Biophore    (nr.occ.= 2):

               N    -CH3

     *** 12 out of the known  15 molecules ( 80%) containing such Biophore
         have an average MTD value in rats of  80. (conf.level= 99%)

                                                        Constant is     19.1
     ** The following Modulators are also present:
     ( 1) CO -OH                                         Inactivating  -41.3
***************************************************************************
         Electronegativity    = -0.20 ;             Its contribution is    -6.6

** The probability that this molecule will induce systemic toxicity is 76.5% **

     ** However, the MTD value is expected to be negligible (10) **
```

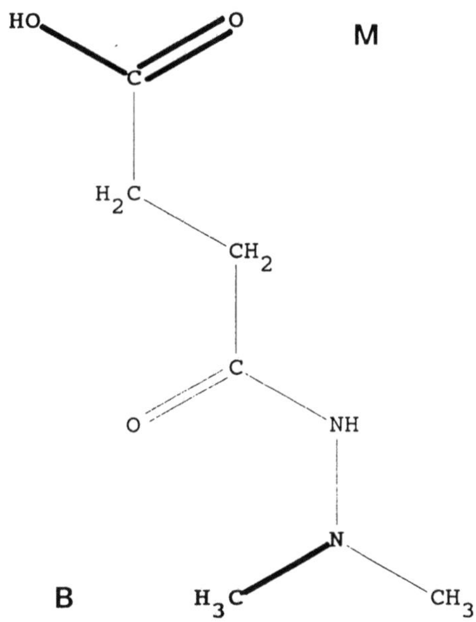

FIGURE 4: Prediction of the negligible toxicity of succinic acid-2,2-dimethylhydrazine.

```
**********************************************************************************
                                 PHENYLBUTAZONE
**********************************************************************************

The molecule contains the Biophore   (nr.occ.= 1):

            CH2 -CH2
                    \
                     CH2 -CH

     ***  6 out of the known   6 molecules (100%) containing such Biophore
       are Rodent carcinogens with an average activity of 50 (conf. level = 98%)
                                                       Constant is     52.9

        The molecule also contains the Biophore    :
          CH =CH -C  =CH -CH =                        <3-N  >

     ** The probability that this molecule is a Rodent carcinogen is  87.5%  **
          increased to  92.1% due to the presence of the extra Biophore

     ** The compound is predicted to be VERY active **
     ** The projected Rodent carcinogenicity activity is 53.0 CASE units **
```

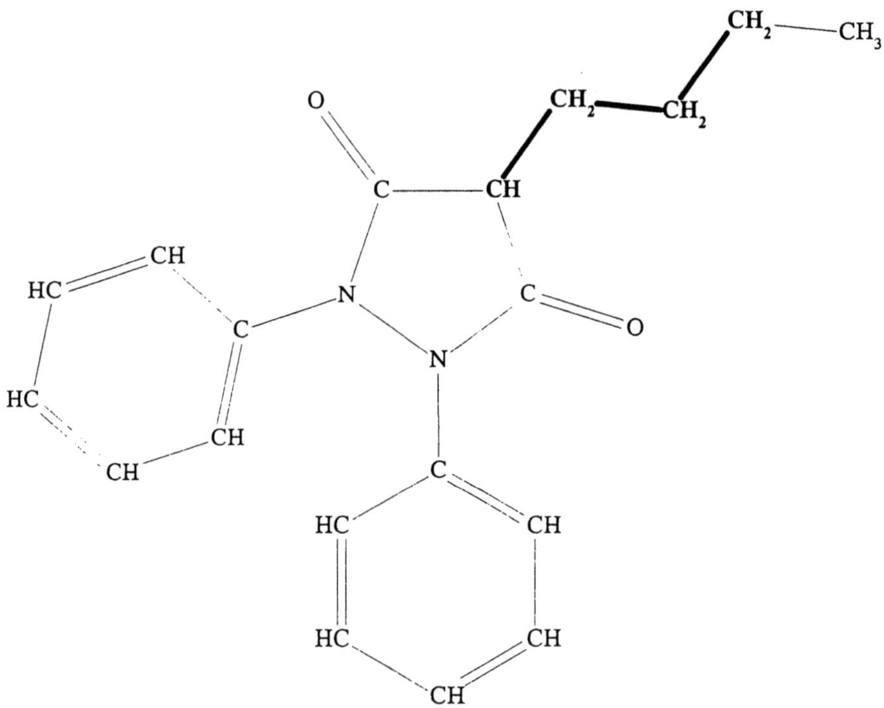

FIGURE 5: Prediction of the carcinogenicity in rodents of phenylbutazone.

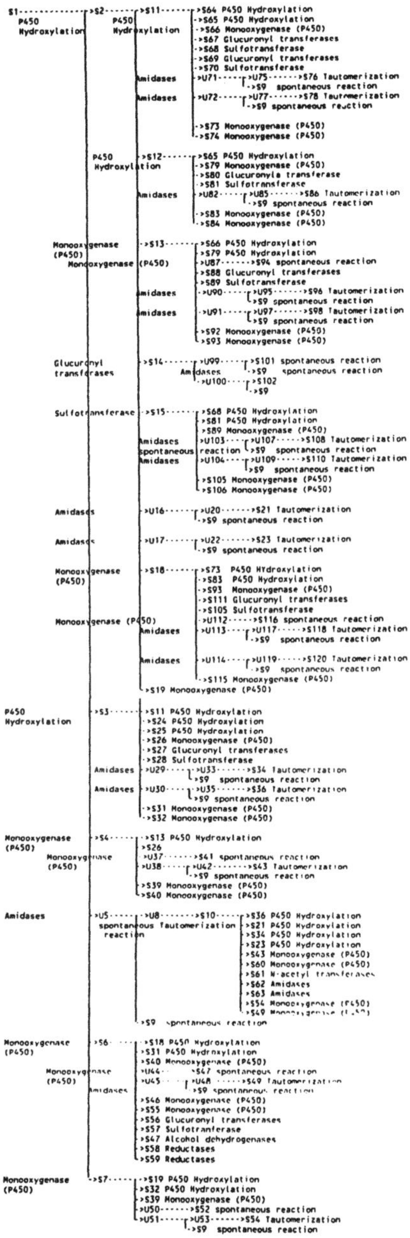

FIGURE 6: Putative biotransformation pathway of phenylbutazone (S1). The META program identifies both the products as well as the enzymes involved. Thus, for example, phenylbutazone is hydroxylated *via* a P450 isozyme to S2 which in turn is glucuronidated *via* glucuronyl transferase to S14 which in turn is hydrolyzed by an amidase to yield U99. The structures of the various metabolites are shown in Figure 7.

META

Metas of Phenylbutazone

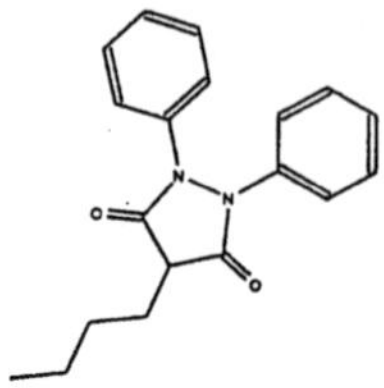

(..) S1. C19H20O2N2

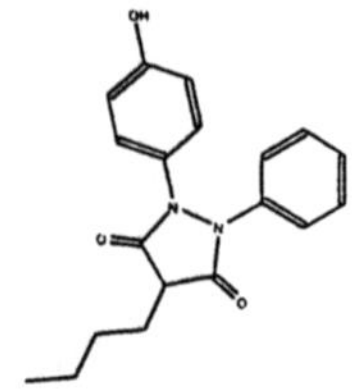

(..) S2. C19H20O3N2

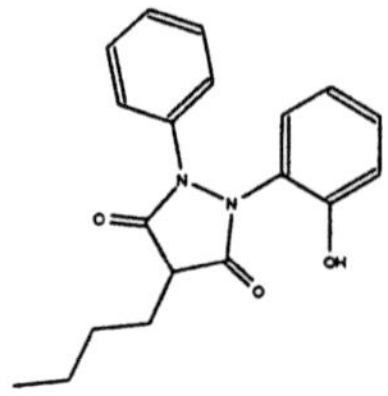

(..) S3. C19H20O3N2

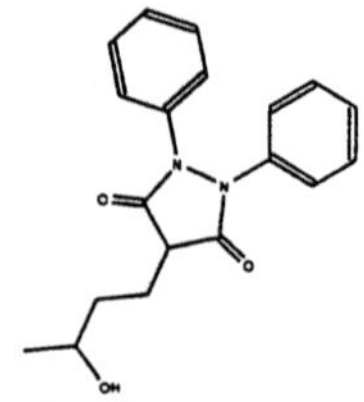

(..) S4. C19H20O3N2

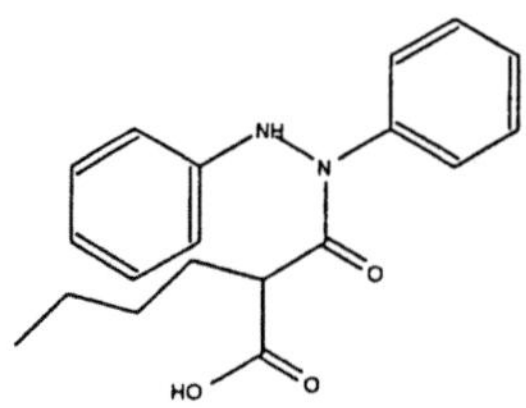

(..) U5. C19H22O3N2

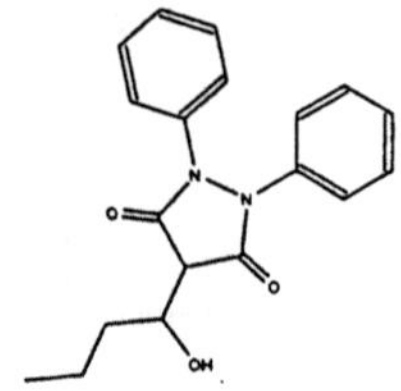

(..) S6. C19H20O3N2

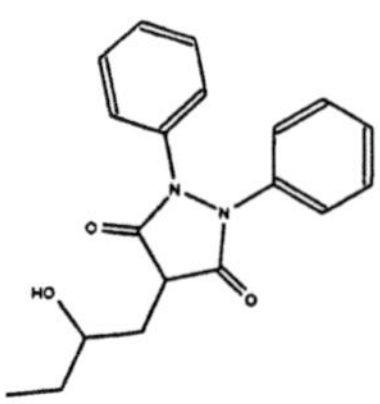

(..) S7. C19H20O3N2

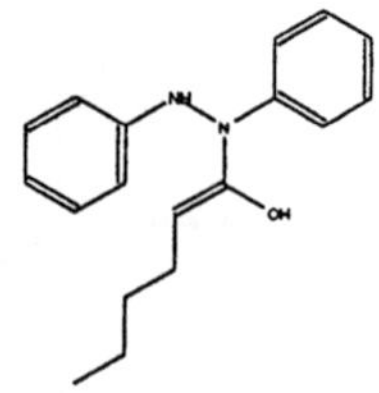

(..) U8. C18H22ON2

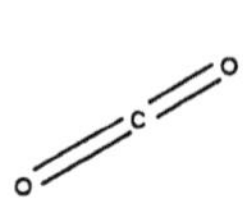

(..) s9. CO2

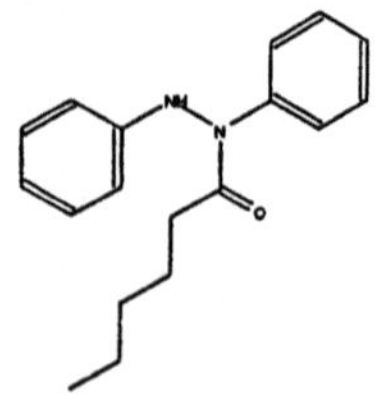

(..) S10. C18H22ON2

FIGURE 7: Structures of the various putative metabolites of phenylbutazone (S1). The designations of the structures refer to pathways shown in Figure 6.

META

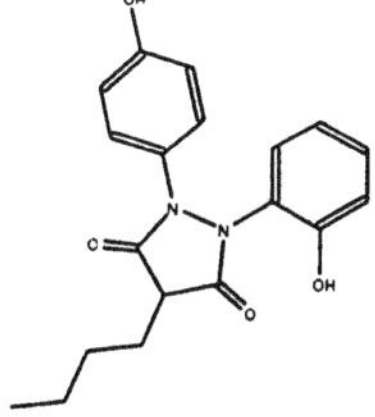

(..) S11. C19H20O4N2

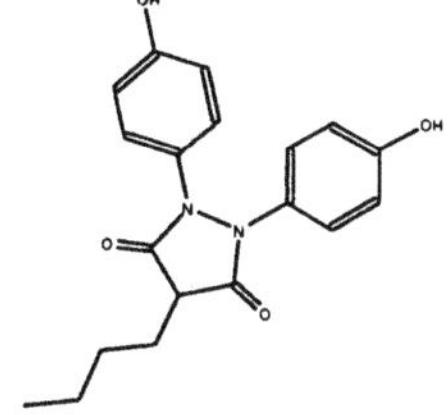

(..) S12. C19H20O4N2

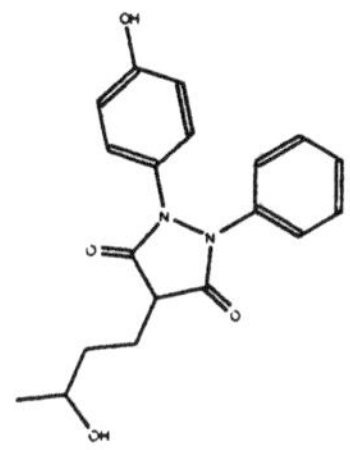

(..) S13. C19H20O4N2

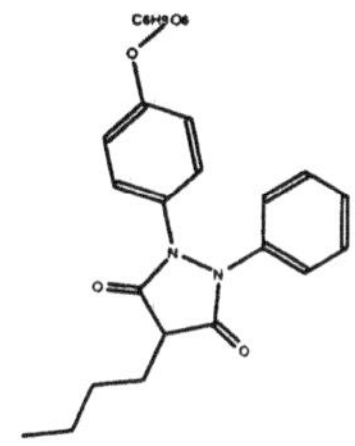

(..) S14. C19H19O3N2C6H9O6

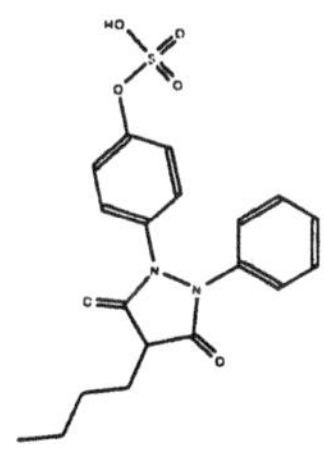

(..) S15. C19H20O6N2S

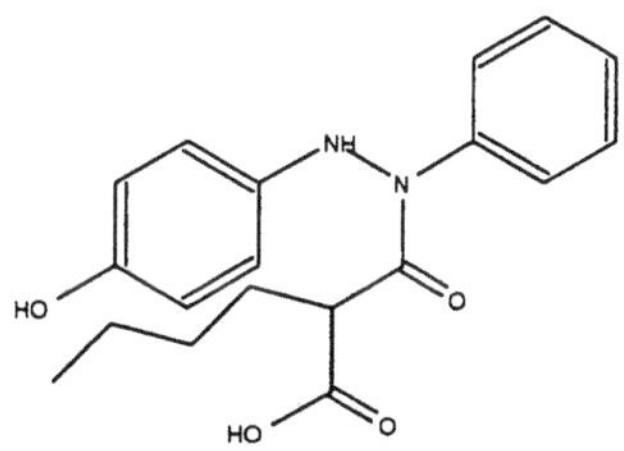

(..) U16. C19H22O4N2

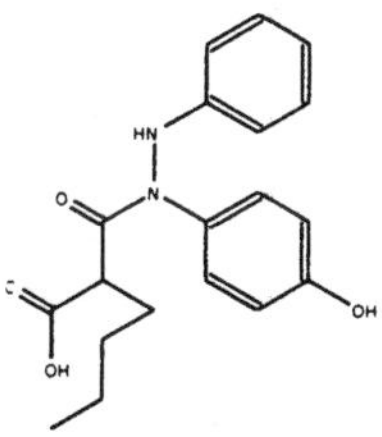

(..) U17. C19H22O4N2

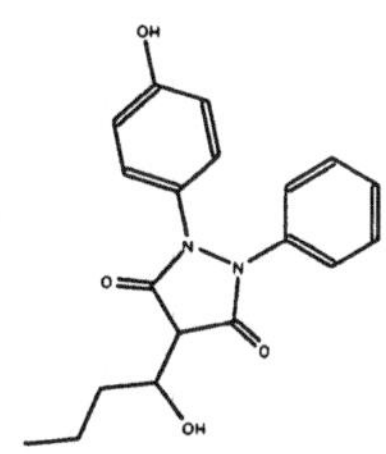

(..) S18. C19H20O4N2

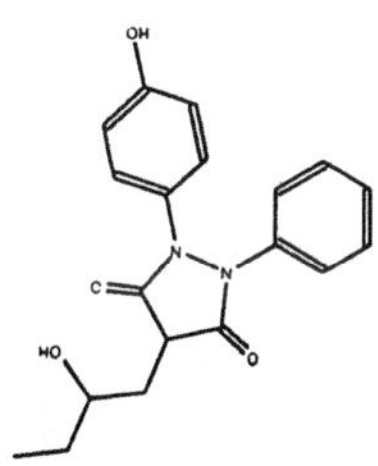

(..) S19. C19H20O4N2

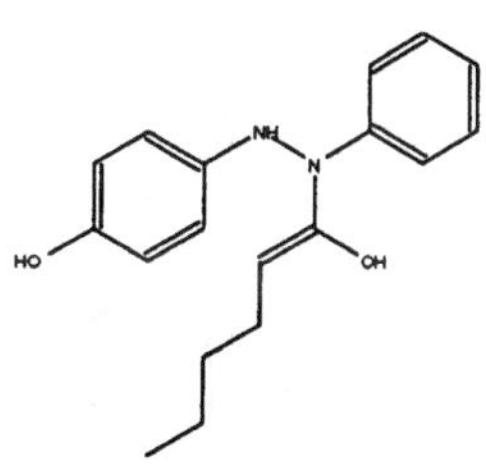

(..) U20. C18H22O2N2

META

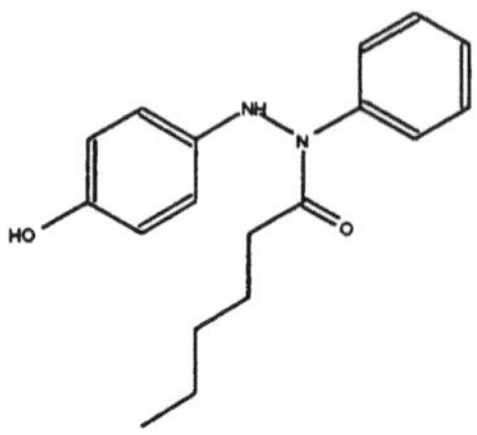

(..) S21. C18H22O2N2

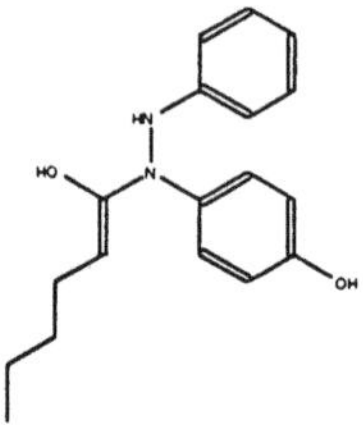

(..) U22. C18H22O2N2

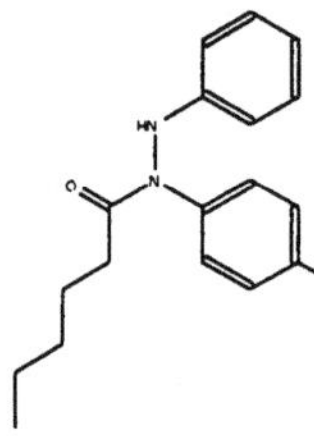

(..) S23. C18H22O2N2

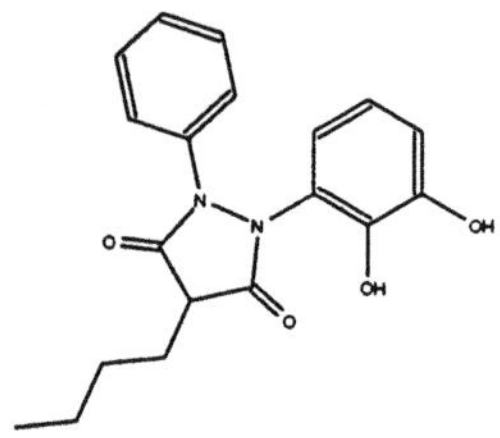

(..) S24. C19H20O4N2

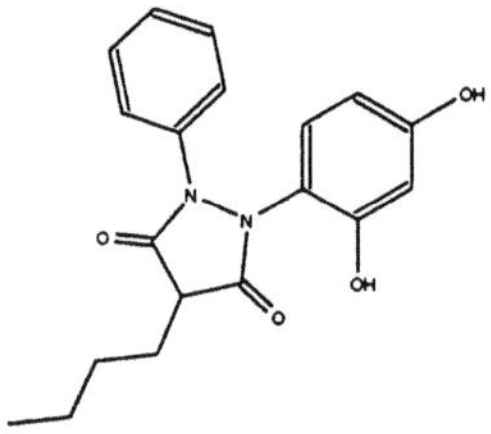

(..) S25. C19H20O4N2

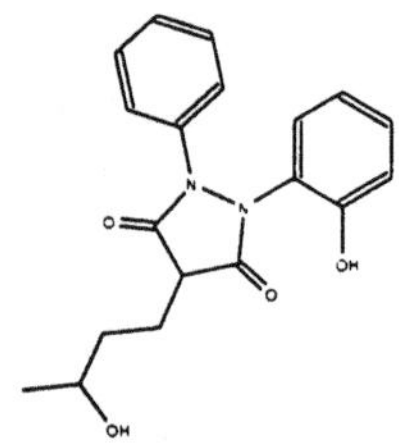

(..) S26. C19H20O4N2

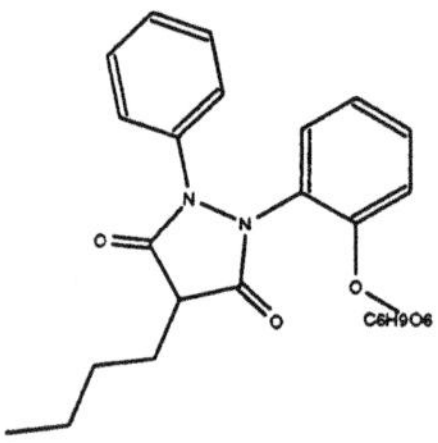

(..) S27. C19H19O3N2C6H9O6

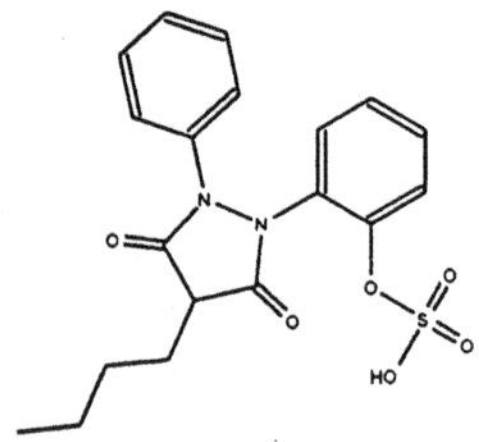

(..) S28. C19H20O6N2S

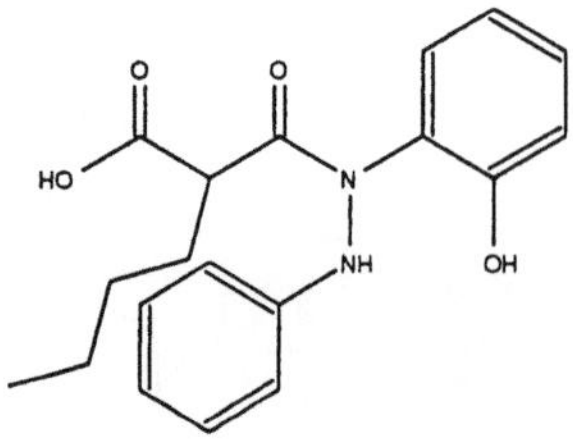

(..) U29. C19H22O4N2

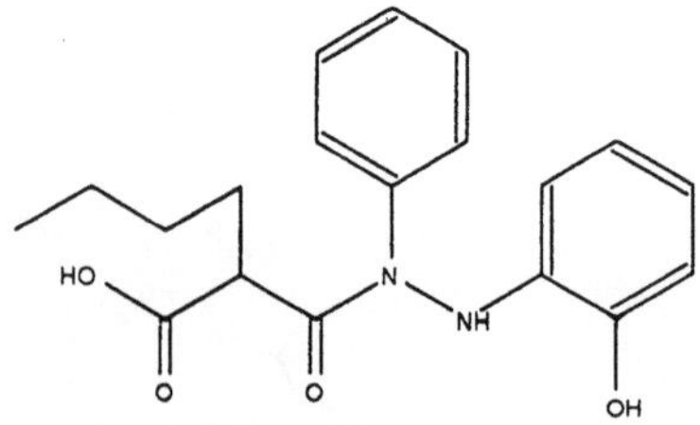

(..) U30. C19H22O4N2

META

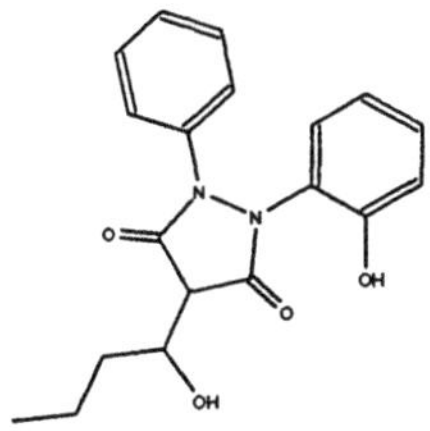

(..) S31. C19H20O4N2

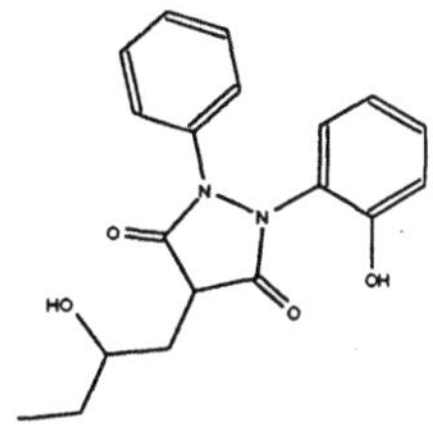

(..) S32. C19H20O4N2

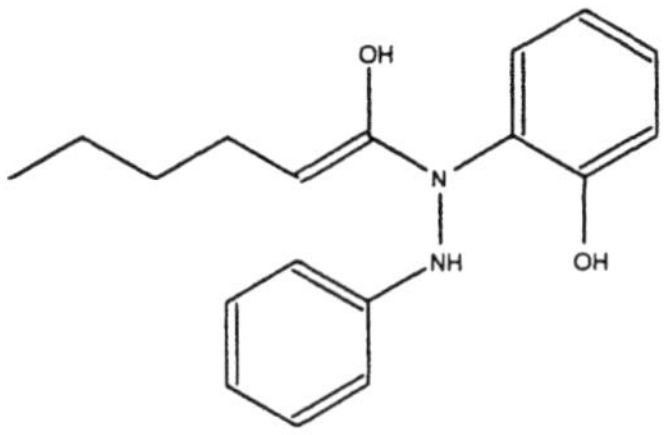

(..) U33. C18H22O2N2

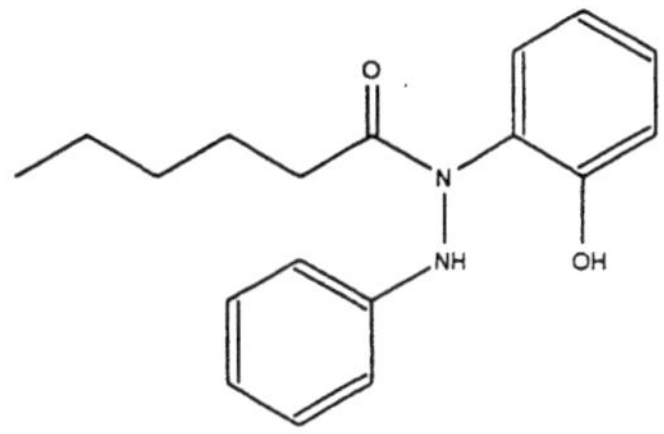

(..) S34. C18H22O2N2

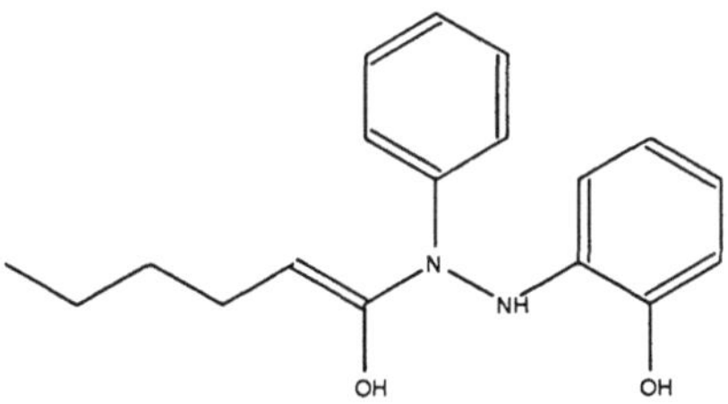

(..) U35. C18H22O2N2

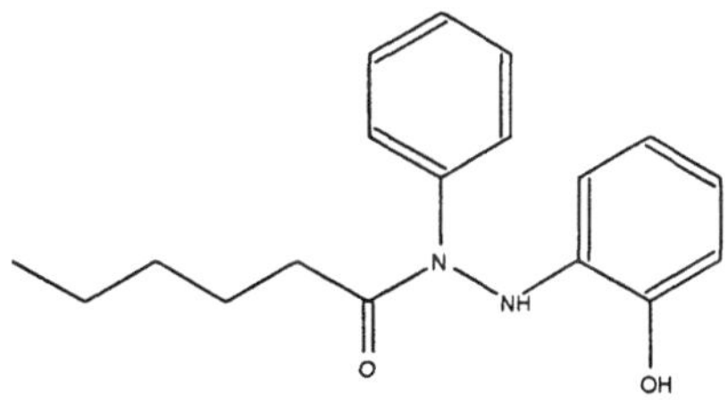

(..) S36. C18H22O2N2

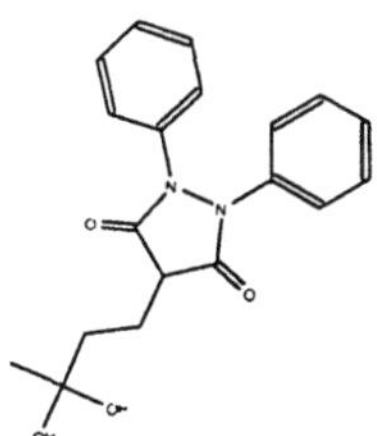

(..) U37. C19H20O4N2

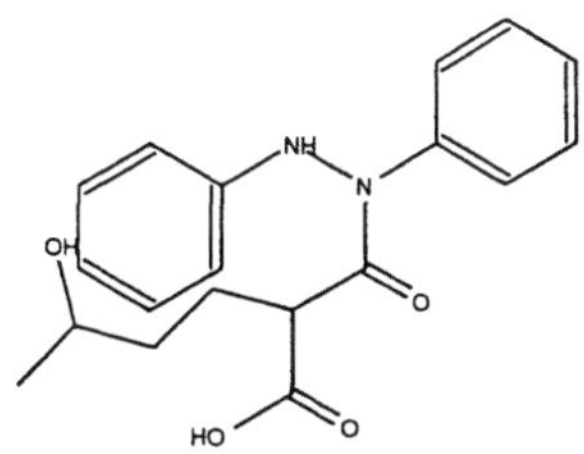

(..) U38. C19H22O4N2

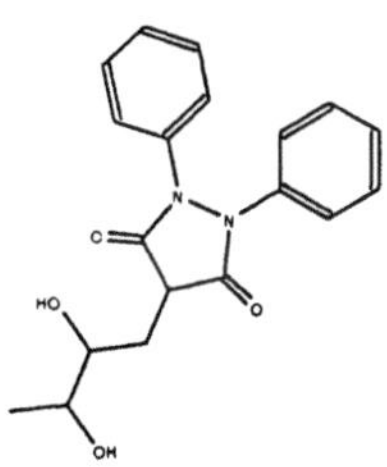

(..) S39. C19H20O4N2

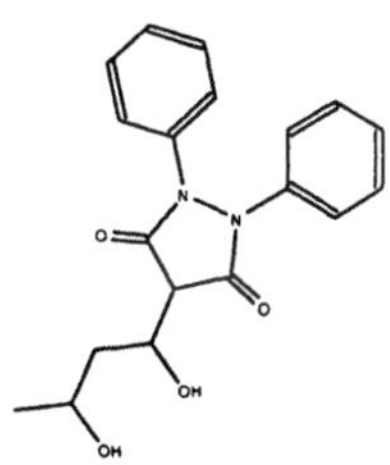

(..) S40. C19H20O4N2

META

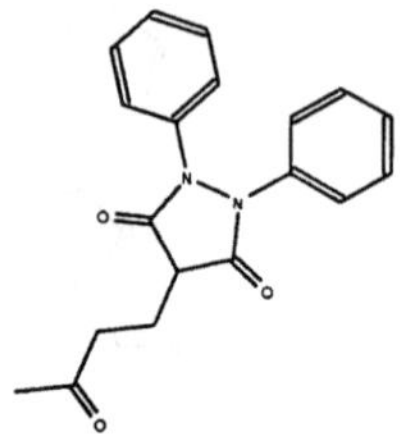

(..) S41. C19H18O3N2

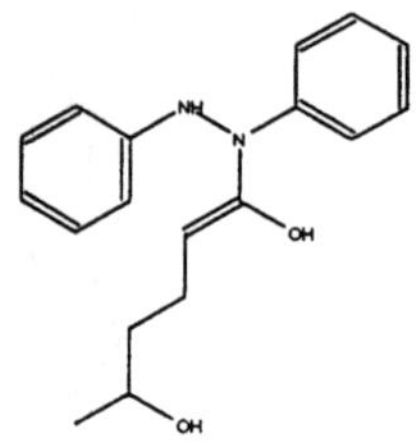

(..) U42. C18H22O2N2

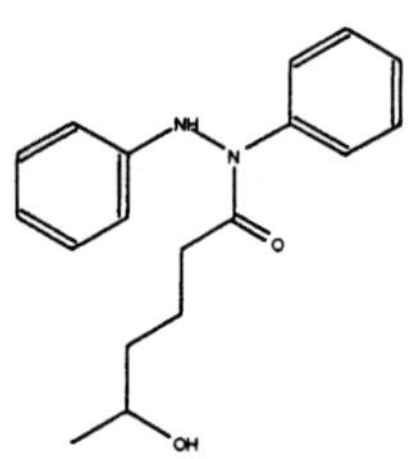

(..) S43. C18H22O2N2

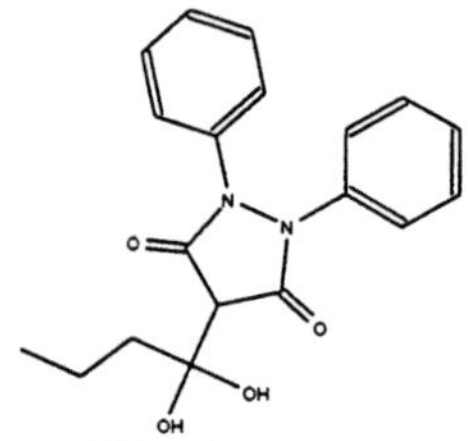

(..) U44. C19H20O4N2

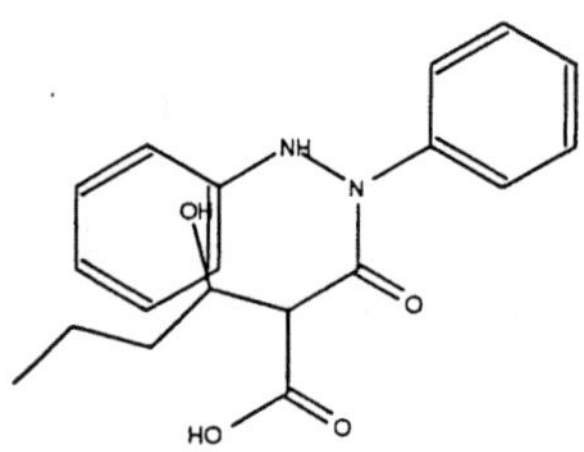

(..) U45. C19H22O4N2

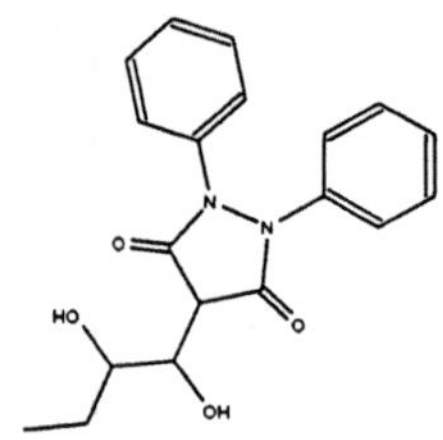

(..) S46. C19H20O4N2

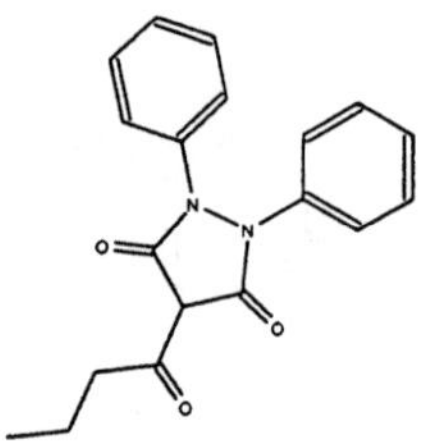

(..) S47. C19H18O3N2

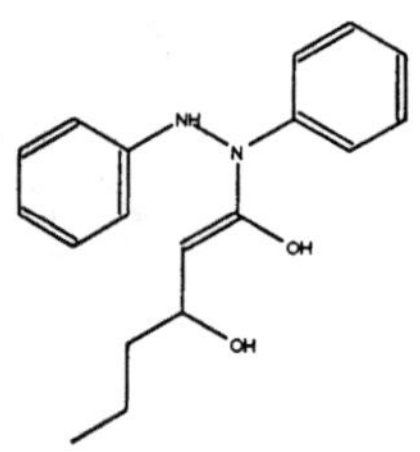

(..) U48. C18H22O2N2

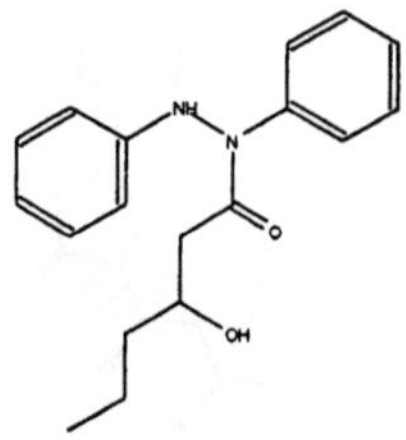

(..) S49. C18H22O2N2

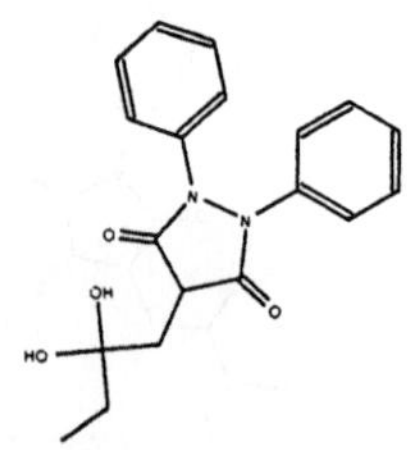

(..) U50. C19H20O4N2

META

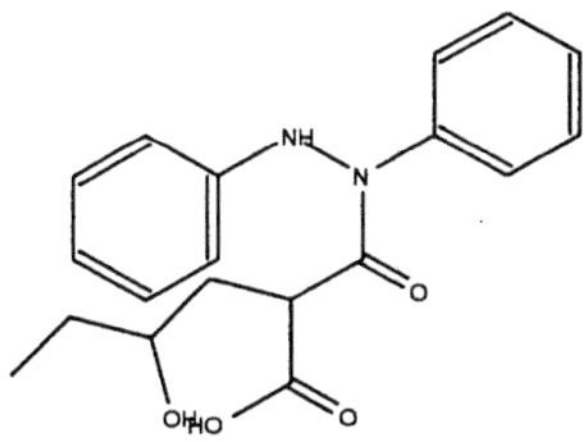

(..) U51. C19H22O4N2

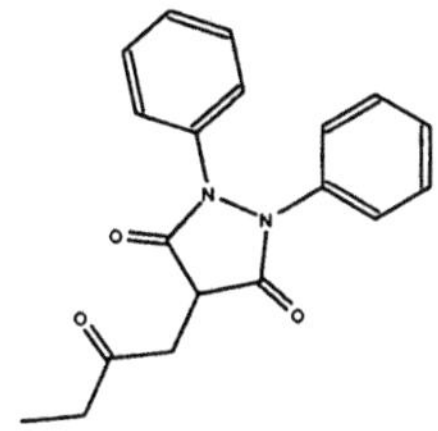

(..) S52. C19H18O3N2

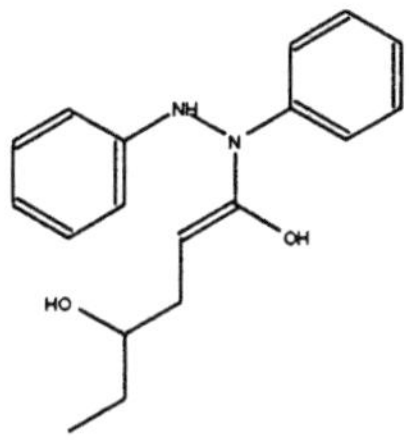

(..) U53. C18H22O2N2

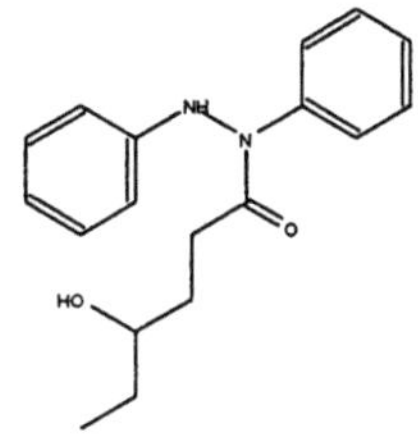

(..) S54. C18H22O2N2

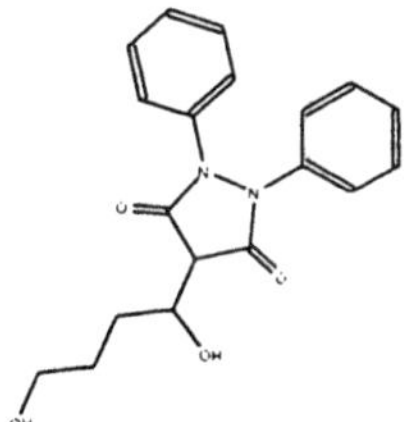

(..) S55. C19H20O4N2

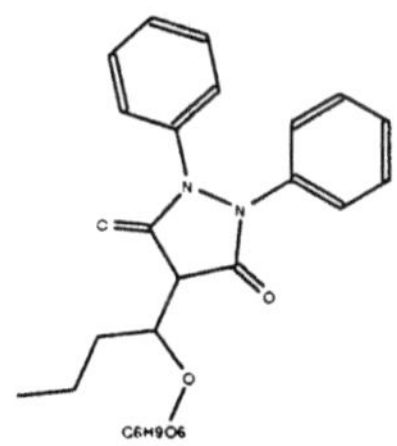

(..) S56. C19H19O3N2C6H9O6

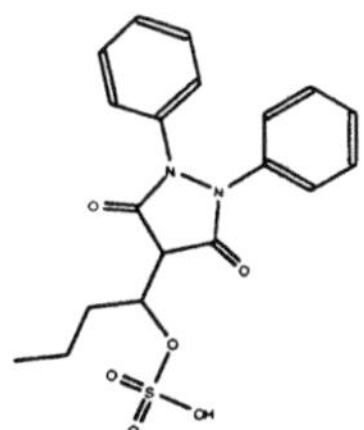

(..) S57. C19H20O6N2S

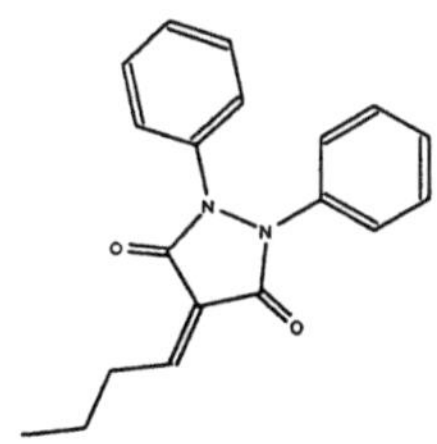

(..) S58. C19H18O2N2

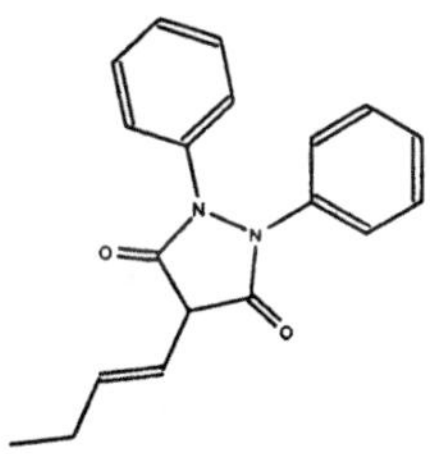

(..) S59. C19H18O2N2

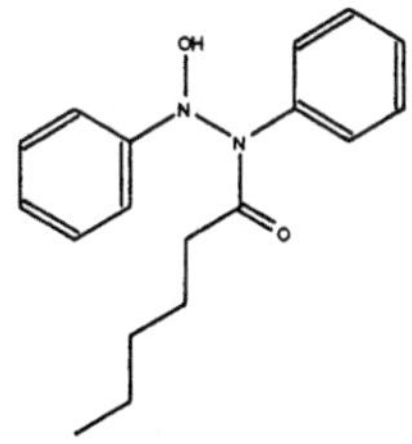

(..) S60. C18H22O2N2

META

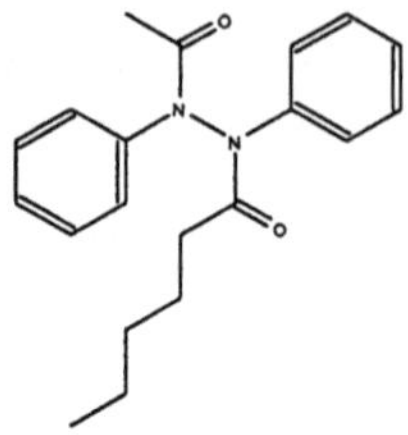

(..) S61. C20H24O2N2

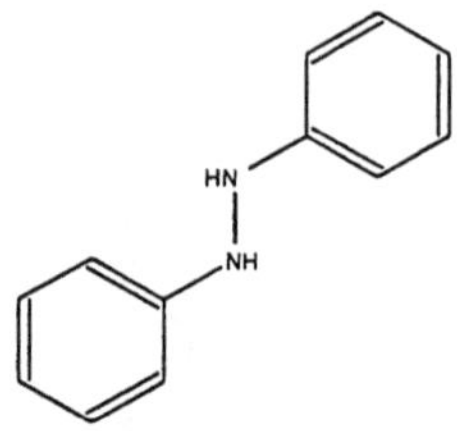

(..) S62. C12H12N2

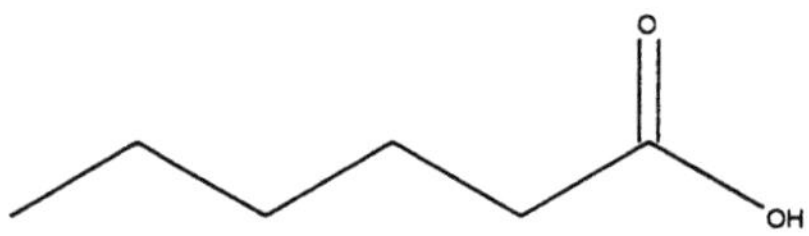

(..) S63. C6H12O2

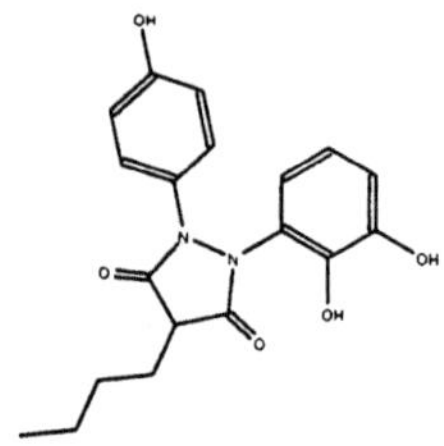

(..) S64. C19H20O5N2

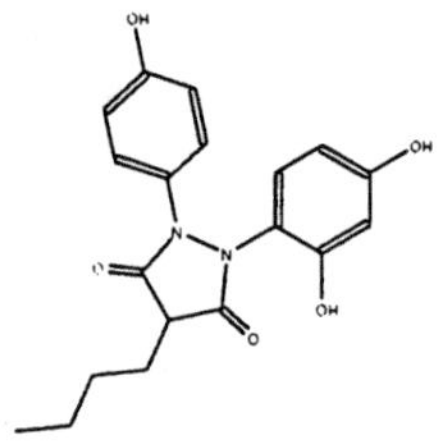

(..) S65. C19H20O5N2

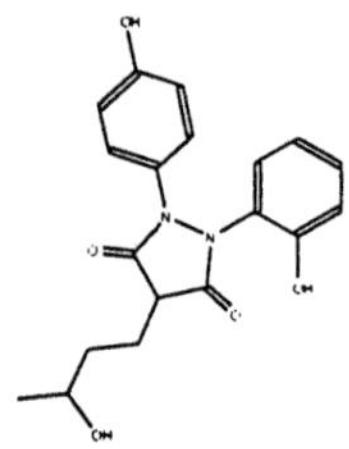

(..) S66. C19H20O5N2

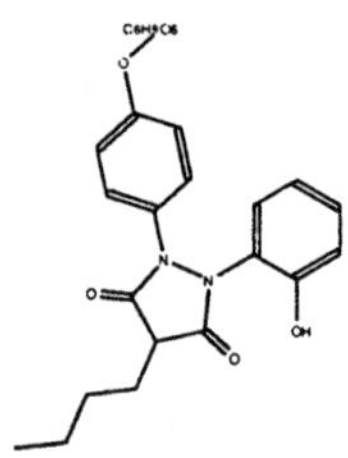

(..) S67. C19H19O4N2C6H9O6

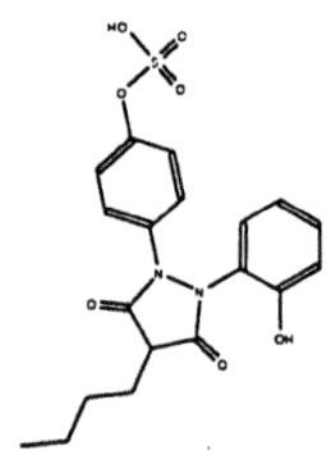

(..) S68. C19H20O7N2S

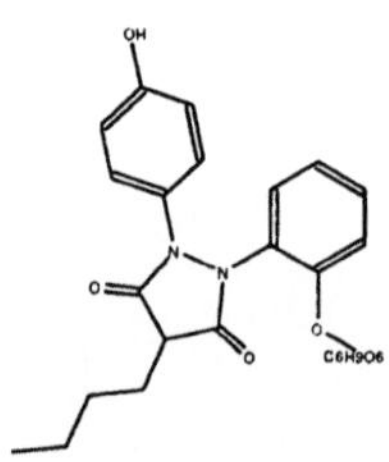

(..) S69. C19H19O4N2C6H9O6

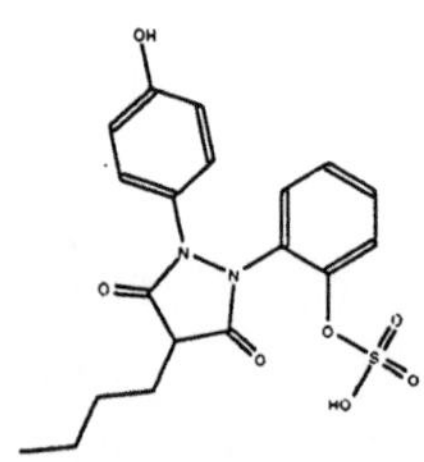

(..) S70. C19H20O7N2S

META

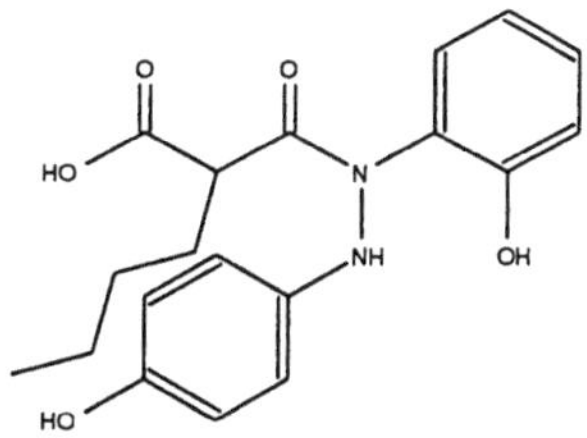

(..) U71. C19H22O5N2

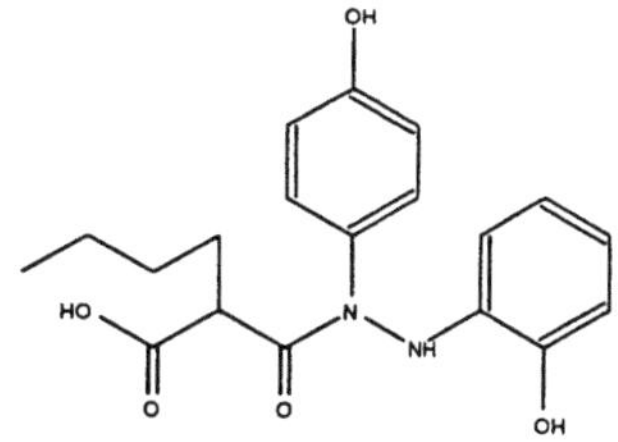

(..) U72. C19H22O5N2

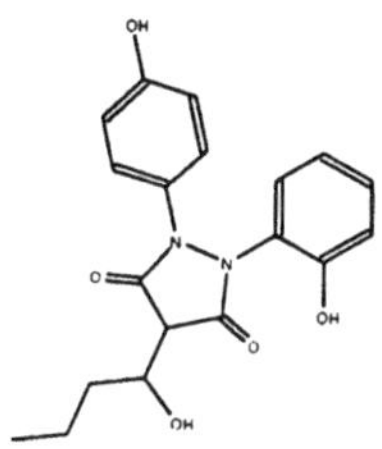

(..) S73. C19H20O5N2

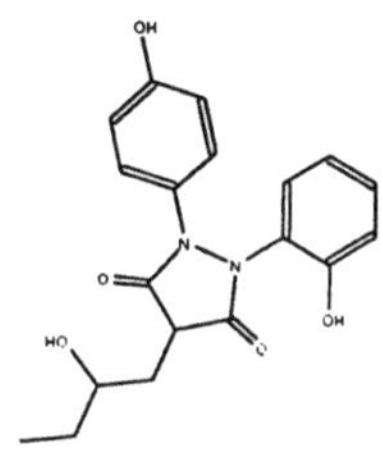

(..) S74. C19H20O5N2

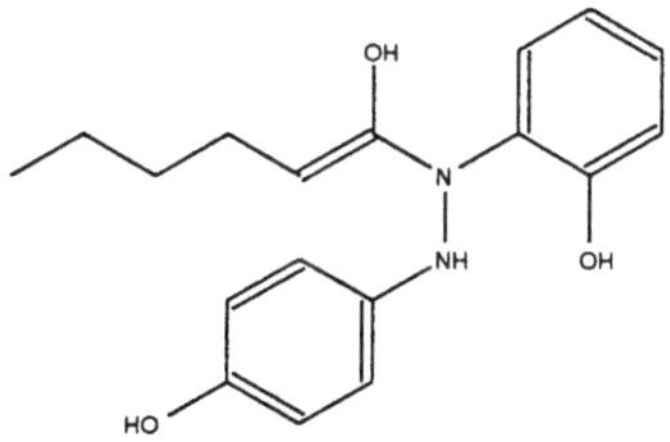

(..) U75. C18H22O3N2

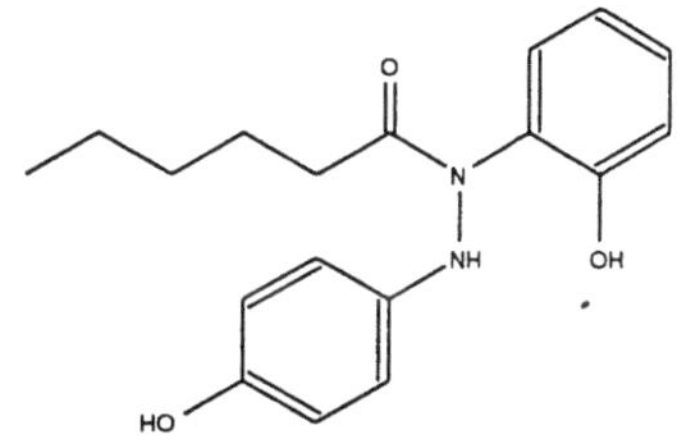

(..) S76. C18H22O3N2

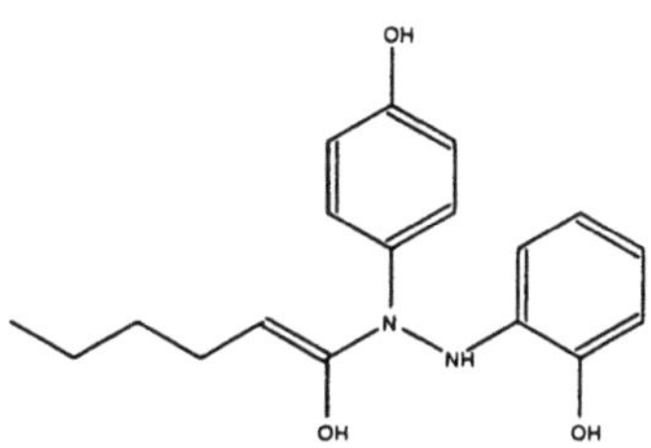

(..) U77. C18H22O3N2

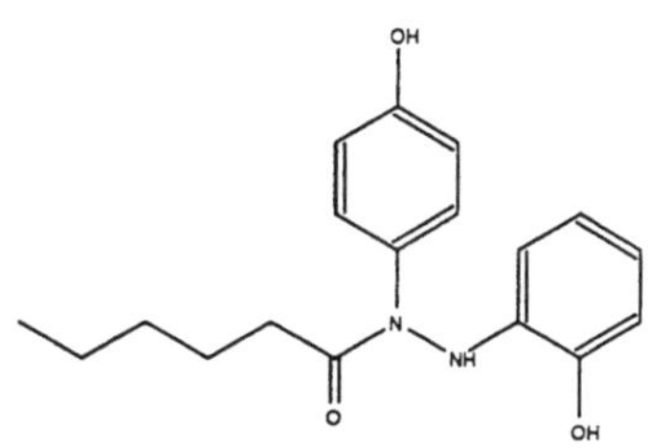

(..) S78. C18H22O3N2

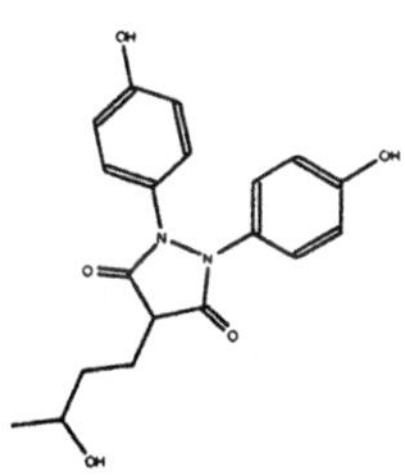

(..) S79. C19H20O5N2

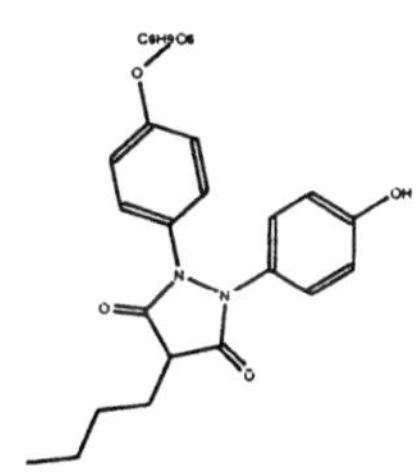

(..) S80. C19H19O4N2C6H9O6

META

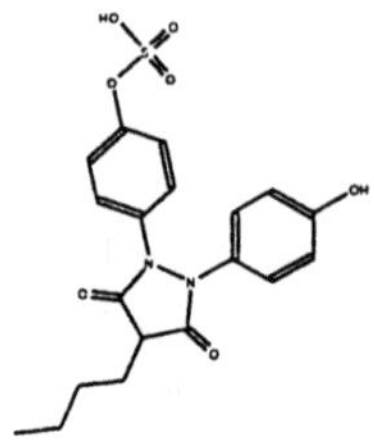

(..) S81. C19H20O7N2S

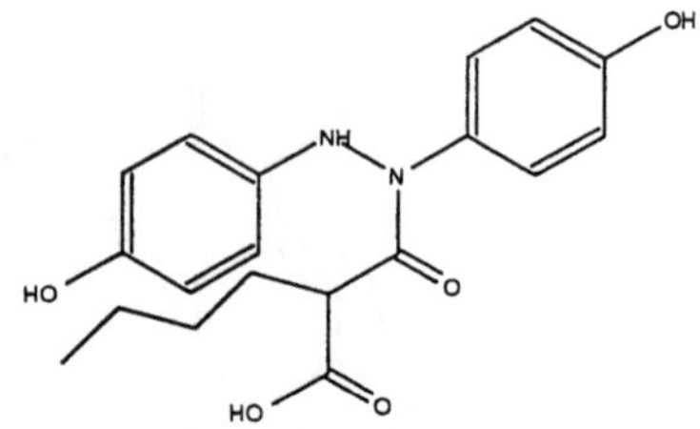

(..) U82. C19H22O5N2

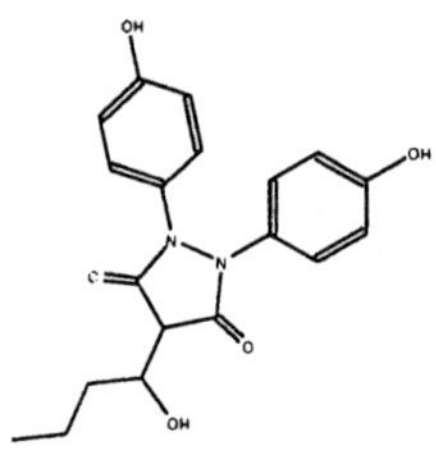

(..) S83. C19H20O5N2

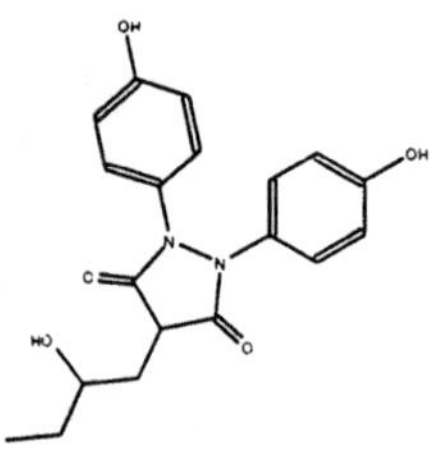

(..) S84. C19H20O5N2

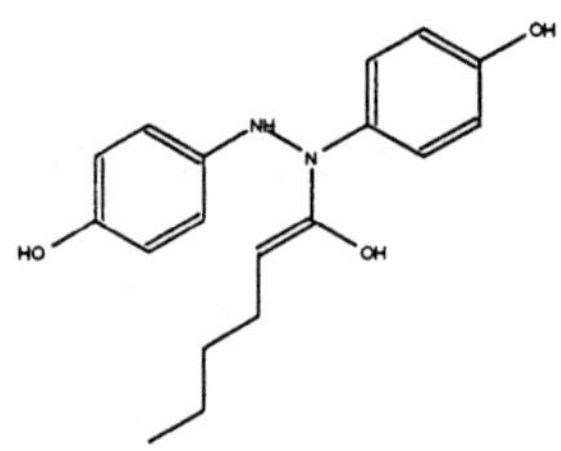

(..) U85. C18H22O3N2

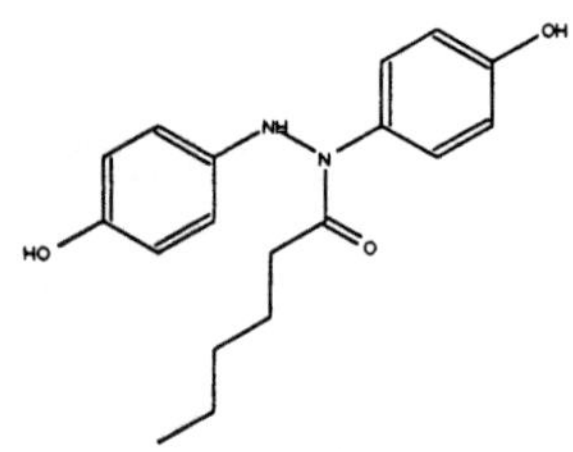

(..) S86. C18H22O3N2

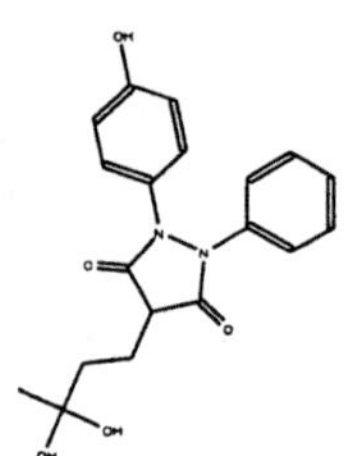

(..) U87. C19H20O5N2

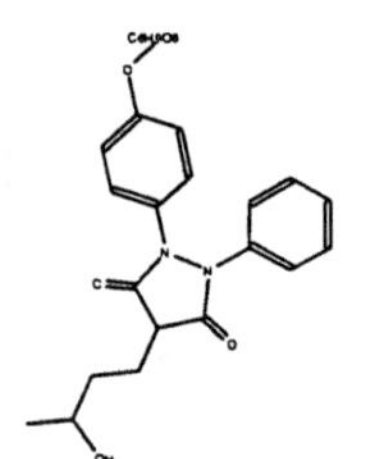

(..) S88. C19H19O4N2C6H9O6

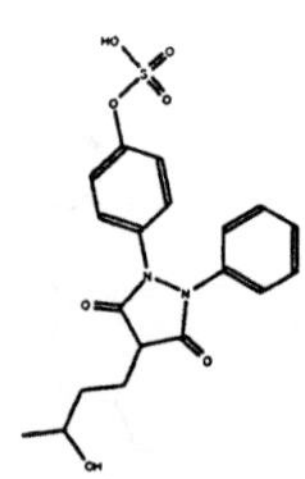

(..) S89. C19H20O7N2S

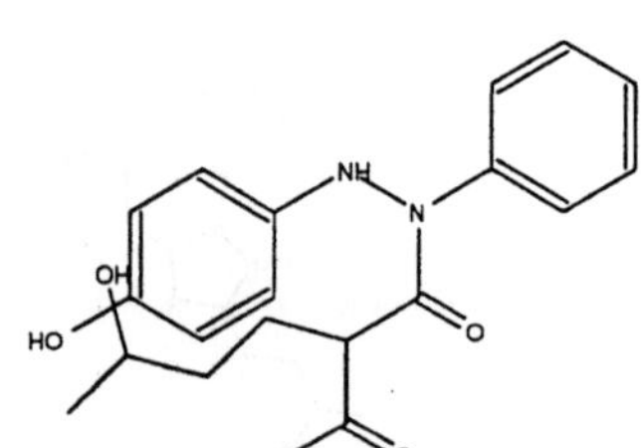

(..) U90. C19H22O5N2

META

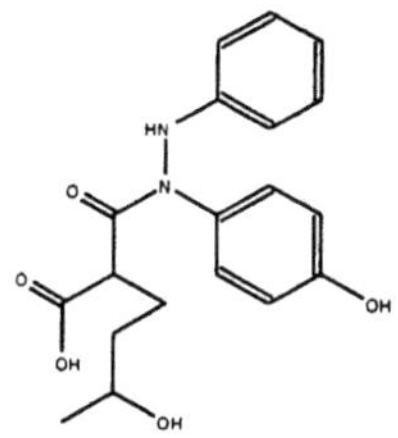

(..) U91. C19H22O5N2

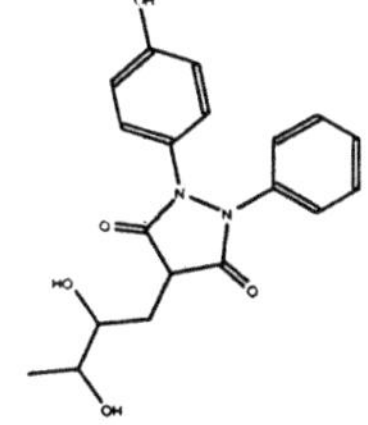

(..) S92. C19H20O5N2

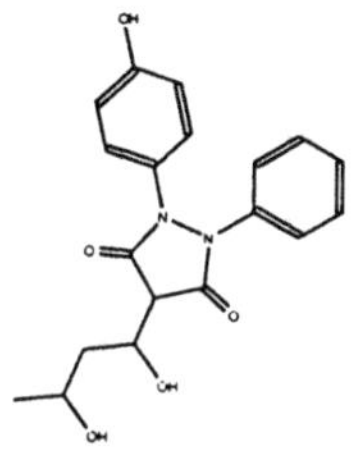

(..) S93. C19H20O5N2

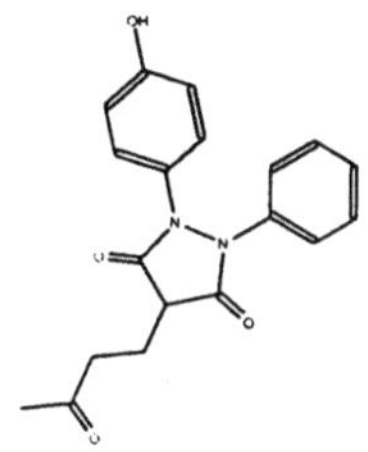

(..) S94. C19H18O4N2

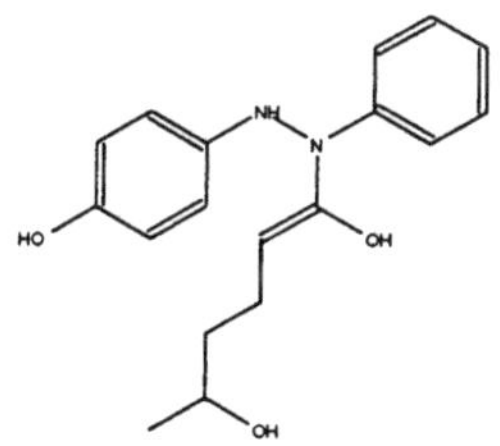

(..) U95. C18H22O3N2

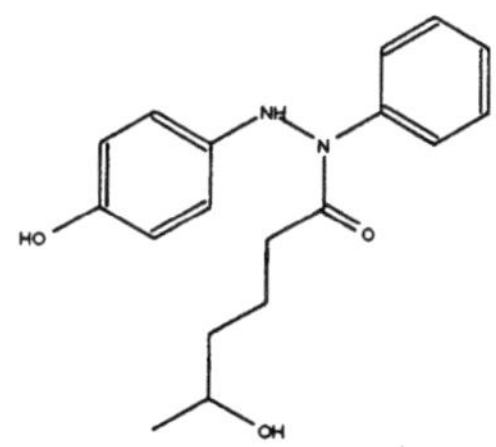

(..) S96. C18H22O3N2

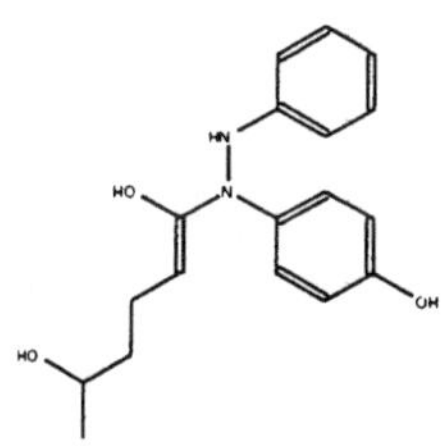

(..) U97. C18H22O3N2

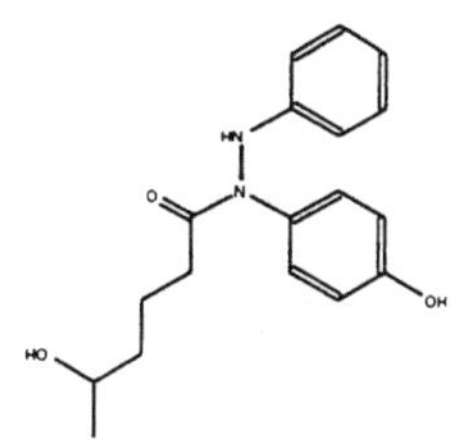

(..) S98. C18H22O3N2

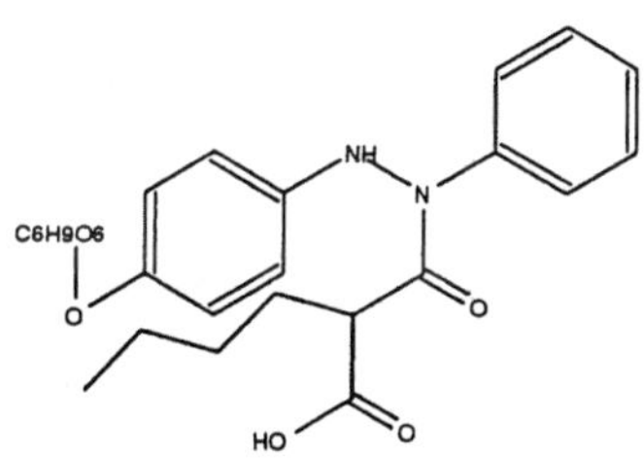

(..) U99. C19H21O4N2C6H9O6

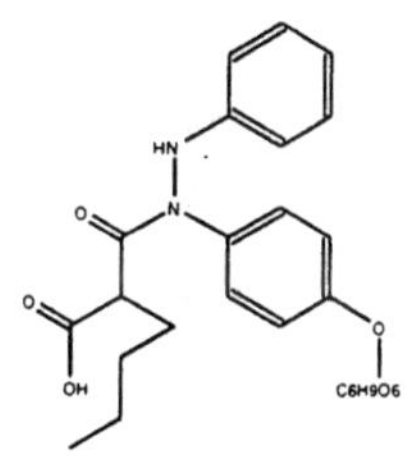

(..) U100. C19H21O4N2C6H9O6

META

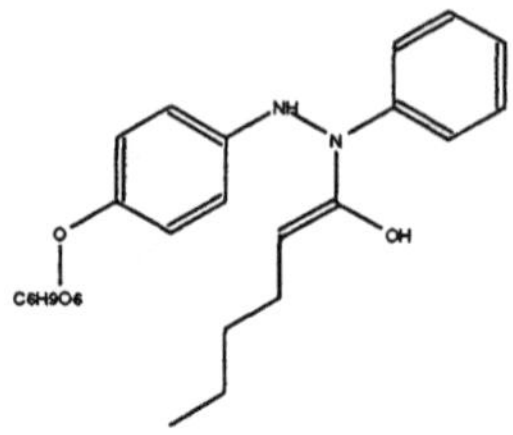

(..) S101. C18H21O2N2C6H9O6

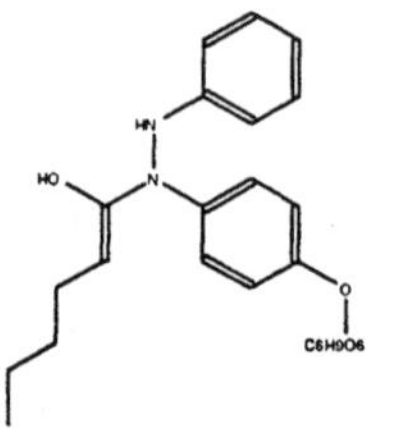

(..) S102. C18H21O2N2C6H9O6

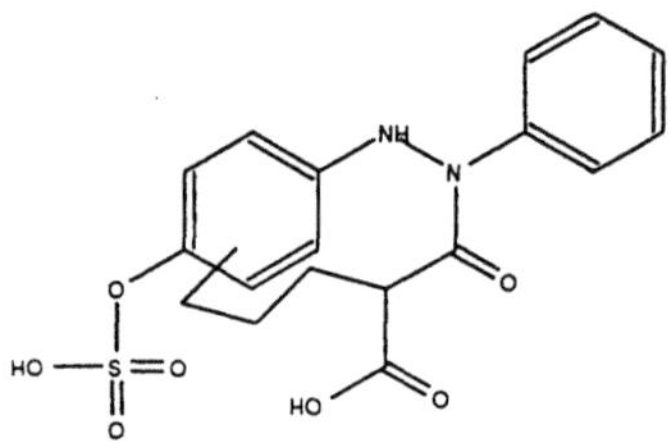

(..) U103. C19H22O7N2S

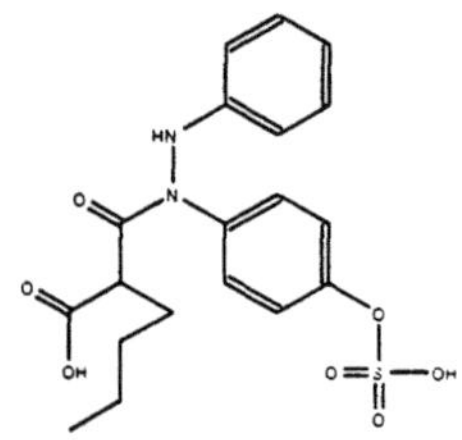

(:.) U104. C19H22O7N2S

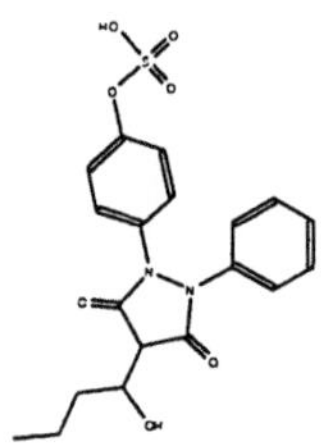

(..) S105. C19H20O7N2S

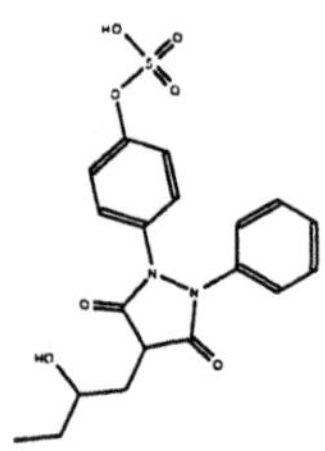

(..) S106. C19H20O7N2S

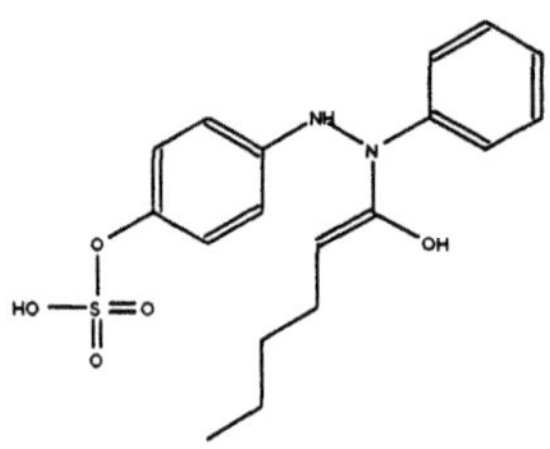

(..) U107. C18H22O5N2S

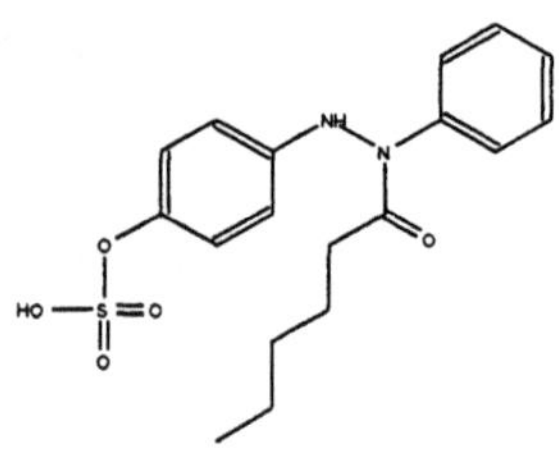

(..) S108. C18H22O5N2S

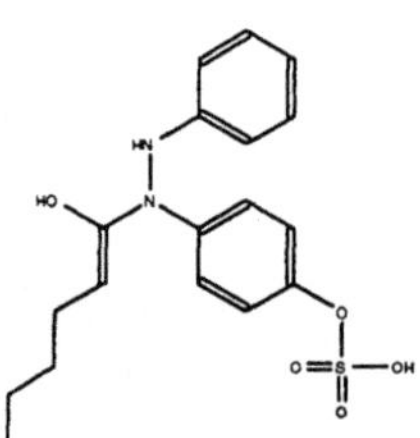

(..) U109. C18H22O5N2S

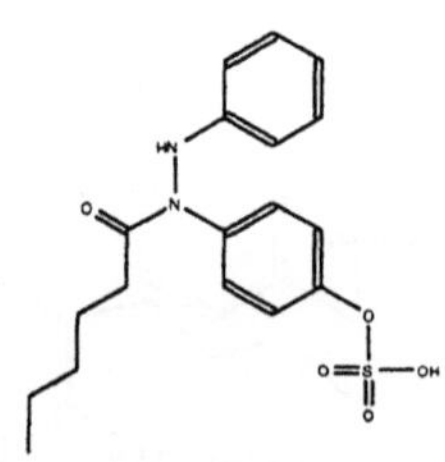

(..) S110. C18H22O5N2S

META

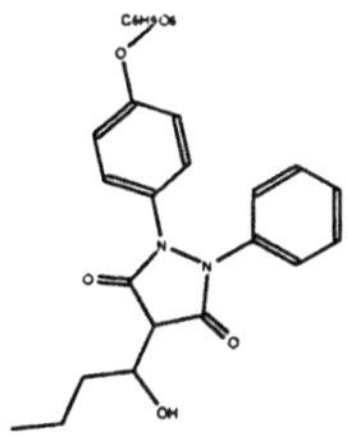

(..) S111. C19H19O4N2C6H9O6

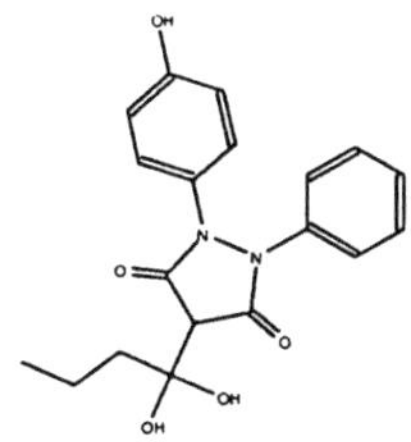

(..) U112. C19H20O5N2

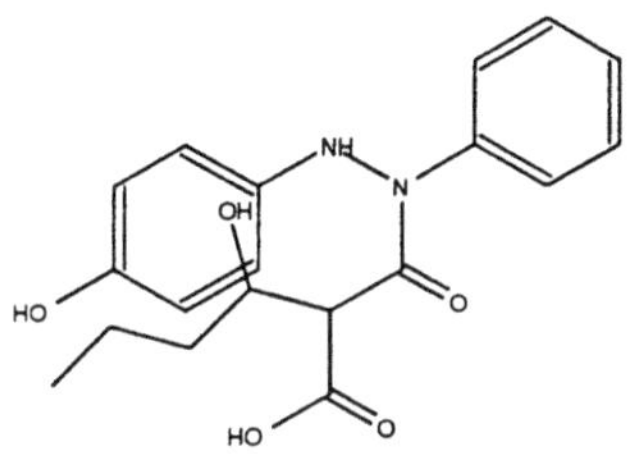

(..) U113. C19H22O5N2

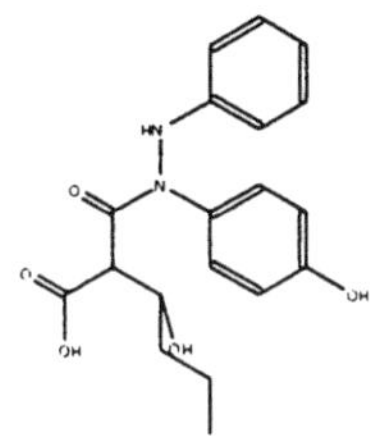

(..) U114. C19H22O5N2

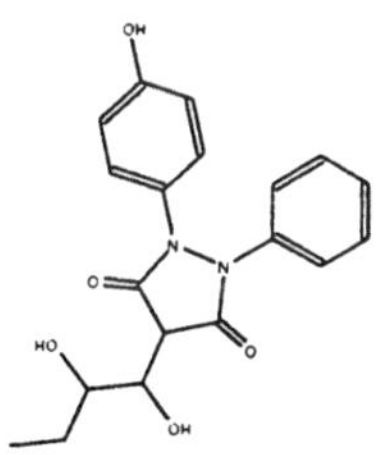

(..) S115. C19H20O5N2

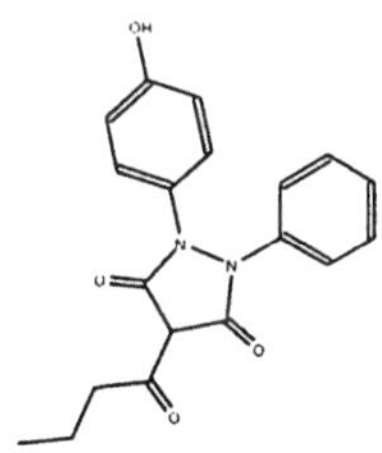

(..) S116. C19H18O4N2

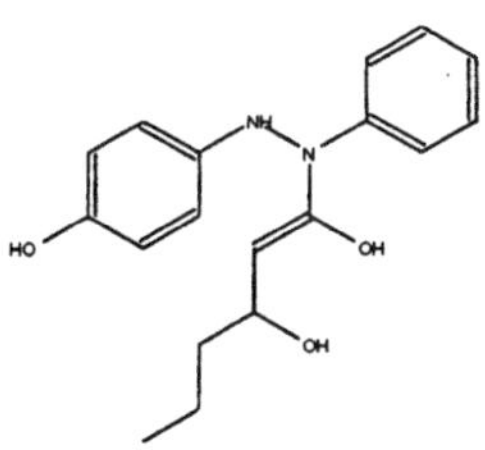

(..) U117. C18H22O3N2

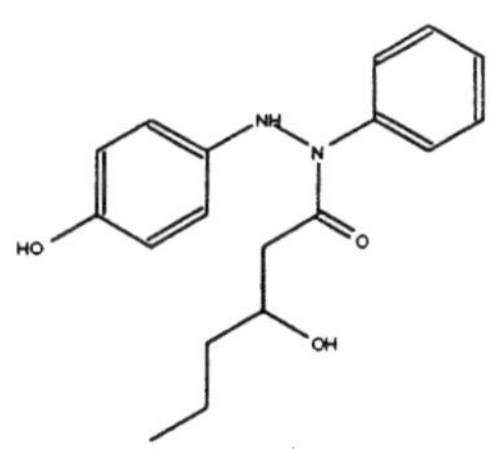

(..) S118. C18H22O3N2

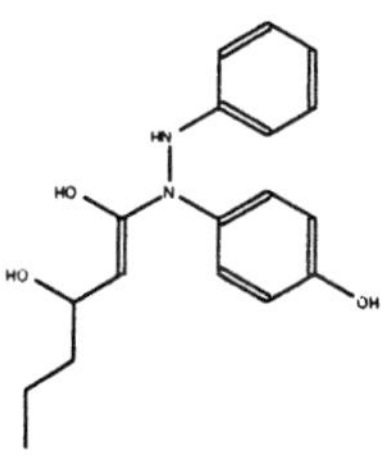

(..) U119. C18H22O3N2

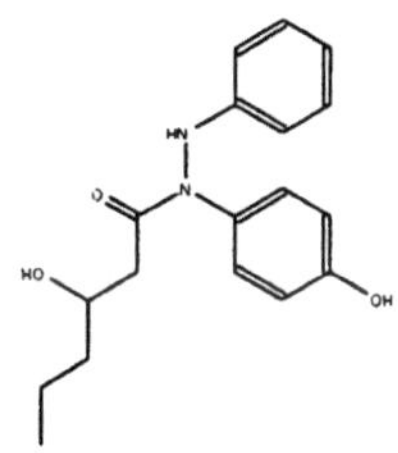

(..) S120. C18H22O3N2

```
*******************************************************************************
                                  SACCHARIN
*******************************************************************************

 *** WARNING *** The following functionalities are UNKNOWN to me :
      *** NH -CO -C. =

         ** The molecule does not contain any known Biophore     **
                 it is therefore presumed to be INACTIVE

      ** The results are INCONCLUSIVE due to the presence of
              UNKNOWN functionalities **
```

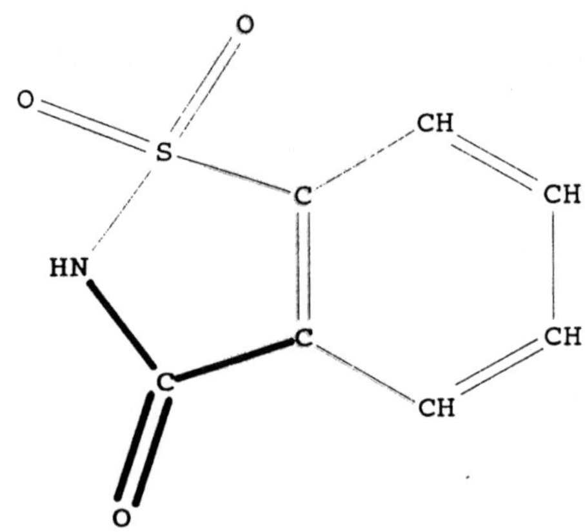

FIGURE 8: Prediction of the lack of toxicity of saccharin. This prediction is based upon the fact that saccharin contains none of the biophores associated with toxicity (see Table 3). However, saccharin contains a fragment (shown in bold) which is "unknown" to the MTD SAR model. Since this fragment could be a biophore, this introduces an element of uncertainty in the prediction.

```
********************************************************************************
                                  METRONIDAZOLE
********************************************************************************

 *** WARNING *** The following functionalities are UNKNOWN to me :
      *** N  -C" -CH3
      *** NO2-C" -N  -

The molecule contains the expanded Biophore   (nr.occ.= 1):

              N   -CH2

     *** 14 out of the known  20 molecules ( 70%) containing such Biophore
         have an average MTD in rats of  69. (conf.level= 98%)
                                                       Constant is     19.1
     ** The following Modulators are also present:
    ( 1) OH -CH2-CH2-                                   Inactivating  -22.5
*************************************************************************
         Electronegativity   =  0.11 ;             Its contribution is    3.6

** The probability that this molecule will induce systemic toxicity is 71.4% **

     ** However, the activity is expected to be negligible (10 CASE units) **
```

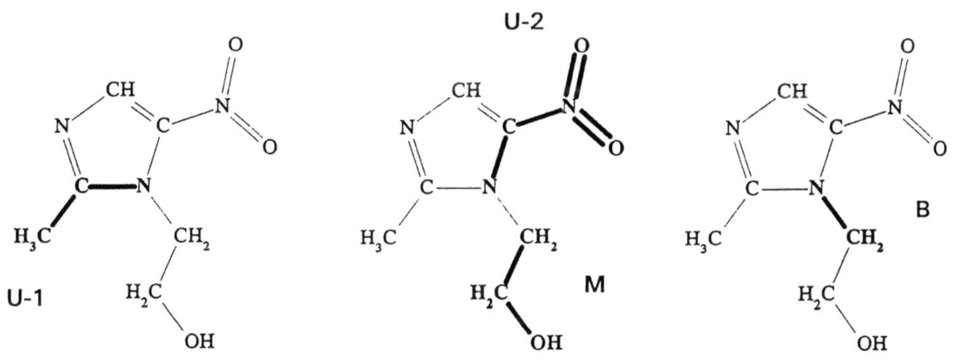

FIGURE 9: Prediction of the negligible toxicity of metronidazole. However, this molecule contains two fragments (U–1 and U–2, shown in bold) which are unknown to the MTD SAR model. Because these fragments could be activating modulators, this introduces an element of uncertainty in the prediction.

STRUCTURAL FEATURES CONTRIBUTING TO BIODEGRADABILITY

Herbert S. Rosenkranz[1], Ying Ping Zhang[1] and Gilles Klopman[2]

[1]Department of Environmental and Occupational Health
University of Pittsburgh
Pittsburgh, PA 15238

and

[2]Department of Chemistry
Case Western Reserve University
Cleveland, OH 44106

NATO ASI Series, Vol. H 93
Modulation of Cellular Responses in Toxicity
Edited by C. L. Galli, A. M. Goldberg, M. Marinovich

SUMMARY

In this review it is shown that biodegradation of organic molecules by mixed bacterial cultures can be modelled successfully by an expert structure–activity relational system.

The model so obtained can be used to predict the biodegradability of as yet untested molecules. This can be applied to the selection of pre–treatment of recalcitrant molecules.

Knowledge of biodegradative pathways can be coupled to the identification of processes which have the least potential for endangering the environmental flora and fauna as well as for causing harm to exposed humans.

INTRODUCTION

The structure–activity modeling (SAR) of biodegradative processes poses a number of problems, both of a toxicological as well as of a methodological nature. Some of these include:

(a) The fact that unlike mammalian metabolism which has an overall unity, albeit it may be complex, environmental biodegradative processes are carried out by a plethora of microbial species present in mixtures of changing compositions which may act simultaneously on a single substrate or consecutively on a series of metabolites.

(b) Depending upon the oxygen tension of the biodegradative process, the metabolism of some microorganisms may be altered from an oxidative one to a fermentative one, thus further increasing the complexity of the system.

(c) Biodegradative processes are often carried out on mixtures of chemicals and different substrates may greatly affect the metabolism of individual chemicals present therein. Thus, growth of a *Pseudomonas sp.* on toluene results in the accumulation of metabolites of chlorobenzene different from those that are generated when growth is in the absence of toluene (Spain *et al.*, 1991).

(d) In view of the fact that biodegradative processes are primarily engineering ones, the emphasis is on the measurement of decreases in mass of the xenobiotic or the extent of its biodegradation as measured by oxygen consumption. The emphasis, however, is not necessarily on the nature of the products that are generated. This is frustrating to the toxicologist as it is conceivable that the chemicals produced or their intermediates may be toxic to the environmental flora and fauna and to humans who may be secondarily exposed to them. This is especially true when in the consideration of a candidate biodegradative process, accumulation of intermediates is not a concern of the engineering component. Thus, for example, toluene, a non–carcinogen and an aquatic non–toxicant, can be converted to benzaldehyde which is a carcinogen as well as an aquatic toxicant.

(e) Finally, biodegradative processes often take place at fixed sites such as landfills that are not readily manipulated. This, in turn, brings into consideration that such sites should not be "poisoned" by bactericidal agents which will either selectively kill some or all of the bacterial components of the biodegradation process (Pangrekar *et al.*, 1994). Thus, agents with such properties should not be processed for biodegradation under such conditions as this may result in the subsequent inability of the bioremediation site to function properly.

Thus, if advanced structure–activity expert systems (such as MULTICASE and META)

(Klopman, 1992; Klopman and Rosenkranz, 1994a,b) are to be applied to biodegradative processes, the above listed considerations may complicate and even present formidable obstacles to such a study.

In the present report, we demonstrate that in spite of these different philosophies with respect to the possible toxicants produced during biodegradative processes, the biodegradative process is amenable to SAR analysis and in fact an SAR model can be developed to predict the biodegradability of yet untested chemicals.

Moreover, the SAR principles developed can be applied to strategies to modify recalcitrant chemicals such that they become biodegradable. On the other hand, it is feasible to develop an expert system similar to META to model biotransformations which takes advantage of the diversity of metabolic processes of individual bacterial species such as to devise chemical-specific biodegradative processes. However, the absence of a concerted effort in this direction and the fact that current experimental and pilot plant processes do not deal with the details of the biodegradative process make the realization of that goal a distant as well as a labor intensive one.

RESULTS AND DISCUSSION

Effects of Molecular Parameters of Biodegradability:

Before embarking upon a full-fledged intensive SAR analysis, we wanted to determine whether overall some physical chemical parameters are, in fact, associated with biodegradability or lack thereof. Thus, lipophilicity has been reported as influencing biodegradability (Pitter and Chudoba, 1990). Accordingly, we have examined the extent of the biodegradability of chemicals in the data base as a function of the calculated log P (where P is the octanol:water partition coefficient; Figure 1). Similarly, we have also examined biodegradability as a function of solubility in water (Figure 2). Neither of these parameters was significantly associated with the extent of biodegradability.

Molecular weights have been used as measures of substituent steric effects as well as contributors to molar refraction (Pitter and Chudoba, 1990), all of which are descriptors which have been included in QSAR analyses of biodegradability (Pitter and Chudoba, 1990). An analysis of the relationship between molecular weight and extent of biodegradability (Figure 3) did not reveal any significant association. Similarly, electronegativity has been suggested as an electronic parameter which would influence the extent of biodegradability (Pitter and Chudoba, 1990). However, our analysis of the relationship of LUMO (Lowest Unoccupied Molecular Orbital) energy, a measure of electrophilicity, did not show such an association (Figure 4). Finally, analyses of the possible relationship of nucleophilicity as expressed numerically by the HOMO (Highest Occupied Molecular Orbital) energy and chemical reactivity as determined by HOMO–LUMO/2, *i.e.* the "hard/soft" ratio (Pearson, 1986), did not reveal any association between these parameters and the extent of biodegradability among the chemicals in the data base (Figures 5 and 6).

However, when we analyzed the differences between the means of log P, LUMO, water solubility and molecular weights for the biodegradable and non–biodegradable molecules, we found highly significant differences (Table 1). This is in accord with earlier observations of an association between these parameters and biodegradability (Pitter and Chudoba, 1990). On the other hand, nucleophilicity (HOMO) did not show a significant difference. Chemical reactivity ("hard/soft" ratio) was significantly different for the two groups. This presumably reflects the significant contribution of LUMO.

These results suggest that presumably log P, water solubility, LUMO and molecular weights affect the biodegradability of subsets of chemicals in the data base. If such is the case, then these parameters should appear as "modulators" in the MULTICASE SAR analyses of the data base.

Data Base:

The data base used for this study consists of a compilation (Pitter and Chudoba, 1990) of 510 organic chemicals for which BOD (biochemical oxygen demand) values had been determined with the aid of mixed cultures. These BOD values were obtained after a 5–day incubation period. Results are expressed as BODx100/TOD values, where TOD represents the theoretical oxygen demand. For the purpose of the present SAR analyses, BODx100/TOD values in excess of 30 were considered to represent biodegradable chemicals. Agents with values above 40 are classified as rapidly degradable (Pitter and Chudoba, 1990). Chemicals with BODx100/TOD values between 20 and 30 were designated as marginally degradable and BODx100/TOD values below 20 were taken to indicate lack of biodegradability. For the MULTICASE SAR analyses BODx100/TOD values were used directly as "CASE Activity units" except that BODx100/TOD values of less than 10 were arbitrarily assigned 10 CASE units of activity.

According to the above classification there are 289 biodegradable, 39 marginally biodegradable and 182 non–biodegradable molecules in the data base.

MULTICASE SAR Analyses:

The MULTICASE structure–activity relational system has been described on a number of occasions (Klopman, 1992; Klopman and Rosenkranz, 1994a,b; Rosenkranz *et al.*, 1993) including the present proceedings (Rosenkranz and Klopman, 1994c). Basically MULTICASE identifies the structural moieties (biophores) associated with the potential for biodegradability. It then searches for the modulators (which may be structural, physical chemical or quantum mechanical) which affect the expression of the potential for biodegradability, *i.e.* the extent or rate of biodegradability. When faced with a new molecule it will determine whether a previously identified biophore is present and if so whether any of the modulators associated with that biophore are present. It will use this information to project a potential for biodegradability (expressed as a percentage) as well as the predicted extent of biodegradability (*i.e.* CASE units) which can readily be converted into BODx100/TOD values.

Analysis of the biodegradation data base with MULTICASE resulted in the identification of a series of major biophores (Table 2) associated with the potential for biodegradability. Thus, biophore 1 (Table 2) which is present in 80 molecules, of which 59 are biodegradable (Table 3), is associated with a 74% probability (Figures 7 and 8) of being biodegradable ($p < 0.001$). Whether and to what extent this potential is realized depends upon the effect of the modulators associated with this biophore (Table 4). Thus, the potential for biodegradability will be affected negatively by high log P values. In addition, the potential for the extent of biodegradability will be augmented or decreased by 4 and 11 structural modulators, respectively (Table 4).

Thus, 3–methylaminopentanedioic acid (Figure 7) contains 2 copies of biophore 1 which endows it with a 74% potential of being biodegradable. The presence of this biophore carries with it 55.1 CASE units of activity. In addition, this molecule also contains the associated modulator No. 1 (Table 4), each copy of which is associated with 19.4 CASE units. The calculated log P of this chemical is –0.76 (Figure 7). Since the formulation of the modulator is $-0.4 \times (\log P)^2$, this assumes a value of –0.2 CASE units. Thus, overall the projected activity of that molecule is 74.3 CASE units (or a BODx100/TOD value of 74), *i.e.* this molecule is predicted to be highly biodegradable. However, NH_2–CH_2 (modulator 1) is also a biophore (No. 3, Table 2; $p < 0.001$). The presence of this second biophore increases the overall probability that this molecule is biodegradable to 90.9% (Figure 7).

On the other hand, 4–methyl–4–propoxy–2–pentanone (Figure 8) also contains biophore 1 (*i.e.* it is associated with a 74% probability of being biodegradable). However, this molecule contains one copy of inactivating modulator No. 2 (Table 4) which endows it with –4.0 CASE units and two copies of inactivating modulator No. 5 (Table 4), *i.e.* 2 x –11.6 = –23.2 CASE units (Figure 8). In addition, the contribution of $-0.4\ (\log P)^2$ is –2.2 for an overall projected activity of 25.7 CASE units which indicates only marginal activity.

Thus in accordance with the expectation of Table 1, we see that log P is associated with the extent of biodegradability of a subset of the data base, *i.e.* those molecules containing biophore 1.

Figure 9 illustrates that the hard/soft index (*i.e.* LUMO energy) is associated with another subset of molecules, *i.e.* those containing biophore 5 (Table 2), and Figure 10 indicates that molecular weight is associated with another group of molecules, *i.e.* those containing biophore 10 (Table 2).

Strain–Specific Metabolism:

Recently, through elective culture procedures and genetic engineering methods, bacterial strains capable of specific and divergent biodegradative pathways have been obtained. Use of these strains has allowed the biodegradation of specific substrates, including those that are recalcitrant to the usual mixed culture approach. Obviously, such methods are highly desirable. However, the use of such pure cultures has also revealed details of specific biodegradative pathways which might actually generate and accumulate products with a potential for health effects. Thus in the degradative schemes shown in Figure 11, while for pathway A neither toluene, m–cresol or the intermediate 1–methylcatechol are predicted to be a carcinogen, methylmuconic acid (Figure 12) is predicted to be a rodent carcinogen. The same is true for 2–chlorocatechol which accumulates in pathway C.

Similarly for pathway B, while p–chlorotoluene is not predicted to be a sensory irritant, the metabolite 1–chloro–4–methylcatechol, which accumulates, is predicted to be a sensory irritant (Figure 13).

These observations suggest that the construction of an expert system to predict species–specific xenobiotic biodegradative metabolism, similar to our recently developed META program for mammalian metabolism (Klopman and Rosenkranz, 1994a,b), coupled to our capabilities for predicting toxicological profiles using MULTICASE may be an effective approach to augmenting biodegradation without the consequent generation of potentially toxic products.

Strategy for Non–Biodegradable Chemicals:

The existence of non–biodegradable substances is one that is of continuous environmental and ecological concern. One of the possible approaches would be to pre–treat such substances prior to biodegradation. As part of our studies using MULTICASE, we explored the possibility that oxidation (*e.g.* exposure to ozone, bleach, etc.) of such recalcitrant chemicals could make them biodegradable.

Thus, we found that while oxidative deamination of benzidine (Figure 14) (*e.g.* as by nitrous acid) increased the potential for biodegradation (*i.e.* 70%) but still would make the rate negligible (10 CASE units), ring oxidation had the potential of greatly increasing biodegradability (*i.e.* 50 and 75 CASE units). Similar results were obtained with a variety of other non–biodegradable agents, *e.g.* dodecane (Figure 15), hexamethylbenzene (Figure 16), dibutyl phthalate (Figure 17). This suggests the possibility of coupling the predictive capability of MULTICASE to developing cost-effective bioremediation processes.

CONCLUSIONS

The present studies indicate that biodegradative processes can be modelled with respect to structural features. Moreover, elucidation of the structural features associated with biodegradability can be used to predict the degradability of molecules as yet untested. The preliminary analyses presented herein also indicate that knowledge of the metabolites generated by degradative processes can be coupled to a determination of their toxicological profiles. This, in turn, can be used to select among alternate processes those with the greatest potential for being environmentally benign and devoid of health risks.

ACKNOWLEDGMENTS

This investigation was supported by the U.S. Environmental Protection Agency (R818275) and Concurrent Technologies Corporation/National Defense Center for Environmental Excellence in support of the U.S. Department of Defense (Contract No. DAAA21-93-C-0046).

REFERENCES

Ashby, J. and R.W. Tennant (1991) Definitive relationships among chemical structure, carcinogenicity and mutagenicity for 301 chemicals tested by the U.S. National Toxicology Program. Mutation Research, 257: 229-306.

Houser, J.J. and G. Klopman (1988) A new tool for the rapid estimation of charge distribution. Journal of Computational Chemistry, 9: 893-904.

Klopman, G., K. Namboodiri and M. Schochet (1985c) Simple method of computing the partition coefficient. Journal of Computational Chemistry, 6: 28–38.

Klopman, G. (1992) MULTICASE 1. A hierarchical Computer Automated Structure Evaluation program. Quantitative Structure–Activity Relationships, 11: 176–184.

Klopman, G., S. Wang and D.M. Balthasar (1992) Estimation of aqueous solubility of organic molecules by the group contribution approach. Application to the study of biodegradation. Journal of Chemical Information and Computer Sciences, 32: 474–482.

Klopman, G. and H.S. Rosenkranz (1994a) Toxicity estimation by chemical substructure analysis: The Tox II program. In: "Decision Support Methodologies for Human Risk Assessment of Toxic Substances".

Klopman, G. and H.S. Rosenkranz (1994b) Prediction of carcinogenicity/mutagenicity using MULTI–CASE. Mutation Research, 305: 33–46.

Pangrekar, J., G. Klopman and H.S. Rosenkranz (1994) An expert system comparison of the structural determinants of chemical toxicity to environmental bacteria. Environmental Toxicology and Chemistry, in press.

Pearson, R.G. (1986) Absolute electronegativity and hardness correlated with molecular orbital theory. Proc. Natl. Acad. Sci. USA, 83: 8440–8441.

Pitter, P. and J. Chudoba (1990) Biodegradability of Organic Substances in the Aquatic Environment. CRC Press, Boca Raton.

Rosenkranz, H.S. and G. Klopman (1990) Structural basis of carcinogenicity in rodents of genotoxicants and non-genotoxicants. Mutation Research, 228: 105-124.

Rosenkranz, H.S., J. Pangrekar and G. Klopman (1993) Similarities in the mechanisms of antibacterial activity (Microtox® Assay) and toxicity to vertebrates. ATLA, 21: 489–500.

Rosenkranz, H.S. and G. Klopman (1994c) Perspective on the use of structure–activity expert systems in toxicology. These Proceedings.

Schaper, M. (1993) Development of a database for sensory irritants and its use in establishing occupational exposure limits. American Industrial Hygiene Association Journal, 54: 488–544.

Spain, J.C., C.A. Pettigrew and B.E. Haigler (1991) Biodegradation of mixed solvents by a strain of *Pseudomonas*. In: "Environmental Biotechnology for Waste Treatment" (G.S. Sayler, R. Fox and J.W. Blackburn, eds.) Environmental Science Research, vol. 41. Plenum Press, New York, pp. 175–184.

TABLE 1

Physical Chemical Properties of Biodegradable Chemicals

	Non-Biodegradable	Biodegradable	*p* Values
HOMO	1.222	1.240	0.4
LUMO	0.970	0.871	0.008
Log P	1.975	0.817	< 0.00001
Water Solubility	0.580	2.267	< 0.00001
Molecular Weight	161.9	127.4	< 0.00001
HARD / SOFT	0.126	0.185	0.009

N = 510

Chemicals with BOD x 100 / TOD values of less than 20 were designated as "non-biodegradable" and chemicals with values in excess of 40 were considered "biodegradable".

TABLE 2

Major Biophores Associated with Biodegradability

Nr.	Fragment 1---2---3---4---5---6---7---8---9---10		Total	Inactive	Marginal	Active
List of BIOPHORE Attributes						
1	CO -CH2-		80	14	7	59
2	OH -CH2-		79	18	6	55
3	NH -CH2-		14	2	2	10
4	NH2-CH2-		13	0	1	12
5	CH =C -CH =	<2-OH >	19	1	1	17
6	CO -O -		23	2	2	19
7	CO -NH2		5	0	0	5
8	N ≡C -		10	0	0	10
9	CH2-CH2-CH -CH2-		9	0	1	8
10	OH -CO -CH -CH3		6	0	0	6
11	OH -SO2-C =CH -CH =C -CH2-		6	0	0	6
12	CH =CH -C =C -CH =	<3-OH >	6	0	0	6
13	OH -CH -CH3		14	2	1	11
14	OH -CH -CH -		11	0	1	10
15	CH3-C =CH -CH =CH -		9	0	1	8
16	CH =CH -CH =CH -C. =CH -CH =CH -		5	0	1	4
17	OH -CO -C =CH -CH =		9	1	1	7
18	CO -CH =		9	1	0	8
19	CH2-CH2-CH2-CH2-CH -		10	1	3	6
20	CH3-CH -C =CH -CH =CH -		5	0	1	4
21	CH2-CH2-C =CH -CH =CH -		6	0	2	4
22	CH3-C =CH2-	<2-CO >	6	0	0	6

<2-OH > indicates a hydroxy substituent on the second non-hydrogen atom from the left.

C. indicates a carbon atom shared by two rings.

Biophore 1 is shown embedded in molecules in Figures 7 and 8; biophore 5 in Figure 9; and biophores 10 and 13 in Figure 10.

The derivation of biophore 1 from among the molecules in the data base is shown in Table 3.

TABLE 3

Derivation of Biophore 1 (CO -CH2) from Among the Molecules in the Data Base

No. of Copies		CASE Activity
1	4-Hydroxy-4-methyl-2-pentanone	46
1	Methyl heptyl ketone	10
1	Methyl hexyl ketone	48
1	Methyl nonanyl ketone	21
2	1,4-Naphthoquinone	38
1	Methyl undecyl ketone	19
1	Butyric acid	48
1	Capriic acid	61
1	Caproic acid	72
1	Caprylic acid	53
1	3,3-Dimethylbutyric acid	10
1	2,3-Dimethylbutyric acid	57
1	2,2-Dimethylglutaric acid	21
1	2,2-Dimethylsuccinic acid	10
1	4,4-Dimethylvaleric acid	10
1	3-n-Dodecenylsuccinic anhydride	72
1	3-Methylbutyric acid	66
1	3-Methylvaleric acid	70
1	4-Methylvaleric acid	67
1	Palmitic acid	37
1	Oenanthic acid	60
1	Pelargonic acid	23
2	Pimelic acid	10
1	4-Phenylbutyric acid	52
1	Valeric acid	60
1	Acetylglycine	29
1	Beta-Alanine	84
1	3-Aminobutyric acid	50
1	4-Aminobutyric acid	56
2	Dibutyl malonate	60
4	Ethylenediaminetetraacetic acid	10
1	4-Chlorophenoxyacetic acid	32
1	3-Bromopropionic acid	50
1	5-Bromovaleric acid	57
1	3-Chlorobutyric acid	46
1	4-Chlorobutyric acid	55
1	3-Chloropionic acid	45
2	Diisobutyl ketone	46
2	Dipropyl ketone	10
2	Ethyl amyl ketone	76
1	Glycolic acid	28
1	Asparagine	53
1	Beta-n-Nonenylsuccinic anhydride	55
1	6-Bromocaproic acid	55
1	3-Bromobutyric acid	54
1	8-Bromocaprylic acid	60
1	5-Chlorovaleric acid	62
1	2-Methyl-4-chlorophenoxy aceti	66
1	Monochloroacetic acid	37
1	Sorbose	65
1	7-Bromoenanthoic acid	53

No. of Copies		CASE Activity
1	Propionic acid	61
2	Glutathione	34
1	Glutamine	42
2	Glutaric acid	57
2	Malonic acid	58
2	Succinic acid	65
1	Methyl octyl ketone	48
1	Methyl amyl ketone	17
1	Methyl butyl ketone	61
1	Methyl propyl ketone	10
2	Diethyl Ketone	71
2	Dibutyl Ketone	10
2	3,3-Dimethylglutaric acid	24
1	e-Caprolactam	84
1	Glycine	78
2	citric acid	53
2	Malic acid	63
1	Glutamic acid	51
1	Aspartic acid	69
1	Methyl isobutyl ketone	55
2	Cyclohexanone	57
1	4-Methoxy-4-methyl-2-pentanone	10
1	Oleic Acid	75
1	Stearic acid	27
1	Methyl ethyl ketone	59
1	2,4,5-Trichlorophenoxyacetic acid	19
1	2,4-Dichlorophenoxyacetic acid	57
2	Adipic acid	10
3	Nitrilotriacetic acid	61

These are the molecules in the data base which lead MULTICASE to identify CO -CH2 as a biophore.

A CASE Activity value > 30 units indicates biodegradability.

Table 3
continued

TABLE 4

MODULATORS related to BIOPHORE 1: CASE Units

CO -CH2- Constant= **55.1**

Fragment

1---2---3---4---5---6---7---8---9---10--------------------------------------

Fragment			Nr.
NH2-CH2-		19.4	1
CO -CH3		-4.0	2
NH -CH2-CH2-		28.9	3
CO -CH2-CH3		9.9	4
CH3-C -CH2-	<2-CH3>	-11.6	5
CH2-N -CH2-CH2-		-11.1	6
Cl -C =C -Cl		-31.4	7
CH2-CH2-CH2-CH =CH -		21.1	8
CH2-CH2-CO -CH2-CH2-		-43.8	9
CH3-CO -CH2-CH2-CH3		-40.7	10
CH3-CO -CH2-CH2-CH2-CH2-CH3		-32.7	11
CO -CH2-CH2-CH2-CH2-CO -OH		-22.5	12
CO -CH2-CH2-CH2-CH2-CH2-CO -OH		-22.4	13
CH3-CO -CH2-CH2-CH2-CH2-CH2-CH2-CH3		-35.9	14
CH3-CO -CH2-CH2-CH2-CH2-CH2-CH2-CH2-CH2-		-21.4	15
(Log P)**2		-0.4	

The presence of the biophore CO -CH2- is associated with 55.1 CASE units of activity. This potential will be increased or decreased by the presence of the modulators associated with this biophore (see text and Figures 7 and 8).

Modulator 1 is shown embedded in a molecule in Figure 7 and modulators 2 and 5 are shown in Figure 8.

Log P was calculated as described previously (Klopman <u>et al.</u>, 1985).

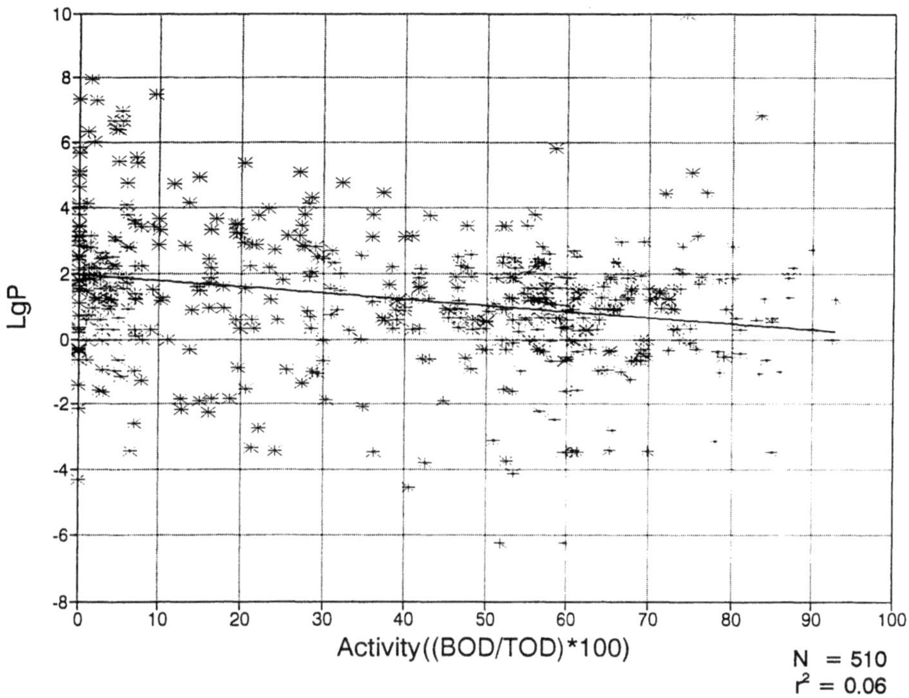

FIGURE 1: Relationship between biodegradability and the octanol:water partition coefficient (log P). Log P was calculated as described previously (Klopman *et al.*, 1985).

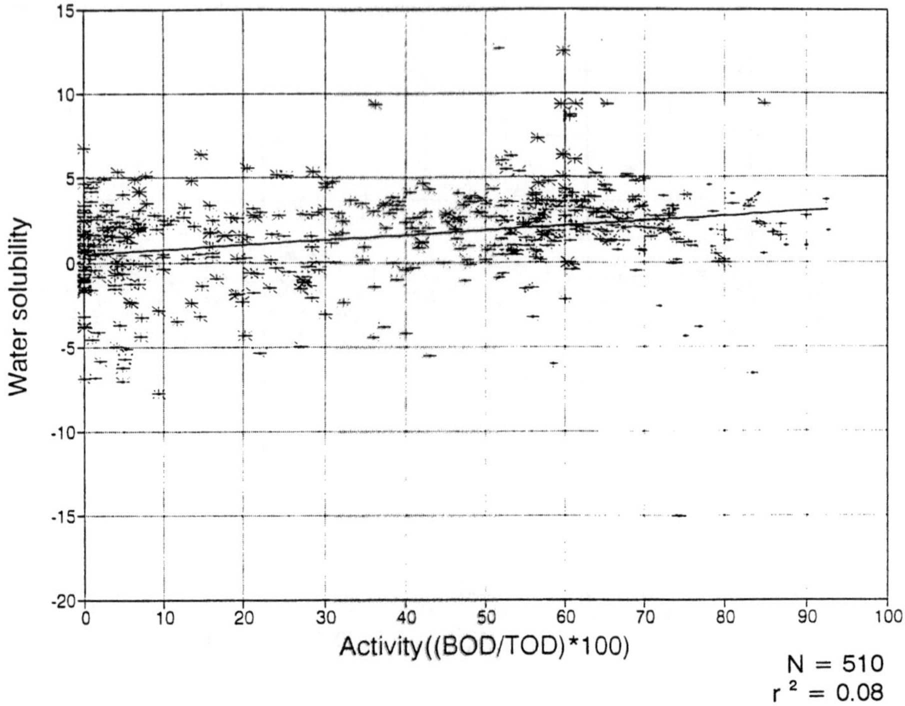

FIGURE 2: Relationship between biodegradability and solubility in water. Solubility was calculated as described previously (Klopman *et al.*, 1992).

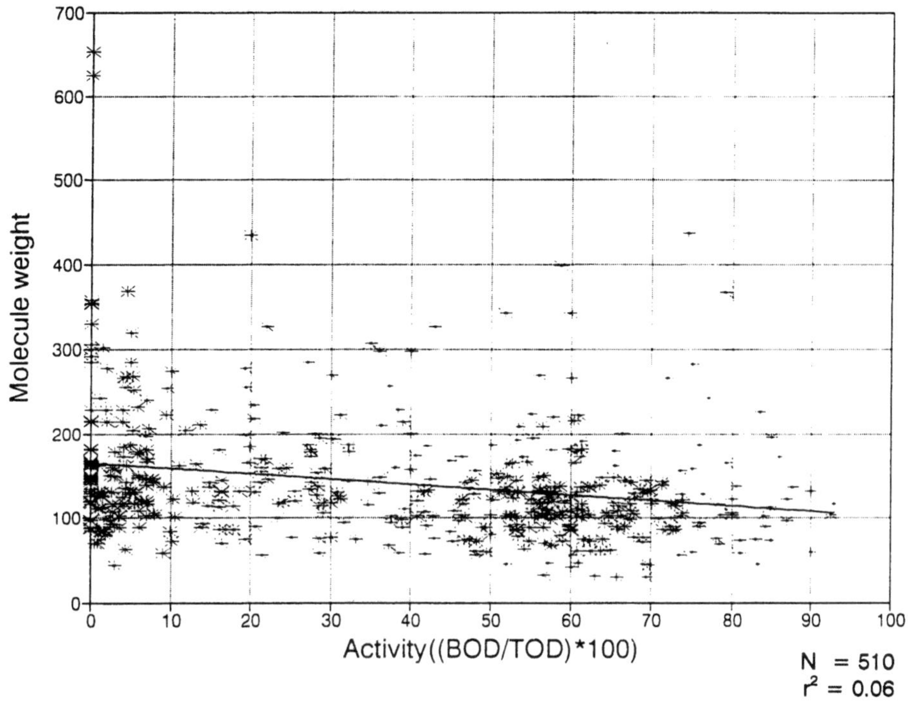

FIGURE 3: Relationship between biodegradability and molecular weight.

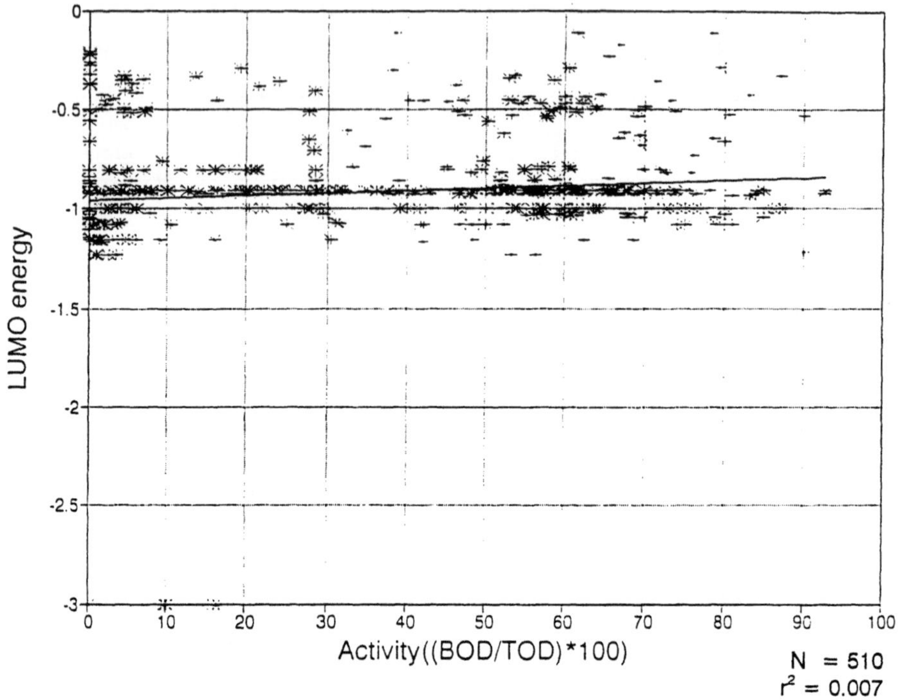

FIGURE 4: Relationship between LUMO (Lowest Unoccupied Molecular Orbital) energy, a quantitative measure of electrophilicity, and biodegradability. LUMO energy was calculated as described previously (Houser and Klopman, 1988).

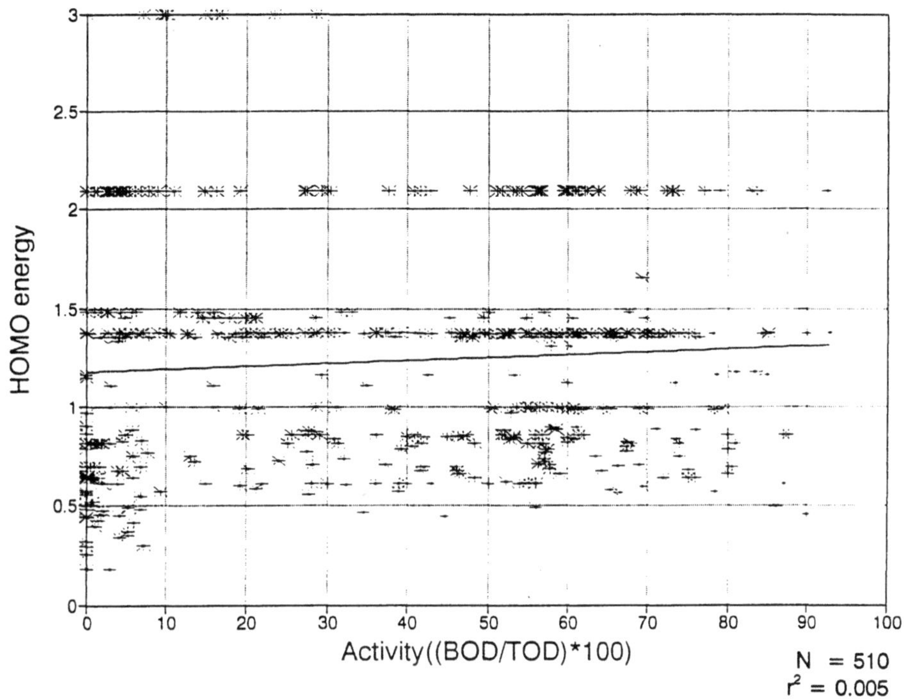

FIGURE 5: Relationship between HOMO (Highest Occupied Molecular Orbital) energy, a measure of nucleophilicity, and biodegradability. HOMO energy was calculated as described previously (Houser and Klopman, 1988).

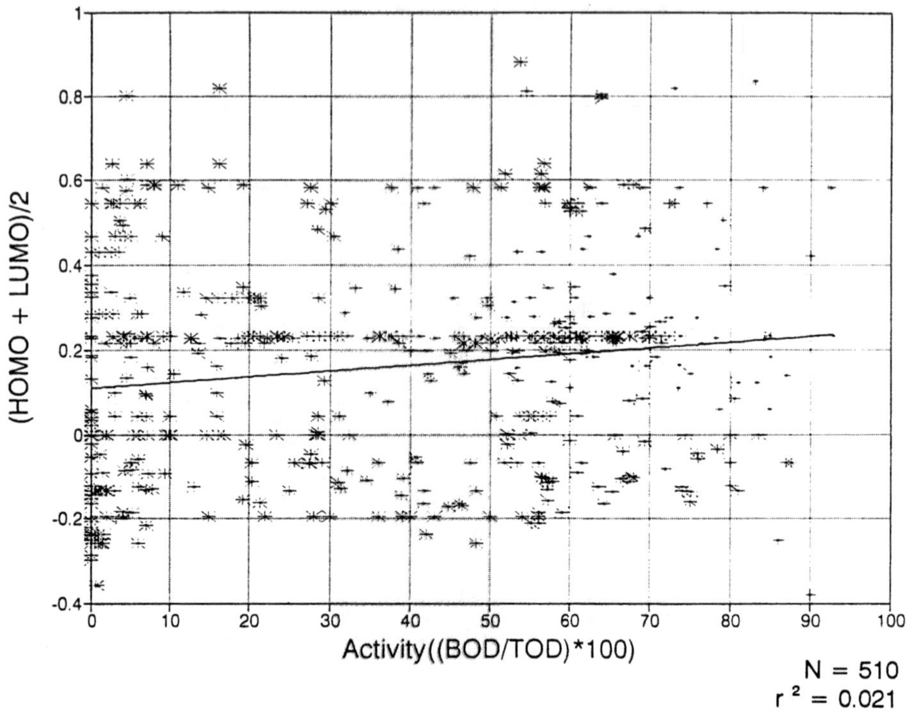

FIGURE 6: Relationship between hard/soft ((HOMO + LUMO) / 2) coefficient (Pearson, 1986) and biodegradability.

```
*******************************************************************************
                        3-METHYLAMINOPENTANEDIOIC ACID
*******************************************************************************

The molecule contains the Biophore   (nr.occ.= 2):

               CO   -CH2

     *** 59 out of the known  80 molecules ( 74%) containing such Biophore
         are Biodegradable with an average activity of  46. (conf.level=100%)
                                                         Constant is     55.1
     ** The following Modulators are also present:
    ( 1) NH2-CH2-                                             Activating   19.4
         Log partition coeff.= -0.76 ;             LogP contribution is   -0.2

        The molecule also contains the Biophore    :
           NH2-CH2-

     ** The probability that this molecule is Biodegradable is  80.0%  **
           increased to  90.9% due to the presence of the extra Biophore

     ** The compound is predicted to be EXTREMELY Biodegradable **
     ** The projected Biodegradability is 74.3 CASE units **
```

B1

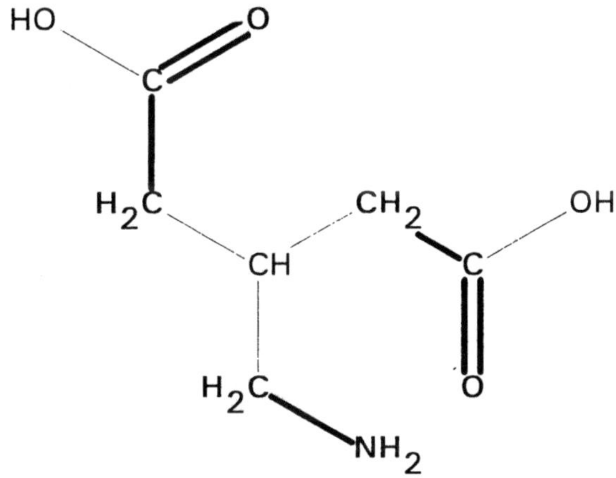

B1

M1

FIGURE 7: MULTICASE prediction of the biodegradability of 3-methylaminopentanedioic acid. This molecule contains two copies of biophore 1 (Table 2) which is present in 80 molecules of the learning set, 59 of which are biodegradable (Table 3). The presence of biophore 1 is associated with 55.1 CASE units. The molecule also contains one copy of the activating modulator 1 (M1, shown in bold) which contributes 19.4 CASE units. -0.4 (log P)^2 is also a modulator (Table 3). The calculated log P of the molecule is -0.76, hence the overall contribution of this modulator is -0.2 CASE units. [Log P is calculated by MULTICASE as described previously (Klopman et al., 1985)].

```
*******************************************************************************
                        4-METHYL-4-PROPOXY-2-PENTANONE
*******************************************************************************

The molecule contains the Biophore   (nr.occ.= 1):

              CO  -CH2

    *** 59 out of the known  80 molecules ( 74%) containing such Biophore
        are Biodegradable with an average activity of  46. (conf.level=100%)
                                                         Constant is     55.1
    ** The following Modulators are also present:
   ( 1) CO -CH3                                              Inactivating   -4.0
   ( 2) CH3-C  -CH2-                                 <2-CH3> Inactivating  -23.2
        Log partition coeff.=  2.32 ;                LogP contribution is   -2.2

    ** The probability that this molecule is Biodegradable is  61.5%  **

    ** The compound is predicted to be MARGINALLY Biodegradable **
    ** The projected Biodegradability is 25.7 CASE units **
```

B1

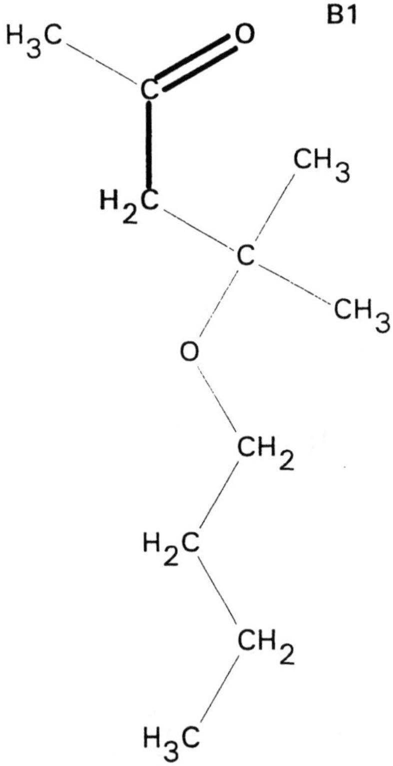

M2

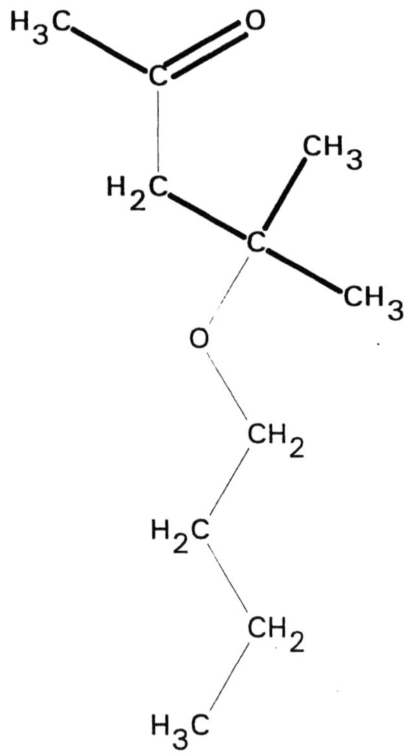

5

FIGURE 8: MULTICASE prediction of the marginal biodegradability of 4-methyl-4-propoxy-2-pentanone. For interpretation of the results, see legend to Figure 8. [Log P is calculated by MULTICASE as described previously (Klopman et al., 1985)].

```
********************************************************************************
                                  4-CHLOROPHENOL
********************************************************************************

The molecule contains the Biophore (nr. occ. = 1):

             OH
            /
      CH =C
            \
             CH"

***   17 out of the known 19 molecules (89%) containing such Biophore are
      Biodegradable with an average activity of 52.  (conf. level = 100%)
                                                          Constant is 116.0
********************************************************************************
      Log partition coeff. = 2.12;                 Log P contribution is 7.3
      Hard/Soft index is   = 0.87;                 Its contribution is -75.4

** The probability that this molecule is Biodegradable is 90.0% **

** The compound is predicted to be VERY Biodegradable **
** The projected Biodegradability is 48.0 CASE units **
```

A

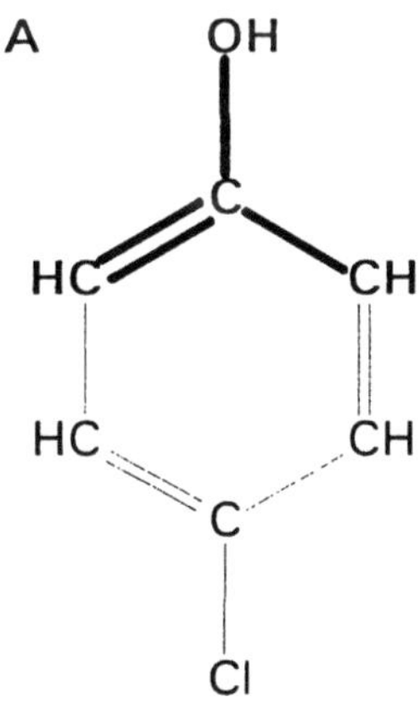

FIGURE 9: MULTICASE prediction of the biodegradability of 4-chlorophenol. This example illustrates that the biophore associated with the potential for activity is modulated by log P (11.2 log P and −3.7 $(\log P)^2$) and the hard/soft index (−87.0 (HOMO–LUMO)/2). Log P, LUMO and HOMO energies are calculated as described previously (Houser and Klopman, 1988).

```
*********************************************************************************
                               LACTIC ACID
*********************************************************************************
```

The molecule contains the Biophore (nr. occ. = 1):

```
(A)    OH —CO
             \
              CH —CH3
```

```
***    6 out of the known 6 molecules (100%) containing such Biophore are
       Biodegradable with an average activity of 59.  (conf. level = 98%)
                                                            Constant is -68.5
       Log partition coeff. = -0.84;                Log P contribution is 0.5
       Ln Nr. Bi/Mol. Wt.   = -4.50;           Nr. Bioph/MW contrib. is 127.6
```

The molecule also contains the Biophore:

```
(B)    OH —CH —CH3
```

```
**     The probability that this molecule is Biodegradable is 87.5% **
       increased to 92.9% due to the presence of the extra Biophore
```

```
** The compound is predicted to be EXTREMELY Biodegradable **
** The projected Biodegradability is 60.0 CASE units **
```

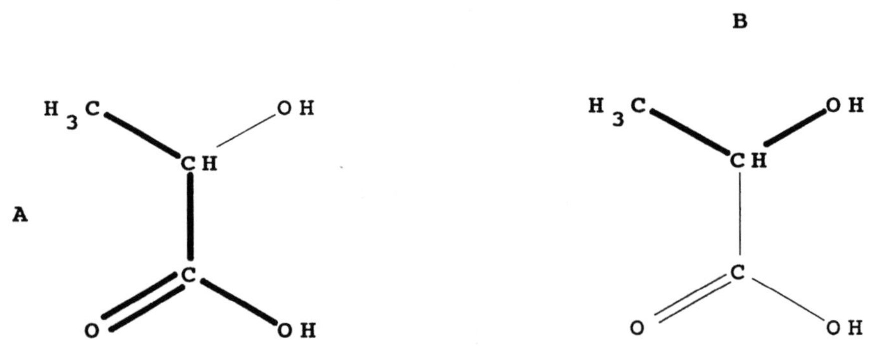

<u>FIGURE 10:</u> MULTICASE prediction of the biodegradability of lactic acid. This example illustrates that the biophore is modulated by log P (0.8 $(\log P)^2$) and molecular weight (−28.3 ln No. of biophores/mol. wt., where ln is the natural logarithm).

Toluene

↘

(A) 1-Methylcatechol → Methylmuconic Acid →

↗

m-Cresol

(B) p-Chlorotoluene → 1-Chloro-4-methylcatechol ↛

4-Chlorophenol

↘

(C) 2-Chlorocatechol ↛

↗

3-Chlorophenol

FIGURE 11: Metabolism of substituted aromatic compounds by a *Pseudomonas sp.* (Spain *et al.*, 1991).

```
********************************************************************************
                              METHYLMUCONIC ACID
********************************************************************************

The molecule contains the Biophore   (nr.occ.= 1):

            CO  -CH"

     ***  3 out of the known   3 molecules (100%) containing such Biophore
       are Rodent carcinogens with an average activity of 46 (conf. level = 87%)
                                                          Constant is     61.2
         Log partition coeff.=  0.35 ;             LogP contribution is   -0.7

     ** The probability that this molecule is a Rodent carcinogen is  80.0%  **

     ** The compound is predicted to be EXTREMELY active **
     ** The projected Rodent carcinogenicity activity is 61.0 CASE units **
```

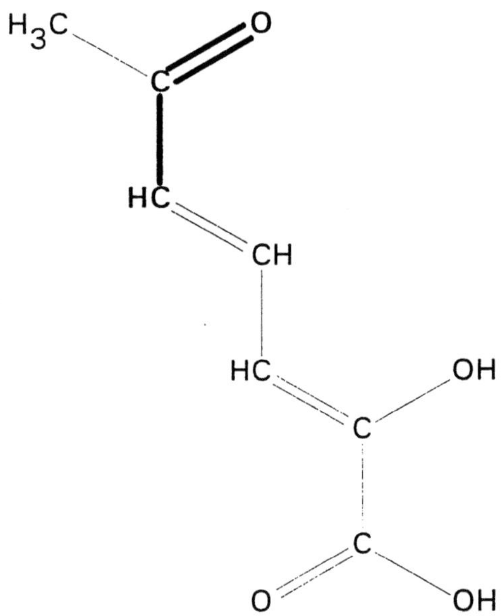

FIGURE 12: MULTICASE prediction of the carcinogenicity in rodents of methylmuconic acid, a biodegradation product of toluene and of m-cresol (Figure 11). The prediction is based upon the rodent carcinogenicity bioassays conducted under the aegis of the U.S. National Toxicology Program (Ashby and Tennant, 1991). A CASE analysis of a subset of that data base has been published (Rosenkranz and Klopman, 1990). An activity of 61 indicates a potential for inducing cancers in male and female mice and rats at multiple sites.

```
********************************************************************************
                          1-CHLORO-4-METHYLCATECHOL
********************************************************************************

The molecule contains the Biophore   (nr.occ.= 2):

              CH   =CH
                      \
                       C
                      //
                 C

    *** 36 out of the known  41 molecules ( 88%) containing such Biophore
      are Sensory irritants with an average activity of 64 (conf. level = 100%)
                                                          Constant is     33.8
       Log partition coeff.=  2.39 ;            LogP contribution is    11.9

    ** The probability that this molecule is a Sensory irritant is  90.2%  **

    ** The compound is predicted to be VERY active **
    ** The projected Sensory irritation activity is   46.0     CASE units **
```

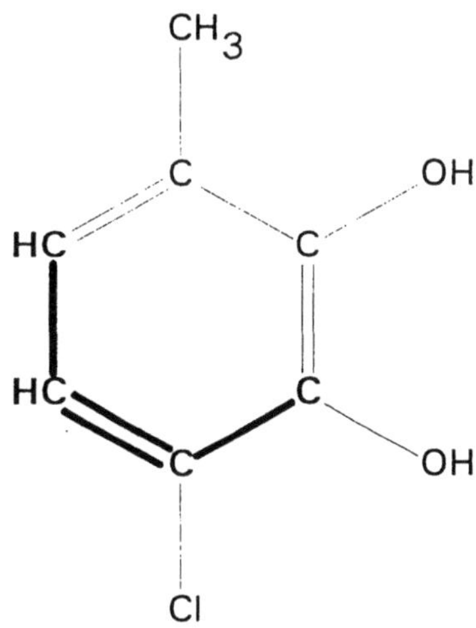

FIGURE 13: MULTICASE prediction of the ability of 1-chloro-4-methylcatechol, a biodegradation product of p-chlorotoluene, to induce sensory irritation in mice. An activity of 46 CASE units corresponds to an RD_{50} value of 13.7 ppm. The sensory irritation data base is taken from the compilation of Schaper (1993).

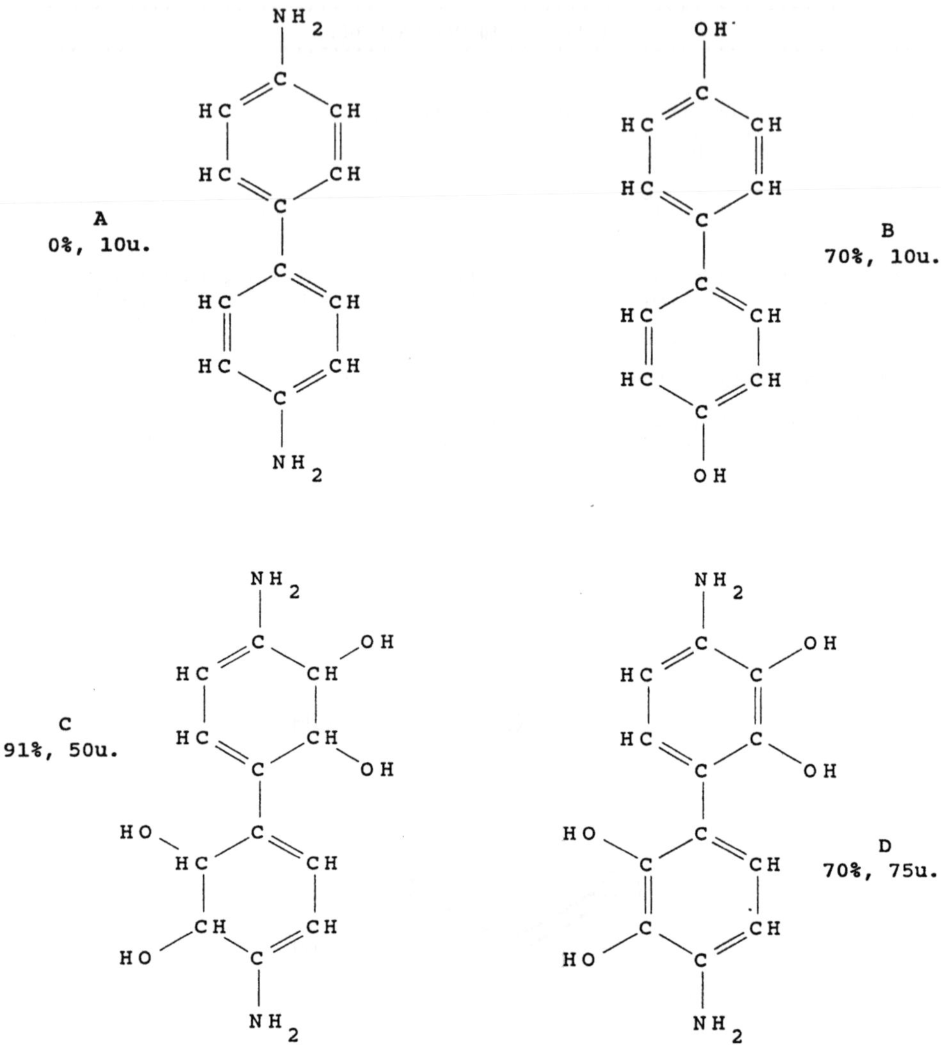

FIGURE 14: Effect of pre-oxidation on the predicted biodegradability of benzidine (A). Benzidine is non-biodegradable (probability 0%, 10 CASE units). Oxidative deamination to B results in an increased probability of biodegradability (70%) but still no appreciable degradability (10 CASE units). On the other hand, ring oxidation to C or D results in predictions of greatly increased biodegradability (50 and 75 CASE units).

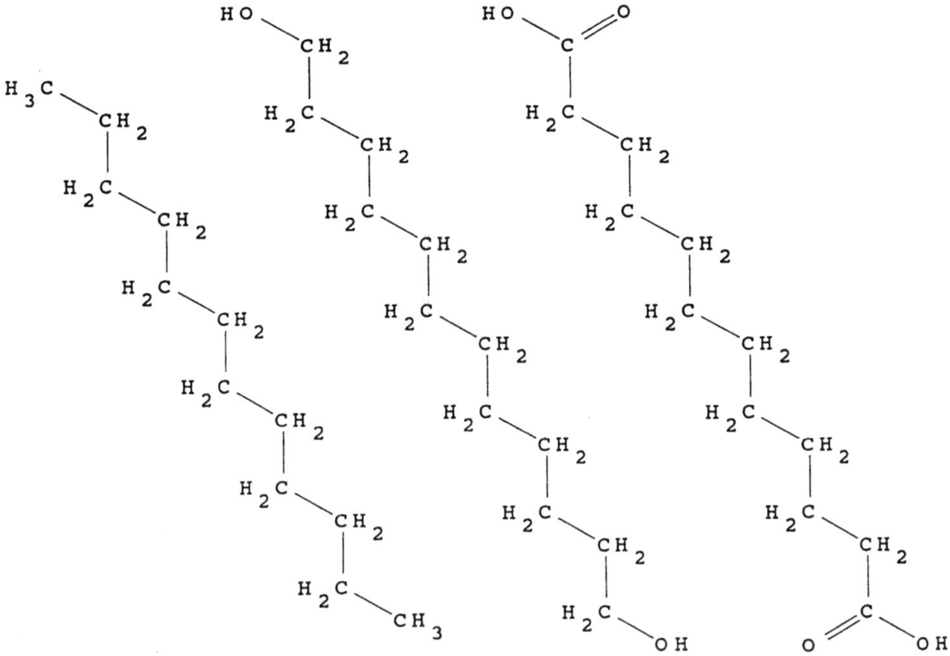

A
0%, 10u.

B
74%, 43u.

C
80%, 51u.

FIGURE 15: Effect of pre-oxidation on the predicted biodegradability of dodecane. Dodecane (A) is predicted to be non-biodegradable (0%, 10 CASE units). Oxidation to the dialcohol (B) or dicarboxylic acid (C) results in predictions of greatly increased biodegradability (B: 74%, 43 CASE units; C: 80%, 51 CASE units).

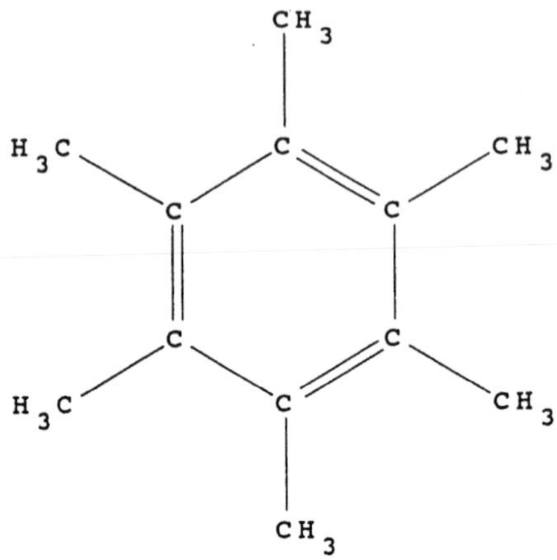

A
0%, 10u.

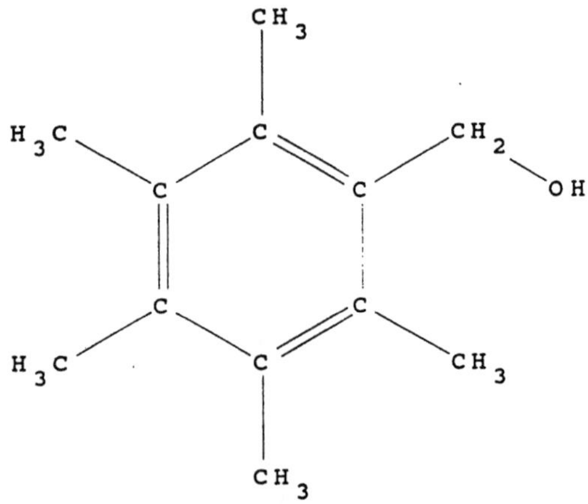

B
74.2%, 43u.

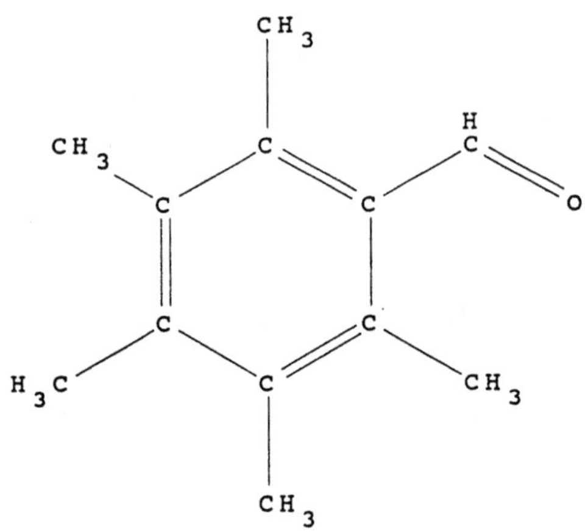

C
83.0%, 66u.

FIGURE 16: Effect of pre-oxidation on the predicted biodegradability of hexamethylbenzene (A). While A is non-biodegradable (0%, 10 CASE units), oxidation of a methyl substituent to either an alcohol (B) or an aldehyde (C) results in predictions of greatly increased biodegradability.

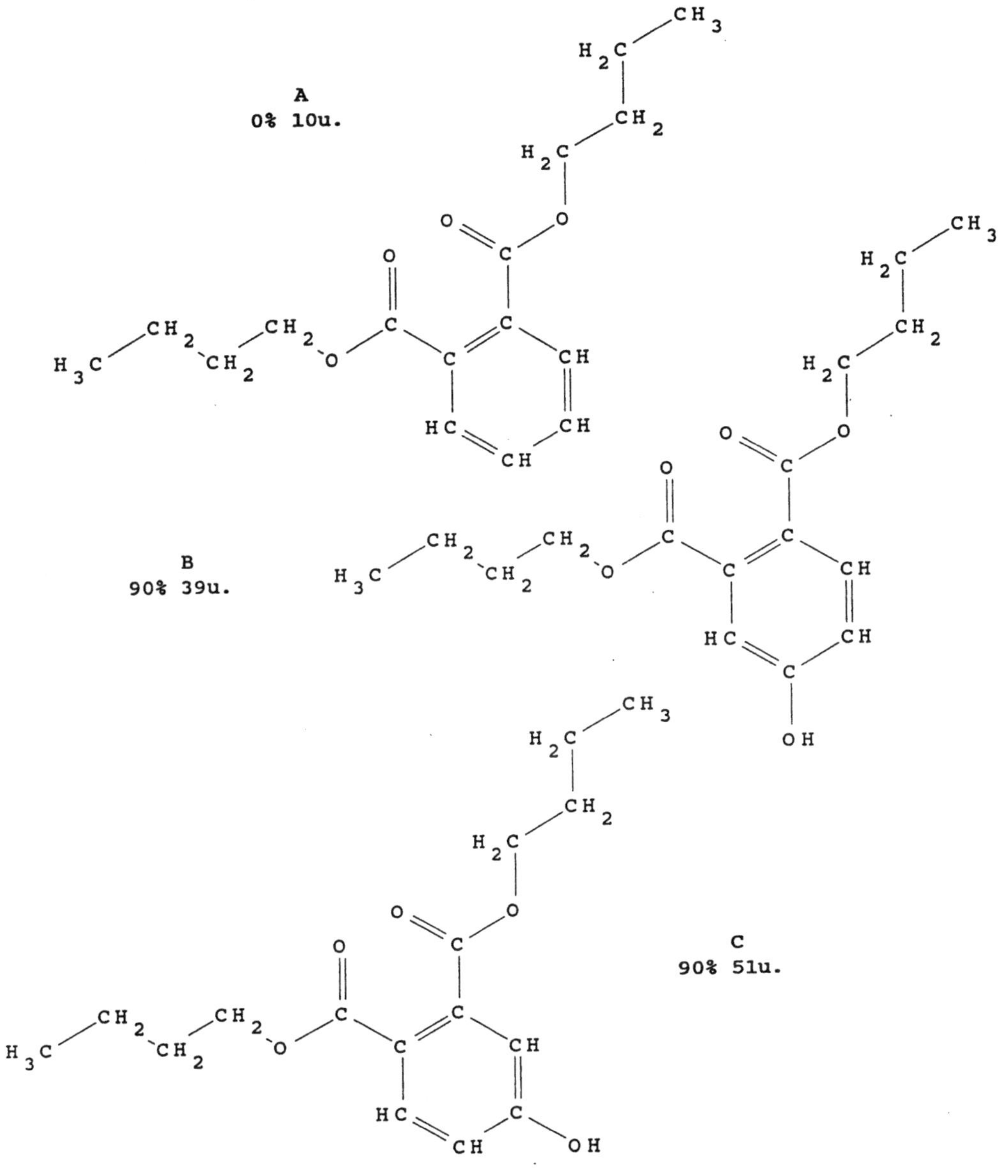

FIGURE 17: Effect of pre-oxidation on the predicted biodegradability of dibutylphthalate (A). While A is non-biodegradable (0%, 10 CASE units), the monohydroxylated derivatives B and C are predicted to be highly biodegradable.

PHARMACOKINETIC CONSIDERATIONS IN THE TOXICOLOGIC EVALUATION OF XENOBIOTICS

Roger K. Verbeeck
Pharmacokinetics and Drug Metabolism Laboratory
School of Pharmacy
Catholic University of Louvain
1200 Brussels
Belgium

1. INTRODUCTION

Toxicokinetics can be defined as the kinetics of absorption, distribution and elimination (i.e. excretion and metabolism) of toxic substances, including therapeutic agents, in animal species used in experimental toxicology. Toxicokinetics is different from pharmacokinetics in that it describes the rates of absorption, distribution and elimination of xenobiotics at relatively high doses often associated with toxic effects. At these high doses the reversible binding of the xenobiotic to plasma and/or tissue proteins may become saturated, as well as the active transport systems and metabolizing enzymes. Consequently, in toxicokinetics one frequently encounters nonlinearity in these rate processes leading to dose-dependent kinetics. This is in contrast to the pharmacokinetic behavior of most drugs following administration of therapeutic doses which usually can be described in terms of linear kinetics. Another important factor which can account or contribute to dose-dependent toxicokinetic behavior of xenobiotics is the potential interaction of the xenobiotic and/or its metabolites with physiological processes such as regional blood flow, urine pH, gastric emptying, etc (Table 1). Such alterations in the normal functioning of the system induced by high doses of the xenobiotic will further complicate the interpretation of toxicokinetic data.

NATO ASI Series, Vol. H 93
Modulation of Cellular Responses in Toxicity
Edited by C. L. Galli, A. M. Goldberg, M. Marinovich

TABLE 1: POSSIBLE CAUSES FOR DOSE- AND TIME-DEPENDENT KINETICS

A. CAPACITY-LIMITED PROCESSES

- carrier-mediated transport across biological membranes, for instance:
  + intestinal absorption
  + transport across the blood brain barrier
  + renal tubular secretion
  + biliary secretion
- reversible binding to plasma proteins and tissue macromolecules
- enzyme-catalyzed metabolic processes

B. EFFECTS OF THE XENOBIOTIC ON THE BODY, e.g.:

- alteration of renal, hepatic, cardiac, etc function
- modification of urine pH and urine flow
- alterations in organ perfusion rates
- depletion of cosubstrates for metabolic processes
- enzyme autoinduction
- enzyme inhibition by metabolites (product inhibition)

It is no longer sufficient to simply administer large doses of a xenobiotic and to describe its toxic effects. It is, however, essential to understand the pharmacokinetic processes (absorption, distribution, excretion and metabolism) which will determine the concentrations of xenobiotic in blood and target tissues to correctly interpret toxic effects of xenobiotics and to extrapolate toxicity data from one species to another. Figure 1 illustrates why dose-toxic response relationships are more difficult to interpret than blood concentration-toxic response relationships. After administration of large doses of a xenobiotic the rates of absorption, distribution, and elimination (i.e. excretion and metabolism) often are no longer linear. This means that blood and target tissue concentrations of the xenobiotic do no longer change proportionately with the administered dose. To understand and correctly interpret the dynamics of the toxicicological response of an animal to a xenobiotic, toxicokinetic parameters based on blood concentrations of the xenobiotic will have to be integrated with the toxicodynamic parameters. Just as pharmacokinetic-pharmacodynamic (PK/PD) integration has gained widespread use even during the development of a new therapeutic agent (Levy, 1993; Peck, 1993), the integrated study of toxicokinetic-toxicodynamic relations, at least from a pharmaceutical industrial standpoint, will become extremely important in the future (Welling, 1993).

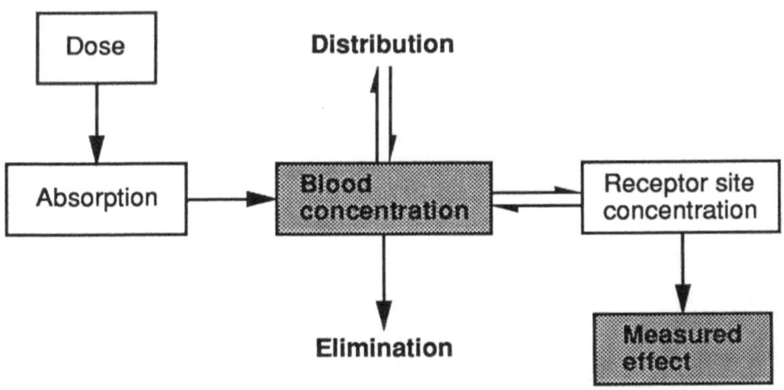

FIG. 1. Schematic diagram illustrating that the DOSE-EFFECT relationship is influenced by the pharmacokinetic processes of absorption, distribution and elimination. A more direct relationship, therefore, exists between xenobiotic BLOOD CONCENTRATIONS, as compared to administered dose, and toxic EFFECT.

In addition, interspecies differences in the susceptibililty to toxic effects of xenobiotics may be explained on the basis of differences in the pharmacokinetic behavior of the compound tested in several species of laboratory animals. The intent of this chapter is to provide a brief review of the processes of absorption, distribution, metabolism and excretion of xenobiotics and to illustrate their effect on the interpretation of toxicity studies in animals and man. For a more detailed description of the pharmacokinetics of drugs and xenobiotics, the reader is referred to a number of excellent text books on this subject (Rowland and Tozer, 1989; Gibaldi and Perrier, 1982; Timbrell, 1991).

2. ABSORPTION

Following oral ingestion most xenobiotics are absorbed by passive diffusion, a process which is not saturable, from the upper gastrointestinal tract. A saturable active transport process is involved in the oral absorption of only a limited number of xenobiotics and oral availability may be drastically reduced after ingestion of large doses of such compounds (Fig 2). It is obvious that the interpretation of toxicity data of such a compound following oral administration may be wrong if blood concentrations are not monitored. Problems of dose-dependent oral bioavailability may also arise because of dissolution considerations, especially if the compound of interest is only sparingly soluble in the aqueous environment of the upper gastrointestinal tract. Above the solubility limit of the xenobiotic in the gastrointestinal juices, the time required for complete dissolution may exceed the intestinal transit time. Consequently, the oral bioavailability will decrease after administration of relatively large doses of a sparingly water soluble xenobiotic. The limited aqueous solubility of 2,3,7,8-tetrabromodibenzo-p-dioxin (TBDD) e.g. may explain its reduced oral bioavailability at high doses. Diliberto et al. (1993)

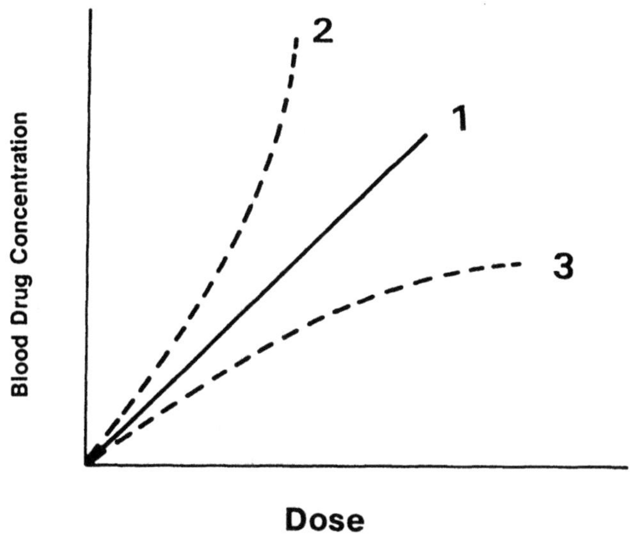

FIG. 2. Graph illustrating that blood concentrations of a xenobiotic are proportionate to the administered dose in case of absorption by passive diffusion (line 1). When the xenobiotic undergoes saturable pre-systemic elimination, blood concentrations increase disproportionately with the administered dose (graph 2). Curve 2 also describes the effect of increasing doses on the blood concentrations of a xenobiotic eliminated by a saturable process (enzymatic metabolism or active tubular secretion). When a xenobiotic is absorbed by a saturable carrier-mediated transport system, or when its solubility at the absorption site is limited the blood concentration-dose curve is nonlinear and shows a maximum (curve 3).

studied the disposition of TBDD in the rat and found a significant decrease in the oral bioavailability of TBDD after administration of higher doses. Oral bioavailability was only 47% following administration of a 500 nmol/kg dose as compared to 80% for the 1 and 10 nmol/kg dose. The oral dosing solution of TBDD was prepared in water/ethanol/Emulphor EL-620 (3/1/1). However, it is likely that TBDD may have precipitated in the gastrointestinal tract, especially at higher doses thus limiting the oral availabiliy.

An interesting example of how a xenobiotic may affect its own absorption has been described with respect to chloroquine (Varga, 1966). Chloroquine causes a dose-dependent inhibition of gastric emptying. Since gastric emptying is an important determinant of the rate of absorption of many xenobiotics because absorption usually takes place in the upper small intestine, the dose-dependent inhibition of gastric emptying following oral administration of chloroquine will eventually slow down its own gastrointestinal absorpton rate. This explains the nearly 18-fold increase in the LD50 of chloroquine in the rat following oral dosing (1080 mg/kg) compared to the intravenous route (60 mg/kg) (Levy, 1968).

Another important phenomenon influencing the quantity of xenobiotic which will reach the systemic circulation following oral exposure is the so-called first-pass effect or presystemic elimination. For most xenobiotics presystemic elimination is most important in the liver, but the intestinal mucosa may also contribute to this so-called first-pass effect (Pond & Tozer, 1984). The lungs are also exposed to the total administered dose and may therefore contribute to the first-pass effect following oral dosing. Contribution of the lungs to the overall presystemic elimination of an orally administered xenobiotic is difficult to evaluate because also following i.v. adminsistration (the reference route) the total dose has to pass across the lungs before reaching the systemic circulation and thus may undergo presystemic elimination. Cassidy and Houston (1980) investigated the role of extrahepatic metabolism in the presystemic conjugative metabolism of phenol in the rat. To estimate the contribution of the small intestine, the liver and the lungs to the overall first-pass effect of phenol in the rat, the experiments were carried out following oral, hepatic portal venous, intravenous and intra-arterial administration of a relatively small dose of 1.5 mg/kg. They found that the small intestine (92% extraction) and the lungs (62% extraction) contributed the most to the first-pass effect of phenol in the rat, whereas the contribution of the liver was only marginal (6% extraction). Following oral administration of large doses of xenobiotics undergoing presystemic elimination, it is possible to saturate the enzymes responsible for the first-pass effect. As a result, bioavailability is nonlinear such that an increase in administered dose produces a disproportionate increase in the systemic blood levels (Fig 2, curve 2). For instance, salicylamide undergoes extensive presystemic gut wall and hepatic elimination in man to inactive glucuronide and sulfate conjugates. Following oral administration of a 1 gram dose, virtually all of the drug is inactivated by presystemic

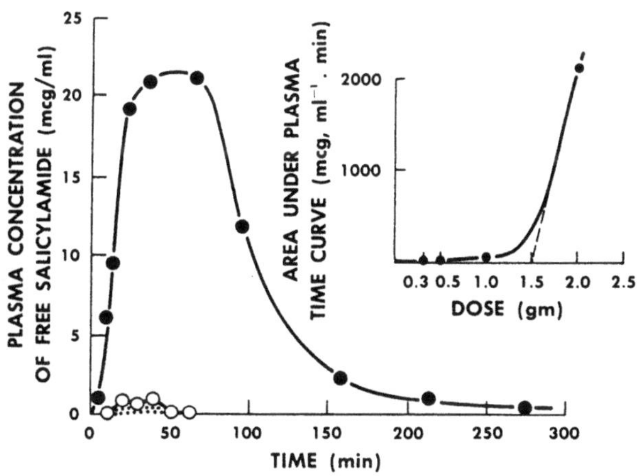

Fig. 3. Plasma concentrations of unchanged (free) salicylamide after oral administration of 1 g (O) or 2 (●) dose. Dotted lines indicate the salicylamide plasma concentrations following oral administration of a 0.3 and 0.5 g dose. The inset shows the corresponding areas under the plasma concentration-time curves for salicylamide as a function of oral dose (from Barr, with permission).

metabolism and the oral bioavailability is extremely low (Barr, 1969). However, following administration of oral salicylamide doses above 1 gram, the area under the salicylamide plasma concentration-time curve increases enormously as a function of the administered dose (Fig. 3). The difference in AUC, and therefore in oral bioavailability, between a 1-g and a 2-g oral dose is spectacular: such a doubling of the dose increases the plasma salcylamide levels several hundred-fold leading to an unexpected rise in therapeutic and toxic effects.

The role of presystemic elimination in explaining route-dependent differences in toxicity is extremely well illustrated by considering the LD50 of nicotine in mice which is 0.3 mg/kg following i.v. and 230 mg/kg following oral administration!

Absorption of xenobiotics through the skin or via the lungs is also toxicologically significant. Lipophilic substances such as the insecticide parathion may cause systemic toxicity in agricultural workers following skin contact. However, lipophilicity is not the only factor influencing percutaneous absorption. In addition, skin tissue contains most of the enzymes capable of metabolizing xenobiotics and biotransformation of xenobiotics during percutaneous absorption may lead to detoxification or, conversely, to the formation of toxic metabolites which could have a local or systemic effect (Kao and Carver, 1990). Pulmonary absorption of foreign substances such as toxic gases, solvent vapors and small particles may be extremely rapid and efficient due to the large alveolar surface area and very high blood flow.

3. DISTRIBUTION

The distribution of a xenobiotic to the different body tissues and organs is affected by many factors such as the rate of diffusion or carrier-mediated transport across biological membranes, blood flow to the various body regions, blood-tissue partitioning and reversible binding to plasma and tissue proteins. Again, following administration of large doses of xenobiotics one or more of these processes may be affected in a way which cannot be predicted based on the distribution characteristics of the xenobiotic at lower doses. For instance, at high doses the xenobiotic may interfere with regional organ and tissue blood flow or binding sites in plasma and/or tissues may be saturated (Table 1). Saturable binding of xenobiotics and its effects on their pharmacokinetic behavior have been well studied (Jusko and Gretch, 1976). Albumin and α_1-acid glycoprotein are the two most important binding proteins for xenobiotics in plasma. Plasma levels of albumin e.g. have been shown to vary between approximately 2 and 5 g/100 ml in eight mammalian species including the mouse, the rat and man (Puigdemont et al., 1989). Assuming one high affinity binding site per albumin molecule, it is evident that when the total plasma concentration of xenobiotic approaches that of the binding protein saturable binding will occur. For a xenobiotic mainly bound at a single site on albumin, saturable plasma protein binding will occur at xenobiotic plasma concentrations of approximately 0.3-0.6 mM (i.e. 75-150 μg/ml for a xenobiotic with a molecular weight of 250 daltons). Such total plasma concentrations will be readily obtained for xenobiotics administered at relatively high doses and having a small distribution volume. For xenobiotics mainly bound to α_1-acid glycoprotein, saturation of the reversible binding will occur at much lower total plasma concentrations of the xenobiotic, because plasma levels of this acute phase protein are 50 to 100 times lower than those of albumin (Tozer, 1984). Saturation of the plasma protein binding sites means that the unbound fraction (fu) of xenobiotic will increase as total plasma concentrations of xenobiotic increase. An increase in fu may have a profound effect on the distribution, elimination and biological activity (toxicity) of the xenobiotic. For instance, administration of doses of the diuretic furosemide above the threshold dose for saturation of plasma protein binding sites in the mouse rapidly causes hepatic damage due to a disproportionate increase in the concentration of the drug in the liver, which is at least partly due to saturation of plasma protein binding sites (Mitchell et al., 1975). Therefore, to understand the mechanisms responsible for unexpected increases in toxicity involving relatively large doses in laboratory animals, it appears essential to determine both total and unbound plasma concentrations of the xenobiotic.

Highly lipophilic xenobiotics such as the chlorinated hydrocarbon insecticides (e.g. DDT) accumulate in adipose tissue. Sequestration of DDT in adipose tissue may be protective, since it may prevent high concentrations of the xenobiotic at the target sites for toxicity. However, mobilization of the fat in the adipose tissue may cause a sudden release of the xenobiotic into the bloodstream with a dramatic rise in concentration in blood and possibly the target tissues. In addition, sequestration of a lipohilic xenobiotic in body fat may also be saturable. Consequently, toxicity may rapidly increase above the threshold dose for saturation of the body fat. Distribution of 2,3,7,8-tetrabromodibenzo-p-dioxin in adipose tissue in the rat decreased as a function of the orally absorbed dose from 24.8 ± 1.9% to 12.7 ± 1.4% at doses of 1 nmol/kg and 500 nmol/kg, respectively (Diliberto et al., 1993). Consequently, relatively more xenobiotic will be available for distribution into other tissues, possibly including the target tissues for toxicity, following administration of high doses of this herbicide.

4. METABOLISM

The biotransformation of xenobiotics is controlled by enzymatic processes that are described by Michaelis-Menten kinetics. This means that the metabolic elimination rate of most xenobiotics following administration of low doses may be described by linear or first-order kinetics. However, when a wider range of doses is administered, as often occurs in toxicity evaluation, the metabolic processes may become saturated. The consequences of such saturable elimination may be profound both in terms of disposition and biological effects: the AUC (area under the blood concentration-time profile) vs. dose relationship is curvilinear (concave-increasing) due to a decrease in total body clearance with increasing dose (Fig 2, curve 2), and the dose-response relationship may show an unusually large increase in pharmacologic (toxic) effects with increasing dose. In addition, saturable metabolism may also be due to depletion of cosubstrate as has clearly been shown for sulfation of acetaminophen in the rat. The capacity-limited factor in the conversion of acetaminophen to its sulfate conjugate is the availability of inorganic sulfate. Galinsky and Levy (1981) demonstrated that the time-dependent increase in acetaminophen plasma concentrations in rats receiving an intra-arterial infusion of the drug for several hours could be prevented if sodium sulfate was simultaneously infused. This experiment showed the important role of normal sulfate body stores on the elimination kinetics of acetaminophen in rats. When sulfate body stores are depleted due to prolonged (several hours) continuous administration of xenobiotics undergoing sulfate conjugation the elimination rate of the xenobiotic may slow down, and as is the case for acetaminophen, an alternative metabolic pathway forming a toxic reactive intermediate may become more important (Levy, 1986).

Enzymes resposible for xenobiotic metabolism are ubiquitous in mammalian species. Although their preponderance in the liver has resulted in a focus of research efforts on hepatic metabolism, observations of organ-specific toxicity of xenobiotics have led to recent growth in research efforts associated with the extrahepatic bioactivation of xenobiotics. Extrahepatic metabolism may take place in many tissues and organs including the small intestine, kidney, lung, brain and skin (Masters et al., 1987; Krishna and Klotz, 1994). The microflora colonizing the digestive tract of mammals may also play a role in the metabolism and possibly bioactivation of xenobiotics. The hepatotoxicity of 2,6-dinitrotoluene is an interesting example of how microfloral metabolism of a xenobiotic is necessary to induce toxicity in a target tissue. 2,4-Dinitrotoluene is metabolized by cytochrome P-450 to 2,6-dinitrobenzylalcohol, which is conjugated to glucuronic acid (Long and Rikkert, 1982). The excretion of this glucuronide shows a clear sex difference: female rats excrete this conjugate predominantly in the urine, while in male rats excretion is mainly via the bile. Following biliary excretion the glucuronide conjugate is hydrolyzed by gut microfloral β-glucuronidase and one or both of the nitro groups of the liberated 2,6-dinitrobenzylalcohol are subsequently reduced to an amino function by microfloral nitroreductase (Mirsalis and Butterworth, 1982). The resulting aminobenzylalcohols (2-amino-6-nitrobenzyl alcohol and 2,6-diaminobenzyl alcohol) are reabsorbed from the small intestine and are metabolically activated in the liver to a reactive genotoxic metabolite (Kedderis et al., 1984). Because of the crucial role of microfloral enzymes in the activation of 2,6-dinitrotoluene, the genotoxic effect of this nitroaromatic compound cannot be demonstrated in vitro using isolated hepatocytes. Enterohepatic cycling also plays an important role in the nephrotoxicity of one of the metabolites of hexachlorobutadiene in a variety of mammalian species (see "Studies on Nephrotoxic Agents in Renal Preparations" by G.G. Gibson).

A class of metabolites which has attracted much attention lately are the acyl glucuronides (Spahn-Langguth and Benet, 1992). Until a few years ago it was generally accepted that glucuronide conjugates of xenobiotics are rapidly excreted in urine and/or bile following their formation in the body and that these conjugates are of no toxicological importance (Mulder 1992). But acyl glucuronides of many drugs have recently been shown to be potentially reactive intermediates which can bind irreversibly to plasma and tissue proteins. Since covalent binding of reactive metabolites to macromolecules may result in serious toxicity, the disposition kinetics of acyl glucuronides are extremely important. Conditions associated with decreased renal or biliary excretion of acyl glucuronides, such as advanced age or co-administration of compounds, such as probenecid, interfering with the renal or biliary excretion of glucuronide conjugates, may significantly prolong the residence time of the acyl glucuronide in the body and therefore enhance the extent of their irreversible binding to plasma and tissue proteins. Certain glucuronide conjugates (especially acyl glucuronides) may also undergo rapid enzymatic (β-glucuronidase, esterases) and/or spontaneous hydrolysis in vitro as well as in vivo and, thus, are interconvertible to their parent xenobiotics (Cheng and Jusko, 1993; Brunelle and Verbeeck, 1993). Metabolic interconversion has been demonstrated in several mammalian species for xenobiotics and their acyl glucuronides, including zomepirac in the guinea pig (Smith et al., 1990), diphenylacetic acid in the rabbit (Sallustio et al., 1989), and clofibric acid in man (Faed and McQueen, 1979; Meffin et al., 1983). The pharmacokinetics of the metabolic interconversion of a xenobiotic and its glucuronide conjugate may be important determinants of the toxicity of this xenobiotic. Indeed, it has been noted that inhibitors of β-glucuronidase may have a profound effect on the carcinogenic potential of certain compounds eliminated by glucuronidation (Walaszek et al., 1984; Walaszek, 1990).

Species-differences in the toxic response to xenobiotics are often related to differences in the metabolism of the xenobiotic. In general, conjugation reactions lead to detoxification of xenobiotics. Species differences in glucuronide conjugation have been described, and the best known example is the deficiency of the cat and other felines to glucuronidate certain (but not all) xenobiotics. When comparing the toxicities of a number of substrates for glucuronide conjugation in the rabbit (a species which is able to form glucuronides quite readily) and the cat a striking difference is observed: substances such as phenol, α-naphthol, β-naphthol and paracetamol have LD50's (expressed as mg/kg) which are 3 to 90 times lower in the cat, the species with a well-documented glucuronidation defect, as compared to the rabbit (Caldwell, 1980).

5. EXCRETION

Many xenobiotics and/or their (toxic) metabolites are removed from the blood by renal excretion. Renal excretion involves one or more of the following mechansims: (1) glomerular filtration, (2) tubular secretion, and (3) tubular reabsorption. Glomerular filtration and tubular reabsorption are essentially non-saturable processes, unlike tubular secretion which is by carrier-mediated transport and thus saturable. A wide range of xenobiotics has been reported to exhibit saturable renal excretion by tubular secretion. The artificial sweetener saccharin is mainly eliminated in the rat by renal excretion of the unmetabolized molecule (Renwick, 1985). Its plasma clearance (which largely corresponds to renal clearance) in rats has been shown to be dose-dependent due to saturation of tubular secretion: a 60% decrease in plasma clearance was found following the I.V. bolus administration of saccharin over a wide range of doses (1-1000 mg/kg) (Sweatman and Renwick, 1980). Because of the saturable renal excretion of saccharin, significant nonlinearity was apparent in the plasma and tissue concentrations of this sweetener as a function of the dose administered (Fig 4). At dietary saccharin concentrations above 3% a disproportionate increase in plasma levels of the sweetener was observed. The concentrations of saccharin in the plasma and tissues of rats given relatively high saccharin diets (>5%) were higher than predicted by linear extrapolation from the concentrations obtained at lower dietary levels. This excessive accumulation of saccharin in plasma and tissues at dietary levels above 3% coincides with a significant increase of the incidence of tumors of the urinary bladder.

The elimination kinetics of the herbicide 2,4,5-trichlorophenoxy acetic acid (2,4,5-T) in the rat are also

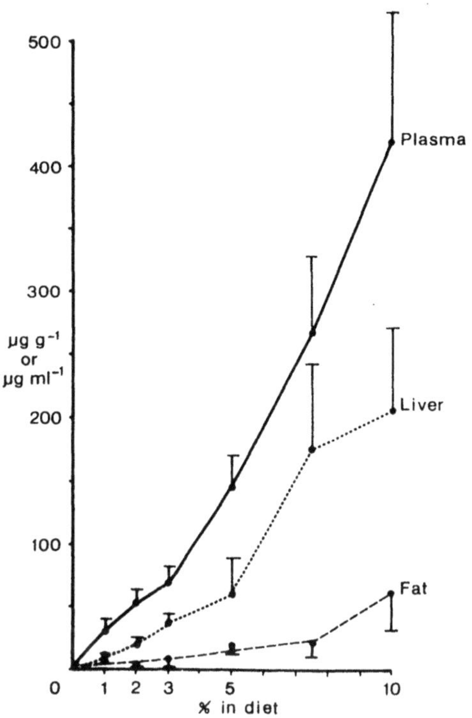

Fig. 4. Concentrations of saccharin in plasma, liver and fat of rats as a function of the percentage of saccharin in the diet. Saccharin containing diets were given ad libitum for 22 days prior to sacrifice (from Sweatman and Renwick, with permission).

nonlinear due to saturation of its elimination by renal tubular secretion (Sauerhoff et al., 1976). However, the saturation of the renal excretion of this xenobiotic may lead to increased elimination by metabolism which may partly compensate for the loss in excretory capacity (Gehring and Young, 1978).

Excretion into the bile is another important route of elimination, especially for polar and amphipathic substances having a relatively high molecular weight. Excretion into bile is usually an active process and various transport systems are involved according to the structure (anionic, cationic, neutral) of the xenobiotic (Klaassen and Watkins, 1984). As a consequence, saturation of the biliary excretion process may occur following administration of relatively large doses of the xenobiotic. Biliary excretion may also lead to the production of toxic metabolites by bacterial enzymes in the gastrointestinal tract. Microfloral nitroreductases are responsible for the formation of aminobenzyl alcohol following the biliary excretion of the 2,6-dinitrobenzyl glucuronide conjugate, a metabolite of 2,4-dinitrotoluene. Aminobenzyl alcohol is subsequently absorbed from the small intestine and is further metabolized in the liver to a hepatocarcinogen by N-hydroxylation followed by sulfation.

Biliary excretion of a xenobiotic and/or its metabolite may be followed by reabsorption in the gut, a process called enterohepatic cycling. Enterohepatic cycling of indomethacin, a nonsteroidal anti-inflammatory drug, may be an important factor contributing to the occurrence of intestinal lesions following the administration of this compound (Duggan et al., 1975). The sensitivity to intestinal lesions following administration of indomethacin correlates remarkably well with the cumulative biliary excretion of this drug and its conjugates in different species. In certain species such as the rat and the dog, enterohepatic cycling is so extensive that the intestine is exposed to more than 100% of the administered dose.

CONCLUSIONS

Pharmacokinetic studies with xenobiotics are needed to provide an understanding of the physiological processes involved in the fate of the xenobiotic in the body and often reveal information as to why exactly toxicity occurs in a given organ or tissue. In addition, due to dose-dependent pharmacokinetic behavior an extremely steep dose-response curve for the toxic effect of the xenobiotic may be observed at doses above a certain threshold. As the severity of toxicity of a xenobiotic is related to two variables, namely the sensitivity of the target organ and xenobiotic concentrations at the site of toxicity, determination of blood and (target) tissue concentrations of the xenobiotic and/or its metabolites is extremely helpful to explain anomalies in the toxic response to a xenobiotic, such as the sudden increase in toxicity at a particular dose level or route of exposure-dependent differences in toxicity. Another important contribution of pharmacokinetics to the toxicologic evaluation of xenobiotics is in the area of risk assessment in man. Comparative pharmacokinetic studies in several animal species form the basis for species-to-species extrapolation and rational risk estimation of xenobiotics in man (Chappell and Mordenti, 1991).

References

Barr WH (1969) Factors involved in the assessment of systemic or biologic availability of drug products. Drug Info Bull 3:27-45.
Brunelle FM and Verbeeck RK (1993) Glucuronidation of diflunisal by rat liver microsomes. Effect of microsomal β-glucuronidase activity. Biochem Pharmacol 46:1953-1958.
Caldwell J (1980): Conjugation reactions. In "Concepts in Drug Metabolism. Part A", Jenner P and Testa B (eds), Marcel Dekker Inc. New York, pp. 211-250.
Cassidy MK and Houston JB (1980) In vivo assessment of extrahepatic conjugative metabolism in first pass effects using the model compound phenol. J Pharm Pharmacol 32:57-59.
Chappell WR and Mordenti J (1991) Extrapolation of toxicological and pharmacological data from animals to humans. In "Advances in Drug Research, Volume 20", Testa B (ed), Academic Press Limited London, pp. 1-116.
Cheng H and Jusko WJ (1993) Pharmacokinetics of reversible metabolic systems. Biopharm Drug Disp 14:721-766.
Diliberto JJ, Kedderis LB, Jackson JA and Birnbaum LS (1993) Effects of dose and routes of exposure on the disposition of 2,3,7,8-[^{3}H]-tetrabromodibenzo-p-dioxin (TBDD) in the rat. Toxicol Appl Pharmacol 120:315-326.
Duggan DE, Hooke KF, Noll RM and Kwan KC (1975): Enterohepatic circulation of indomethacin and its role in intestinal irritation. Biochem Pharmacol 25:1749-1754.
Faed EM and McQueen EG (1979) Plasma half-life of clofibric acid in renal failure. Br J Clin Pharmacol 7:407-410.
Galinsky RE and Lavy G (1981) Dose- and time-dependent elimination of acetaminophen in rats: pharmacokinetic implications of cosubstrate depletion. J Pharmacol Exptl Ther 219:14-20.
Gehring PJ and Young JD (1978) Application of pharmacokinetic principles in practice. In: "Proceedings of the First International Congress on Toxicology. Toxicology as a Predictive Science", Plaa GL and Duncan WAM (eds), Academic Press New York, pp. 119-141.
Gibaldi M and Perrier D (1982) Pharmacokinetics. Second edition, revised and expanded. Marcel Dekker Inc. New York.
Jusko WJ and Gretch M (1976) Plasma and tissue protein binding of drugs in pharmacokinetics. Drug Metab Rev 5:43-140.
Kao J and Carver MP (1990): Cutaneous metabolism of xenobiotics. Drug Metab Rev 22:363-410.
Kedderis GL, Dryoff MC and Rickert DE (1984) Hepatic macromolecular covalent binding of the hepatocarcinogen 2,6-dinitrotoluene and its 2,4-isomer in vivo: modulation by the sulfotransferase inhibitors pentachlorophenol and 2,6-dichloro-4-nitrophenol. Carcinogenesis 5:1199-1204.
Klaassen CD and Watkins III JB (1984) Mechanisms of bile formation, hepatic uptake, and biliary excretion. Pharmacol Rev 36:1-67.

Krishna DR and Klotz U (1994) Extrahepatic metabolism of drugs in humans. Clin Pharmacokinet 26:144-160.
Levy G (1968) Dose dependent effects in pharmacokinetics. In "Importance of Fundamental Principles in Drug Evaluation", DH Tedeschi and RE Tedeschi (eds.), Raven Press New York, pp. 141-172.
Levy G (1986) Sulfate conjugation in drug metabolism: role of inorganic sulfate. Fed Proc 45:2235-2240.
Levy G (1993) The case for preclinical pharmacodynamics. In "Integration of Pharmacokinetics, Pharmacodynamics, and Toxicokinetics in Rational Drug Development", A. Yacobi, JP Skelly, VP Shah and LZ Benet (eds.), Plenum Press New York, pp. 7-13.
Long RM and Rickert DE (1982) Metabolism and excretion 2,6-dinitro-[^{14}C]-toluene in vivo and in isolated perfused rat livers. Drug Metab Disp 10:455-458.
Masters BSS, Muerhoff AS and Okita RT (1987) Enzymology of extrahepatic cytochromes P-450. In "Mammalian Cytochromes P-450", FP Guengerich FP (ed), CRC Press, Boca Raton FL, pp. 107-132.
Meffin PJ, Zilm DM and Veenendaal JR (1983) Reduced clofibric acid clearance in renal dysfunction is due to a futile cycle. J Pharmacol Exptl Ther 227:732-738.
Mirsalis JC and Butterworth BE (1982) Induction of unscheduled DNA synthesis in rat hepatocytes following in vivo treatment with dinitrotoluene. Carcinogenesis 3:241-245.
Mitchell JR, Potter WZ, Hinson JA, Snodgrass WR, Timbrell JA and Gillette JR (1975) Toxic drug reactions. In "Handbook of Experimental Pharmacology Vol 28 Part 3. Concepts of Biochemical Pharmacology", JR Gillette and JR Mitchell (eds.), Springer Verlag, Berlin, pp. 383-419.
Mulder GJ (1982) Pharmacological effects of drug conjugates: is morphine-6-glucuronide an exception? Trends Pharmacol Sci 13:302-304.
Peck CC (1993) Rationale for the effective use of pharmacokinetics and pharmacodynamics in early drug development. In "Integration of Pharmacokinetics, Pharmacodynamics, and Toxicokinetics in Rational Drug Development", A. Yacobi, JP Skelly, VP Shah and LZ Benet (eds.), Plenum Press New York, pp. 1-6.
Pond SM and Tozer TN (1984) First-pass elimination. Basic concepts and clinical consequences. Clin Pharmacokinet 9: 1-25.
Puigdemont A, Arboix M, Gaspari F, Bortolotti A and Bonati M (1989) In vitro plasma protein binding of propafenone and protein profile in eight mammalian species. Res Comm Chem Pathol Pharmacol 64:435-440.
Renwick AG (1985): The disposition of saccharin in animals and man - a review. Food Chem. Toxicol. 23:429-435.
Rowland M and Tozer TN (1989): Clinical Pharmacokinetics. Concepts and Applications. Lea & Febiger Philadelphia.

Sallustio BC, Purdie YJ, Birkett DJ and Meffin PJ (1989) Effect of renal dysfunction on the individual components of the acyl-glucuronide futile cycle. J Pharmacol Exptl Ther 251: 288-294.

Sauerhoff MW, Braun WH, Blau GE and Gehring PJ (1976): The dose-dependent pharmacokinetic profile of 2,4,5-trichlorophenoxy acetic acid following intravenous administration to rats. Toxicol Appl Pharmacol 36:491-501.

Smith PC, McDonagh AF and Benet LZ (1990) Effect of esterase inhibition on the disposition of zomepirac glucuronide and its covalent binding to plasma proteins in the guinea pig. J Pharmacol Exptl Ther 252:218-224.

Spahn-Langguth H and Benet LZ (1992) Acyl glucuronides revisited: is the glucuronidation process a toxification as well as a detoxification mechanism? Drug Metab Rev 24:5-48.

Sweatman TW and Renwick AG (1980) The tissue distribution and pharmacokinetics of saccharin in the rat. Toxicol Appl Pharmacol 55:18-31.

Timbrell JA (1991) Principles of Biochemical Toxicology, Taylor & Francis London.

Tozer TN (1984) Implications of altered plasma protein binding in disease states. In "Pharmacokinetic Basis for Drug Treatment", Benet LZ, Massoud N and Gambertoglio JG (eds), Raven Press New York, pp. 173-193.

Varga F (1969) Intestinal absorption of chloroquine in rats. Arch Int Pharmacodyn Therap 163:38-46.

Walaszek Z (1990) Potential use of D-glucaric acid derivatives in cancer prevention. Cancer Lett 54:1-8.

Walaszek Z, Hanausek-Walaszek M and Webb TE (1984): Inhibition of 7,12-dimethylbenzanthracene-induced rat mammary tumorigenesis by 2,5-di-9-acetyl-D-glucaro-1,4:6,3-dilactone, a β-glucuronidase inhibitor. Carcinogenesis 5: 767-772.

Welling PG (1993) Pharmacokinetic principles: linear and nonlinear. In "Drug Toxicokinetics", Welling PG and de la Iglesia FA (eds), Marcel Dekker Inc. New York, pp. 19-41.

Biosensors in pharmacology and toxicology in vitro

Philippe Catroux, Martine Cottin, André Rougier, Jacques Leclaire
Central Department of Products Safety , L'OREAL laboratory
1, avenue Eugène Schueller,
93600 Aulnay-sous-Bois,
France.

Introduction

The need of disposing of analytical information in real time in many fields of bioanalytical and bioindustries explains the great attention devoted to biosensor development. A biosensor is usually defined as resulting from the combination of a sensitive biological element capable of molecular recognition and a transductor (electrode, optical detector...) which gives a meaningful mostly electrical signal. During the last decade, major development has concerned electrode tipped with enzymes, antibodies or other reagents that interact chemically with analytes -the substances being analysed. Such biosensors are expected to have numerous applications in the field of medicine and biology (Brennan and Krull,1992).

Here, we would like to focus on the recent development of biosensors based on silicon chip technology and their various application in the field of pharmacology and toxicology *in vitro*. Among these, cytosensor (formerly silicon microphysiometer) is a biosensor which detects changes in the physiological state of cultured cells by monitoring on line and accurately the rate at which cells excrete their acidic products of metabolism.(Mc Connell et al,1991).

Experimental apparatus

The cytosensor is mainly formed from eight independent fluidic assemblies that comprise each reagent reservoirs and computer controlled pump head, debubbler block, computer controlled valve, sensor chamber and reference electrode (figure 1). Cells are sandwiched between two porous polycarbonate membranes inside disposable cell capsules. Both adherent and non adherent cells may be accommodated by growing them directly on the lower

NATO ASI Series, Vol. H 93
Modulation of Cellular Responses in Toxicity
Edited by C. L. Galli, A. M. Goldberg, M. Marinovich

lower membrane or by entrapping them within a fibrous protein matrix. The assembled capsule is placed in a sensor chamber at the time of experiment. A plunger creates a microvolume cavity to contain the cells. The cells are supplied with a controlled flow of nutrients, with or without an effector agent of interest.

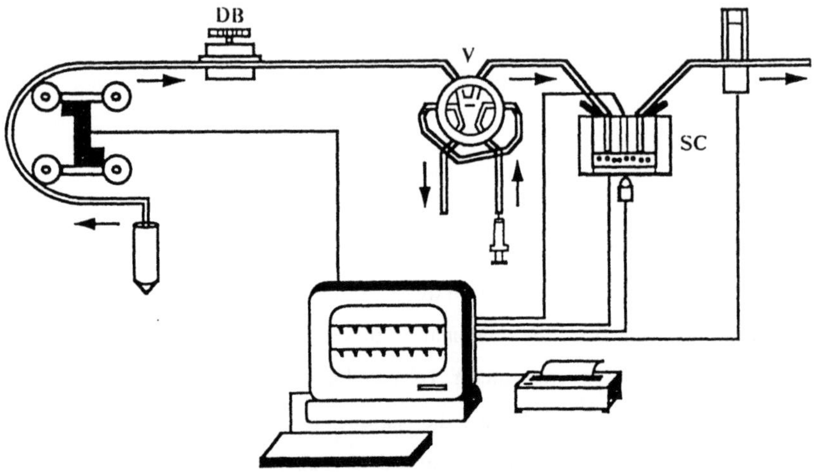

Fig. 1. Schematic diagram of the cytosensor with reagent reservoirs, pump head, debubbler block (DB), valve (V) and sensor chamber (SC).

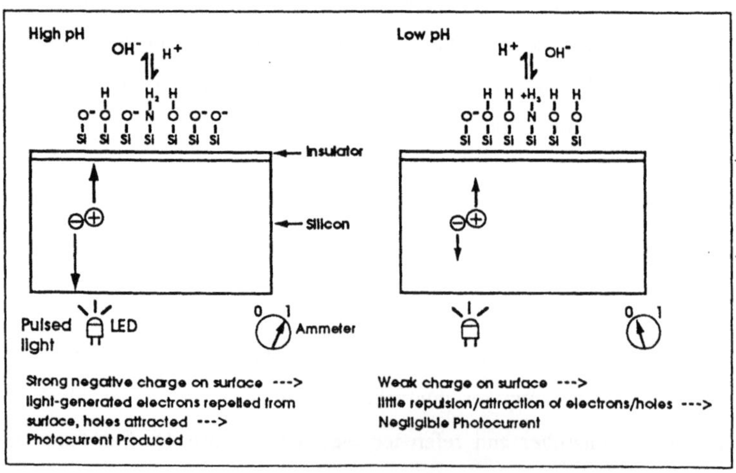

Fig. 2. Schematic diagram of the ligh addressable potentiometric sensor.

Changes in the metabolic activity of the cells result in changes in the rate of release of acidic products, so altering the surface charge of the light addressable potentiometric sensor (LAPS). Figure 2 is a schematic representation of the LAPS. A light emitting diode illuminates the LAPS every second to produce a photomicrocurrent, the potential of which is mainly influenced by the number of protons that binds to the surface of the LAPS. A surface charge dependant voltage parameter, related to pH in a Nernstian manner, is processed by the cytosensor system and reported every second to the computer. Extracellular acidification rates (EAR) are determined as microvolts per second during periods of flow cessation. Rates are recorded and plotted as a function of time.

The chemical and biological factors that determine the performance and applications of the cytosensor has been excellently reviewed by Owicki and Parce (1992). In mammalian cells in culture, the extracellular acidification results mainly from excretion of lactic and carbonic acids formed during the energy metabolism using glucose and glutamine as the principal carbon sources.

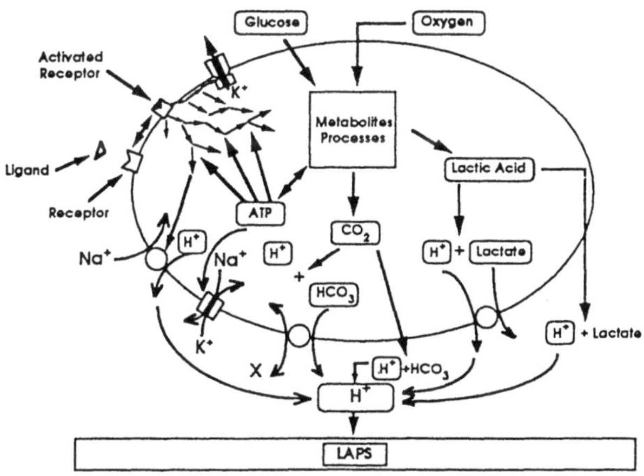

Figure 3 : Schematic representation of cellular metabolism and its relationship with extracellular acidification and physiological processes (adapted from Owicki and Parce, 1992)

When expressed per ATP synthetized, the production of protons is particularly high from glycolysis. Respiration which is energetically more efficient produces less acid for ATP generated. In culture, the energy metabolism of cells depends more on glycolysis than on respiration. This is due mainly to the composition of medium and the condition of relative hypoxia *in vitro*. Extracellular acidification rate is also influenced by the mechanisms of acid excretion. These include the direct transport of protons by Na^+/H^+ antiporter, H^+/K^+ exchanges or H^+ ATPases, for example. Other sources are catalyzed and uncatalyzed transports for weak acids which produces a proton upon dissociation after transport.

Change in extracellular acidification rate is not necessary the consequence of only a change in metabolic process and may result from direct or indirect effect on acid excretion mechanism. For these reasons, microphysiometry should not be considered strictly as a non invasive measurement of the energy metabolism of the cells.

Cytosensor in pharmacology *in vitro*

Applications which include studies on receptor activation and inactivation, signal transduction, agonist and antagonist profile represent the major development of the cytosensor.

• Study of receptor activation and agonist/antagonist profiles

Both a binding assay to assess receptor-ligand affinity and a functional assay to evaluate agonist/antagonist profile are usually required to study receptor activation. Numerous reports have shown that the accurate measurement with cytosensor of changes in EAR may represent a convenient and universal functional assay for such application (Owicki et al,1990). The receptors that has been detected belong to various classes of different receptors which are coupled to all of the classical second-messenger pathways in whole cells. These include tyrosine kinase receptors, G protein linked receptors, ion channel gated receptors and others (table I).

Receptor activation usually results in an increase in EAR as exemplified in figure 4 from various agonists and cell types. The magnitude and kinetic of this effect vary with the receptor and cell types but typically involve a 10-100 % increase in EAR within seconds to minutes after the application of the agonist. Depending on both the nature of test agent and time of exposure (bolus versus continuous perfusion) changes may be sustained along minutes or hours or may be only transient as a burst of extracellular acidification.

Responses have been shown receptor specific in different ways. In many cases, experiments

EFFECTOR AGENTS	AGONISTS	ANTAGONISTS	REFERENCES
Tyrosine kinase receptors			
. EGF			Owicki J.C. et al, PNAS (1990)
. TGFα			
. NGF	recombinant growth factors	antibodies genisteine herbamycine	Pitchford S. et al, Soc. Neurosci (1992)
. PDGF			
. IGF I and II			
Cytokines receptors			
. GMCSF			
. IL1, IL2, IL3, IL4	recombinant cytokines	antibodies	McConnell H.M. et al., Science (1992)
. TNFα			
. PAF			
G-Protein linked receptors			
. Adrenergic drugs	Isoproterenol	Propanolol	Owicki J.C. et al, PNAS (1990)
. Cholinergic drugs	Carbachol	Atropine	Miller D.L. et al, Biophys. J., (1993)
. Dopaminergic drugs (D1, D2, D3, D4)	Dopamine Quinpirole	SCH 23390 Clozapine Epideprine	Neve K.A. et al, J. Biol. Chem. (1992) Rosser M.P. et al, Soc. Neurosci (1991) Bouvier C. et al, Soc. Neurosci (1992)
. Neuropeptides	Bombesine Bradykinine Vasopressine	Antibodies	unpublished
. Cholecystokinin and Gastrin	CCK 4 CCK 8		Denyer J.C. et al, J. Physiol, (1993)
Ion gated channel receptors			
. Excitatory aminoacids	Kainate	Kinurenate	Raley-Susman KM. et al, J. Neurosc. (1992)
Others			
. MHC complexes	MHC I MBP peptide complexes	Anti CD3 Anti TRC	Nag H.G. et al, J. Immunol, (1992)

Table I . Some exemples of effector agents which has been detected with the cytosensor.

were performed on cells transfected with the gene encoding the receptor of interest. These cells responded to the appropriate agonits whereas non transfected cells exhibited no response. The specificity of the effect in many cases was further documented using neutralising antibodies or specific pharmacologic reagents. Figure 5, for example, shows the response of in EAR. Pitchford et al (1992) have shown that this response can be reduced by preincubating the cells with genistein, an inhibitor of the intracellular domain of tyrosine kinase receptors. In chinese hamster ovary (CHO) cells transfected with human β_2 adrenergic receptors, isoproterenol produces a 20-60 % increase in the rate of extracellular acidification, a response that is completely antagonised by the β blocker propanolol (Owicki et al, 1990).

Receptor subtyping also can be studied with the cytosensor. Salon et al (1991) have measured the dopamine responsive metabolic rates of cells transfected with human D_1 or D_2 receptors which are known to couple through G_s and G_i mechanisms respectively. Increase in EAR induced by unspecific (dopamine) or specific (SKF - 38393) D_1 agonists in cells expressing D_1 receptors could be inhibited by specific (Sch 23390) or non specific (flupentixol) antagonists of this receptor subtype. These antagonist were ineffective to inhibit changes in EAR induced by dopamine or quinpirole (specific D_2 agonist) in cells transfected with cDNA for D_2 receptors, in contrast to the D_2 specific antagonist spiperone.

Dose response curves can be carried out for assessing the potency of agonists and antagonists then determining usual pharmacological parameters (EC $_{50}$, IC $_{50}$). Table II is a summary of data comparing cytosensor with different assays that are traditionally used to evaluate receptor triggering and agonist/antagonist profiles (Hirst and Pitchford 1993). The authors showed that EC 50 (the concentration of test product that produces 50 % of the maximal effect) determined from microphysiometry are very similar to those obtained from biochemical, electrical or physiological investigations.

Receptor Cell Type	Signal Transduction	EC50 (assay type)	EC50 Cytosensor	Agonist
Muscarinic M1-CHO	IP3/DAG	3.2μM (IP3)	2.5μM	Carbachol
Adrenergic β2	cAMP (increase)	1nM (cAMP)	4nM	Isoproterenol
Kainic Acid hippocampal	Ligand-gated	100μM (ion current)	140μM	Kainic Acid
GM-CSF TF-1	PKC	4.3pM (proliferation)	3.6pM	GM-CSF
NGF PC-12	Tyrosine Kinase	148 pM (neurite outgrowth)	153pM	NGF

Table II . Comparison of data obtained with cytosensor and traditionnal functional assays (from Hirst and Pitchford, 1993)

◆ **Study of signal transduction pathways**

The role of various second messenger pathways in cell response following receptor triggering has been assessed with the cytosensor. For example, cAMP dependant second messenger pathway in β_2 adrenergic receptor triggering has been shown in CHO cells transfected with human receptor (Owicki et al, 1990). These cells responds to isoproterenol by a large increase in EAR. This effect was mimicked by treating the cells either with the membrane permeable cAMP analog, 8-bromo cAMP or the adenylate cyclase activator, forskolin. The involvement of a Gs protein type was further documented by showing the irreversible activation of this system with cholera toxin, a compound that activates Gs via ADP ribosylation. Recently, the nature of Pkc isoform involved in GM-CSF signal transduction has been determined on TF1 cells, a cell line etablished from the bone marrow of a human erythroleukemia patient (Baxter et al,1992). GM-CSF induced a dose dependant increase in EAR, a response which was suppressed by calphostin c, an inhibitor of Pkc or by Pkc depletion following long lasting treatment with phorbol ester PMA. Using isoenzyme specific antisense oligonucleotides, the autors were able to demonstrate the involvement of a calcium dependent, phospholipid independant kinase (Pkc ε isoform) in this response. Then, using specific pharmacological probes, it is possible with cytosensor to evidence the intracellular mechanisms involved in cell response to receptor activation.

◆ **Changes in EAR : what is measured?**

Because cell response to agonist involves the interplay and cross-talk of a number of cytosolic and membrane bound events, the mechanisms that determine rapid changes in EAR are difficult to etablish. The metabolic cost due to early events such as receptor autophosphorylation or generation of second messengers (generation of cAMP from ATP, generation of phosphatidyl inositol ....) may not represent the overall increase in EAR following receptor activation. Parce et al (1989) for example have etablished that generation of cAMP from ATP following triggering of β receptor may cause no more than 1% of the apparent increase in acidification. The most generally important source of sustained changes in EAR is change in the rate of metabolism. A direct modification of the activity of key regulatory enzymes involved in energy metabolism as for example phosphofructokinase 1 regulatory enzymes involved in energy metabolism as for example phosphofructokinase 1 may occur on receptor activation. However, it is likely that the bulk of changes in EAR in many cases results from energy demand required to reestablish ion gradients that have been dissipated either directly by the receptor or indirectly by a second messenger. Another important mechanism involves direct or indirect activation of sodium/proton antiporter system which causes a transient increase in EAR and concommitant increase in intracellular pH. For example, GM-CSF on TF1 cells causes a rapid increase in extracellular acid production which is dose dependant and competitively inhibited by a specific anti-GM-CSF antibody. The activation of sodium/proton antiporter contributes a short lived minor component of this response in contrast to the activation of glycolysis which appears as a longer lasting major event. (Wada et al, 1993).

◆ Other applications

The cytotoxic effect of chemotherapeutic agents on tumor cell line has been evaluated with the cytosensor (Parce et al, 1993 - Wada et al, 1991). Experiments were performed with human uterine sarcoma line sensitive or rendered resistant to doxorubicin and vincristine. The resistant cells had an enhanced activity of the P-glycoprotein transporter which confers multidrug resistance. The authors showed that the cytotoxic effect of the two chemotherapeutic agents, as manifested by a progressive decrease in EAR was significantly faster in sensitive cells. They suggested that such approach should be advantageous to screen biopsies from cancer patients against panels of chemotherapeutic agents to help predict therapeutic efficacy in individual cases.

The interference of a biological material with cellular functions can also be detected with cytosensor. Wada et al (1991) have reported that the time course of the infection of cultured cell lines by rapidly acting (vesicular stomatis virus) or slower acting (HIV 1) virus could be monitored by EAR measurements. They further reported that efficacy of antiviral drugs such as AZT or ribarivin could be evaluated precisely.

Basically, many physiological functions may be addressed with the cytosensor. It depends mainly on their relative importance to influence directly or indirectly the rate of metabolism and/or ionic exchanges through the membranes. In addition to receptor activation and signal transduction processes, the physiological function the most dependant on energy metabolism are cell growth and biomass, protein secretion, maintenance of ion gradients and motion.

Cytosensor in toxicology *in vitro*

In toxicology, applications of the cytosensor has been mainly directed towards the development of *in vitro* alternatives to rabbit eye irritancy tests *in vivo*. Assessment of effects on the metabolic rate of cultured cells was expected to be a more sensitive and predictive parameter to evaluate the ocular irritancy potential of test products than cellular methods based on the measurement of cytolysis or growth inhibition as endpoints. In cytosensor, cells are exposed sequentially to increasing concentrations of test products until the extracellular acidification cease (figure 6). Time of exposure for each concentration is short, usually comprised between 5 and 12 minutes to mimic *in vitro* the conditions of accidental exposure to the eye *in vivo*. Endpoint which is determined is the MRD_{50}, the concentration of test product that decreases the extracellular acidification rate by 50 %. This value is subsequently compared to maximal average score (MAS) from Draize test in an *in vivo/in vitro* correlation based approach.

Bruner and coll (1992) have investigated 17 materials including different chemicals and formulations (bar soap, shampoos, diswashing liquids...). They found a positive linear correlation ($r = 0.86$) between the *in vivo* irritancy potential of the products and their MRD_{50}. They compared seven alternative *in vitro* methods for testing these products and reported that the microphysiometer was one of the best for predicting rabbit eye irritancy. Close results have been found in a multicentric study in which we were involved (Bagley et al, 1991). The intralaboratory comparison of silicon microphysiometer data yielded a linear correlation of 0.93 and the linear correlation between *in vivo* and *in vitro* results were within a range of 0.8 - 0.9. In a recent study, we have evaluated the effect of 53 products (21 surfactants and 32 surfactants-based formulations) (Catroux et al, 1993). Figure 7 shows the interpolation from a plot of MRD_{50} versus MAS. The linear (Pearson) and rank (Spearman) correlation between *in vivo* and *in vitro* data were 0.91 and 0.89, respectively. Moreover, we have been able to distinguish different classes of products *in vitro* which roughly correspond to slightly irritant, irritant and very irritant products *in vivo*. Using this classification, three was no false negative and only 8 false positive but borderline values.

Notwithstanding these good results, cytosensor must not be considered on an universal method to evaluate ocular irritancy potential for different reasons. First, testing with cytosensor at the present time, has only been addressed to a reduced number of product categories. These include mainly surfactant and surfactant based formulations, alcohols, acids and alkalines, indeed products that usually can be well evaluated by other cellular and non cellular *in vitro* tests (Cottin et al, 1992). Second, testing with cytosensor is limited until now to water soluble products.

Conclusions

Experiments made with the cytosensor or others biosensors (a similar approach has been made recently with a biosensor based on H+-sensitive field effect transistor -Grattarola et al,1993) show that many physiological functions can be assessed by extracellulary measurement of physicochemical parameters. A major interest of these methods is that measurement are on-line and non invasive, so that the effect of a test compound can be monitored kinetically avoiding false negative results for all or nothing phenomenons. Microphysiometry can be coupled to other non invasive methods as quantitative fluorescence imaging to obtain more informations (Miller et al,1993). Another advantage of these type of biosensor is its sensitivity since precise measurement can be obtained from less than ten thousands cells present in the sensor chamber.

The major development of the cytosensor is pharmacological investigations and more precisely to study receptor activation and inactivation and signal transduction pathways. In this field, numerous pharmacological probes are available to characterize the mechanisms by which EAR is changed. In contrast, the use of cytosensor in toxicology remains empiric since the intracellular mechanisms leading to drug- induced decrease in extracellular acidification rate have not been sufficiently investigated until now.

To conclude, the speed, ease of use and versatility of the cytosensor make it a valuable assay tool for science research laboratory.

Acknowledgements : We thank Drs J. Owicki, L. Laxhuber and R. Metzger from Molecular Devices (Palo Alto, USA) for helping discussions and providing us some documents reported here.

Bibliography

Bagley, D.M., Bruner, L.H., de Silva, O., Cottin, M., O'Brien, K.A.F., Uttley, M., Walker, A.P. (1991). An evaluation of five potential *in vitro* alternatives to the *in vivo* rabbit eye irritation test. Toxic. In Vitro, 5, 277-284.

Baxter, G.T., Miller, D.L., Kuo, R.C., Wada, H.G., Owicki, J.C. (1992). PKCε is involved in granulocyte-macrophage colony-stimulating factor signal transduction : evidence from microphysiometry and antisense oligonucleotide experiments. Biochemistry, 31, 10950-10954.

Brennan, J.D., Krull, U.J.(1992). A novel biochemical sensor. Chemtech, 12, 227-235.

Bruner, L.H., Miller, R., Owicki, J.C., Parce, J.W., Muir, V.C. (1992). Testing ocular irritancy *in vitro* with the silicon microphysiometer. Toxic. In Vitro, 6, 272-284.

Catroux, P., Rougier, A., Dossou, K.G., Cottin, M. (1993). The silicon microphysiometer for testing ocular toxicity *in vitro.* Toxic.In Vitro, 7, 465-469.

Cottin, M., Catroux, P., Delabarre, I., Rougier, A., Dossou, K.G. (1992). Complementary of four cytotoxicity methods and their relevance for ocular safety assessment of surfactants. Toxicology Letters, 15, 237.

Grattarola, M., Martinola, S., Melani, M., Tedesco, M., Parodi, M.T.(1993). On-line extraçellular pH measurements in culture as a tool for cell metabolism monitoring. STP. Pharma. Sciences, 3, 31-34.

Hirst, M.A., Pitchford, S.(1993). Use of a single assay system to assess functionnal coupling of a variety of receptors. The Journal of NIH Research, 5, 69.

McConnell, H.M., Rice, P., Wada, G.H., Owicki, J.C., Parce, J.W. (1991). The microphysiometer biosensor. Current opinion in structural biology, 1, 647-652.

Miller, D.L., Olson, J.C., Parce, J.W., Owicki, J.C. (1993). Cholinergic stimulation of the Na^+/K^+ ademosine triphosphatase as revealed by microphysiometry. Biophys. J., 64, 813-823.

Nag, B., Wada, H.G., Fok, K.S., Green, D.J., Sharma, S.D., Clark, B.R., Parce, J.W., McConnell, H.M. (1992). Antigen-specific stimulation of T-cell extracellular acidification by MHC class II - peptide complexes. J. of Immunology, 148, 2040-2044.

Owicki, J.C., Parce, J.W., Kercso, K.M., Sigal, G.B., Muir, V.C., Venter, J.C., Fraser, C.M., McConnell, H.M. (1990). Continuous monitoring of receptor-mediated changes in the metabolic rates of living cells. Proc. Natl. Acad. Sci., 87, 4007-4011.

Owicki, J.C., Parce, J.W. (1992). Biosensors based on the energy metabolism of living cells : the physical chemistry and cell biology of extracellular acidification. Biosensors and bioelectronics, 7, 255-272.

Parce, J., Owicki, J., Kercso, K., Sigal, G., Wada, VC., MUIR, L., Bousse, L., Ross, K., Sikic, B., Mc. Connell, HM.(1989). Detection of cell affecting agents with a silicon biosensor. Sciences (Wash.DC), 240, 1182-1185.

Parce, J.W., Owicki, J.C., Kercso, K.M. (1990). Biosensors for directly measuring cell affecting agents. Ann. Biol. Clin., 48, 639-641.

Pitchford, S., Glaeser, B.S., De Moor, K. Rapid measurements of response by PC 12 cells to nerve growth factor (NGF) using the cytosensor microphysiometer system.(unpublished)

Raley-Susman, K.M., Miller, K.R., Owicki, J.C., Sapolsky, R.M. (1992). Effects of excitotoxin exposure on metabolic role of primary hippocampal cultures : application of silicon microphysiometry to neurobiology. Journal of Neuroscience, 12, 3, 773-780.

Rosser, M.P., Kozlowski, M.R., Neve, R.U., Neve, K.A.(1991) Effects of D_2 and D_3 receptor activation measured by microphysiometry. Soc. Neurosci.(Abs), 17,818.

Salon, J.A., Johnson, R.A., Civelli, O.(1991) Real-time measurements of human D_1 and D_2 receptor activity with a silicon-based biosensor. Soc. Neurosci.(Abs), 17,86.

Wada, H.G., Indelicato, S.R., Meyer, L., Kitamura, T., Miyajima, A., Kirk, G., Muir, V.C., Parce, J.W. (1993). GM-CSF triggers a rapid glucose dependant extracellular acidification by TF1 cells. Evidence for a sodium/proton antiporter and PKC mediated activation of acid production. J. Cell. Physiol, 154, 129-148.

Networks of intercellular communication: Eicosanoids and cytokines as mediators of the response of skin to toxicity

Gerhard Fürstenberger and Karin Müller-Decker
German Cancer Research Center
Research Program Tumor Cell Regulation
D-69120 Heidelberg
Germany

Skin architecture

Skin is not only the largest but also one of the most complex and most reactive organs of the body. As an integument it imparts a distinct character to the individual, as a barrier it protects the body against water loss and prevents toxins and microorganisms from invading, and as an interface between the body and the surroundings it is destined to communicate and to cope with environmental conditions. The skin fulfills all these different functions via a complex tissue structure and a highly sophisticated communication network between the different compartments and cell types.

Skin is a multicellular tissue consisting of three main layers: the subcutaneous tissue, the dermis, and the epidermis (Fig. 1). Overlaying the subcutaneous tissue is the dermis which basically provides the epidermis with physical and nutritional support. The dermis primarily consists of connective tissue with a heterogenous network of blood vessels and nerves. The predominant cells are fibroblasts, endothelial and smooth vascular, mast, and nerve cells.

The epidermis is composed of several distinct layers of keratinocytes which differ in their state of terminal differentiation (Fig. 1). In the basal cell compartment, actively proliferating cells divide and move upwards. As they ascend they lose their capacity to replicate and terminally differentiate into the dead horny cells of the stratum corneum, the uppermost layer of the epidermis. In addition to keratinocytes, which make up 95 % of the epidermal cell mass, Langerhans cells, melanocytes, and Merkel cells populate the epithelium.

NATO ASI Series, Vol. H 93
Modulation of Cellular Responses in Toxicity
Edited by C. L. Galli, A. M. Goldberg, M. Marinovich

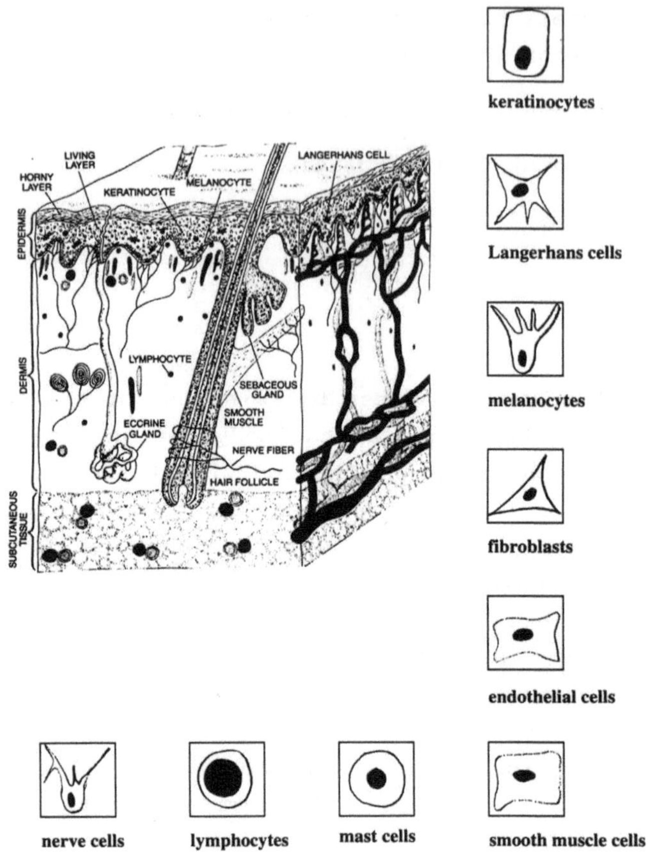

Fig. 1 A cross-section of human skin, showing the different cell layers and appendages and the major resident cell types

<u>Keratinocytes as initiators of hyperplastic transformation, the general response of skin to damage and toxicity</u>

The general response of skin to a wide variety of external stimuli - physical or chemical toxins - is acute inflammation followed by epidermal hyperplasia also termed hyperplastic transformation (Marks, 1990; Marks & Fürstenberger, 1993). Essentially all resident cell types of skin and, in addition, blood-derived cells as well as a plethora of local tissue factors are known to have taken part in the evolvement of this response. Vascular effects such as erythema and edema formation, invasion of inflammatory cells including granulocytes, lymphocytes, and monocytes are symptoms of acute inflammation, and as a common sequel to these inflammatory processes, hypertrophy and hyperplasia of epidermis and scale formation occur.

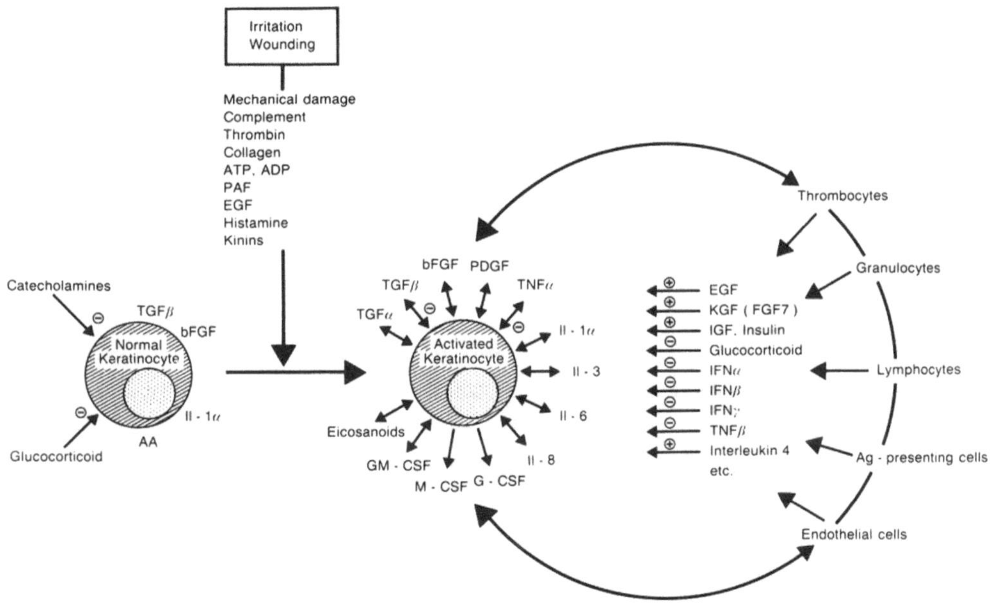

Fig. 2 Initiation of hyperplastic transformation of skin by keratinocyte activation. Upon irritation and injury of skin, keratinocytes generate and release a complex cocktail of growth factors, cytokines and lipid mediators. These factors autoactivate keratinocytes and recruit and activate other cell types to contribute factors controlling keratinocyte proliferation along para- and autocrine mechanisms: (+) stimulation, (-) inhibition.

Its direct contact with the environment and its vicinity to dermal blood vessels and nerves make the integumental epithelium a highly sensitive signaling interface between the environment and the body. Within epidermis, the keratinocytes represent a versatile multipotent cell type (Milstone & Edelson, 1988). Activation of keratinocytes results in the induction or upregulation of genes coding for a wide variety of growth factors, cytokines and their cognate receptors. The activated cells express cell adhesion molecules and release lipid-derived proinflammatory mediators (Fig. 2). These observations led Kupper and others to develop the concept of the "activated keratinocyte" as an initiator of acute inflammation providing the critical trigger for the dramatic alterations that occur following contact between the epidermis and a host of noxious agents (Kupper, 1990 a and b; Nickoloff et al., 1990; Barker et al., 1991). Among the keratinocyte-derived mediators proinflammatory cytokines and eicosanoids form an important part of the "language" facilitating the intercellular communication between the cell types spread over the different compartments of skin and involved in hyperplastic transformation.

The eicosanoid network in hyperplastic transformation

A critical role of eicosanoids in hyperplastic transformation is indicated by several pieces of evidence. A wide variety of exogenous irritant stimuli and many endogenous proinflammatory factors are known to activate eicosanoid biosynthesis. Eicosanoids mediate different aspects of the inflammatory response as well as the induction of an epidermal hyperplasia in mouse skin in vivo. Inhibitors of arachidonic acid metabolism have been shown to be potent anti-inflammatory agents (reviewed in Fürstenberger, 1990; Fürstenberger & Marks, 1990).

Activation and regulation of epidermal eicosanoid biosynthesis

Eicosanoids comprise a large and heterogeneous family of oxygenated C-20 fatty acids (Needleman et al. 1986). They are local hormones which are not stored in cells and tissues, but are synthesized on demand. The biological activity of eicosanoids is primarily determined by their biosynthesis. The common substrates are polyunsaturated fatty acids such as arachidonic acid (5,8,11,14-eicosatetraenoic acid). In cells, they are esterified to phospholipids carrying the fatty acid at the sn-2 position. Thus, release of substrate from these phospholipid stores is an important control point in eicosanoid biosynthesis (Burgoyne & Morgan, 1990).

Several pathways exist for the release of arachidonic acid (Fig. 3; Axelrod, 1990). The most direct pathway involves the

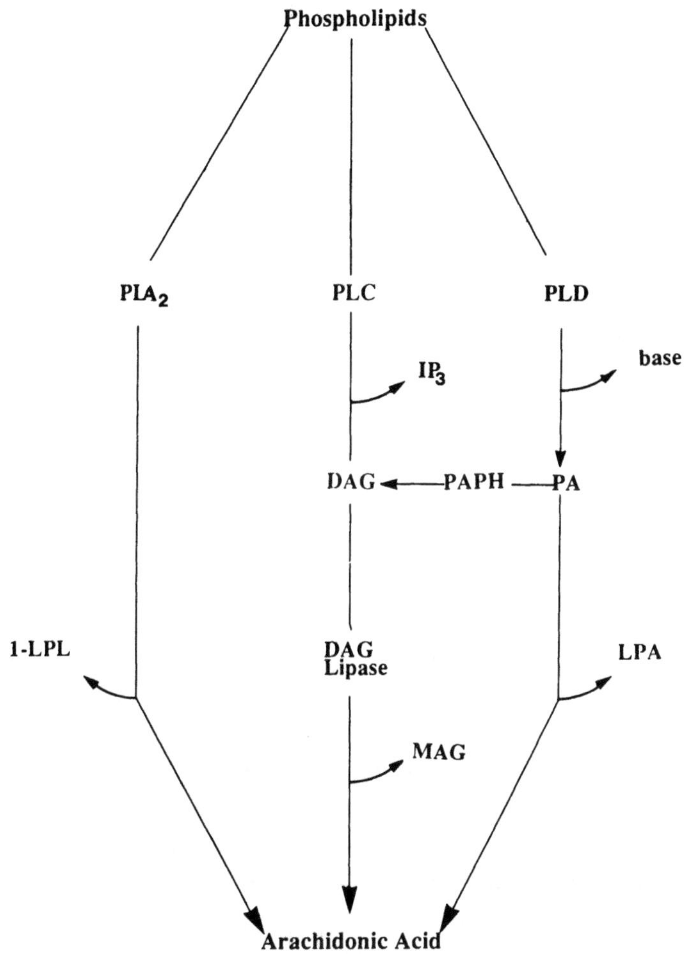

Fig. 3 Pathways of arachidonic release from phospholipids. An increase of free arachidonic acid levels can be produced directly or indirectly by cleavage of phospholipids by phospholipases A_2, C and D. DAG, diacylglycerol; IP_3, inositol-1,4,5-trisphophate; MAG, monoacylglycerol; PA, phosphatidic acid; LPL,lysophospholipid; LPA, lysophosphatidic acid; PAPH, phosphatidic acid phosphohydrolase.

activation of phospholipases A_2 releasing arachidonic acid from the sn-2 position of the phospholipids. Indirect routes have been demonstrated involving the release of arachidonic acid from diacylglycerol by diacylglycerol lipase. Diacylglycerol can be generated by phospholipase C-catalyzed hydrolysis of phosphatidylinositol or -choline or by dephosphorylation of phosphatidic acid, which can be produced from phospholipase D-catalyzed hydrolysis of phosphatidylcholine (Fig. 3). In keratinocytes and epidermis in vivo phospholipase A_2 activities provide the major mechanism involved in the release of arachidonic acid. Phospholipases A_2 form a large family of enzymes (Kudo et al., 1993) and two different members have been shown to be expressed in keratinocytes and in mouse and human epidermis in vivo (Fürstenberger et al., 1994a; unpublished results). They include a high molecular weight cytosolic PLA_2 ($cPLA_2$) which is activated via receptor-G protein-, PKC- and EGF receptor-dependent pathways (Kast et al., 1991, 1993 a,b). Activation of this enzyme involves an increase of the enzyme activity and the Ca^{2+}-dependent association of $cPLA_2$ with the membrane. A transient increase in the cytoplasmic Ca^{2+} concentration appears to be necessary for the translocation of the enzyme and phosphorylation on serine to be important for the increase of $cPLA_2$ activity (Clark et al., 1991; Lin et al., 1992). In cells overexpressing the enzyme, a MAP kinase-catalyzed phosphorylation of serine-505 has indeed been shown to be critical for the rise of $cPLA_2$ activity (Lin et al., 1993). Most probably, activation of $cPLA_2$ is causal for the immediate release of arachidonic acid from keratinocytes observed in response to endogenous proinflammatory mediators such as bradykinin, histamine, or platelet-activating factor, and growth factors such as transforming growth factor-α (TGF-α) but also exogenous chemical and physical stimuli such as phorbol ester tumor promoters, Ca^{2+}-ionophore and UV-light (Kast et al., 1993 a,b; Kang-Rotondo et al., 1993). In addition to activation through posttranslational modification $cPLA_2$ has been found to be regulated at the transcriptional and post-transcriptional level by different proinflammatory cytokines such as IL-1α and TNF-α but also by growth factors such as EGF/TGF-α in different cell types (for review see Glaser et al., 1993; Kramer, 1993).

In keratinocytes, de novo biosynthesis of $cPLA_2$ is induced by IL-1α (unpublished results). Since $cPLA_2$ preferentially liberates arachidonic acid, the activity of this enzyme is thought to play a critical role in the initiation of eicosanoid biosynthesis in keratinocytes as well as in other cell types. Keratinocytes coexpress another member of the PLA_2 family, a low-molecular weight secretory PLA_2 type II ($sPLA_2$ II; unpublished results). Again, de novo synthesis of this enzyme has been shown to be induced in various cell types by proinflammatory cytokines (Glaser et al., 1993), e.g. in keratinocytes by IL-1α (unpublished results), and to be strongly expressed in psoriatic epidermis (Andersen et al., 1994). In contrast to $cPLA_2$, the secretory enzyme does not show specifity for arachidonic acid. Nevertheless, upon release into the extracellular space this enzyme could act as a proinflammatory mediator propagating or amplifying the production of eicosanoids. However, the detection of high affinity binding proteins for $sPLA_2$ on different epithelial cell types points to additional functions of these enzymes including autocrine induction of their own biosynthesis (Kishino et al., 1994; Lambeau et al., 1994). Thus, it is conceivable that regulation of arachidonic acid release by modulation of the activity and expression of PLA_2 is an important control device for the biosynthesis of eicosanoids in keratinocytes. However, free cellular arachidonic acid levels may also be modulated by the activity of enzymes catalyzing the reacylation of the fatty acid into phospholipids (Fuse et al., 1989).

Once released from phospholipids arachidonic acid is metabolized by one of three different oxygenases (Fig. 4). Distinct cytochrome P450-dependent monooxygenases insert an oxygen atom into the double bonds of arachidonic acid yielding regioisomeric epoxy derivatives. In addition, cytochrome P450-catalyzed arachidonic acid oxygenation produces regioisomeric enantiomeric monohydroxy eicosatetraenoic acids (HETEs) and, as a result of ω-oxidation, 19- and 20-HETE (Fitzpatrick & Murphy, 1989). Upon incubation with arachidonic acid freshly isolated human keratinocytes have been shown to produce 5,6 and 8,9-epoxyeicosatetraenoic acids in addition to an enantiomeric mixture of 12-HETE (Holtzman et al., 1989).

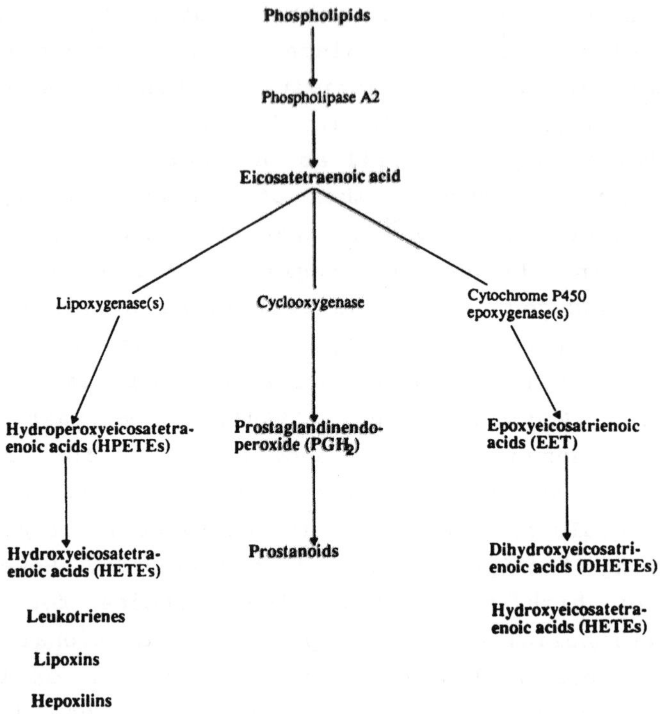

Fig. 4 Pathways of arachidonic acid oxygenation

Lipoxygenases introduce one molecule of oxygen enantiospecifically at carbons 5, 8, 9, 11, 12, or 15 of arachidonic acid to generate a series of regioisomeric hydroperoxy fatty acids (HPETEs), which are reduced to the corresponding hydroxy fatty acids (HETEs; Pace-Asciak & Asotra, 1989; Yamamoto, 1992; Fig. 5). In contrast to the above-mentioned products of the cytochrome P450-catalyzed arachidonic acid oxygenation the lipoxygenase-derived HETEs are pure S-enantiomers. 5-S- 8-S-, 12-S- and 15-S-HETE have been found as products of lipoxygenase-catalyzed arachidonic acid oxygenation in human and murine keratinocytes (Hammarström et al., 1979; Burral et al., 1988; Green, 1989; Fürstenberger et al., 1991; Lehmann et al., 1993; Takahashi et al., 1993; Hussain et al., 1994; Janßen-Timmen et al., 1994).

Accordingly, the expression of the 5-, 8-, and platelet-type 12-lipoxygenases have been detected at the mRNA and protein level in normal and/or hyperproliferative human and mouse epidermis and in keratinocytes (Takahashi et al., 1993; Chen et al., 1994; Hussain et al., 1994; Janßen-Timmen et al., 1994; Löschke et al., 1994).

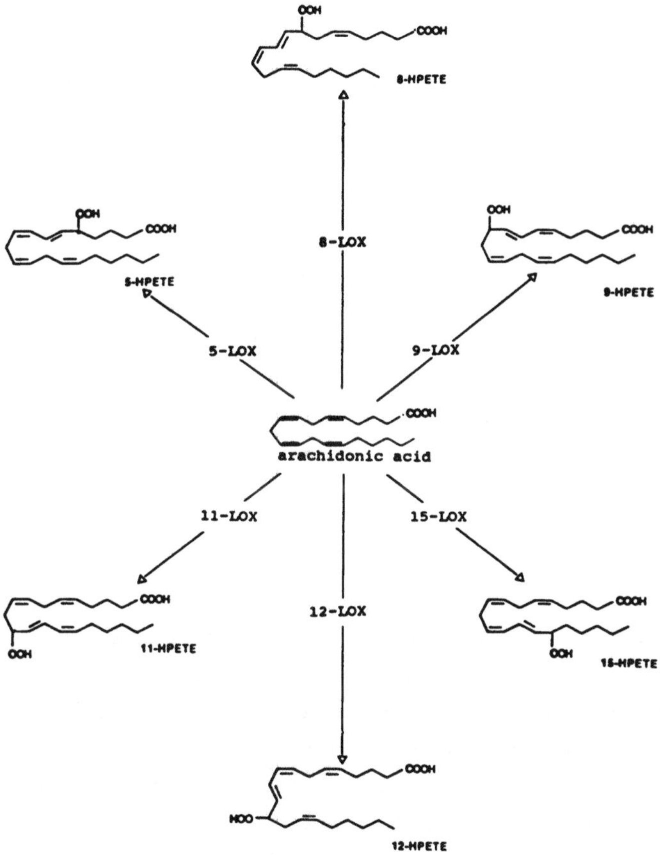

Fig. 5 Lipoxygenase-catalyzed arachidonic acid oxygenation. Mammalian lipoxygenases introduce one molecule of oxygen enantiospecifically into the carbon skeleton of arachidonic acid generating regioisomeric hydroperoxyeicosatetraenoic acids (HPETEs). The 5-, 12-, and 15-lipoxygenases (LOX) from different tissues and species have been characterized at the molecular level while 8-, 9-, and 11-lipoxygenases have not yet been completely characterized.

In addition, the expression of the platelet-type 12-lipoxygenase is restricted to the basal layer of normal human epidermis and strongly increased in psoriasis consistent with a high 12-S-HETE generating activity (Hammarström et al., 1979; Hussain et al., 1994). Psoriatic scales, however, have been shown to contain high levels of enantiomeric 12-HETE which are assumed to be the product of an arachidonic acid-metabolizing P450-dependent monooxygenase in suprabasal keratinocytes (see above; Holtzman et al., 1989). The mechanisms involved in the selective expression or regulation of the epidermal lipoxygenases are widely unknown. Recently, EGF was found to increase the expression of the 12-lipoxygenase in the epidermoid carcinoma cell line A 431 (Chang et al., 1992) and overexpression of the EGF homologue TGF-α is thought to be one factor responsible for the high level of 12-lipoxygenase epxression in psoriatic epidermis (Hussain et al., 1994). Moreover, irritant and hyperplasiogenic phorbol ester tumor promoters strongly induced the epidermal 8-lipoxygenase in mouse epidermis in vivo (Fürstenberger et al. 1991). Although HETEs appear to be the predominant products of the epidermal lipoxygenase pathways, keratinocytes are able to transform 5-HPETE into leukotrienes (Fig. 6). In addition to the insertion of molecular oxygen at C-5 of arachidonic acid, the 5-lipoxygenase catalyzes the dehydration of the intermediately formed 5-HPETE yielding an 5,6 epoxide of arachidonic acid, leukotriene A_4 (LTA_4). Upon generation LTA_4 is immediately transformed along two divergent enzymatic pathways leading to leukotriene B_4 by addition of one molecule of water and to the cysteinyl leukotriene C_4 (LTC_4) by addition of a glutathionyl group at C-6 and simultaneous opening of the 5,6-epoxide. The cysteinyl leukotrienes LTD_4 and LTE_4 are formed by subsequent enzymatic cleavage of glutamic acid and glycine (Fig. 6). Murine epidermis in vivo as well as human keratinocytes transform LTA_4 into LTB_4 through the activity of an epidermal LTA_4 hydrolase and generate cysteinyl leukotrienes (Iversen et al., 1993 and 1994). These authors assume that the epidermal leukotrienes are products of transcellular processing of leukocyte-released LTA_4 rather than true keratinocyte-derived products. However, recent data demonstrate a functional 5-

lipoxygenase in human keratinocytes. The enzyme activity and the generation of 5-HETE, LTB_4, and LTC_4 were found to be upregulated upon terminal differentiation or challenge of the cells with the ionophore A 23187 (Janßen-Timmen et al., 1994). Moreover, treatment of mouse skin in vivo with phorbol esters rapidly increased the epidermal content of cysteinyl leukotrienes (Fürstenberger et al., 1994b). Taken together, these data present evidence that skin keratinocytes are capable to express 5-, 8, 12 and 15-lipoxygenases which are induced by exogenous stimuli and by endogenous factors modulating growth and differentiation of the epithelium.

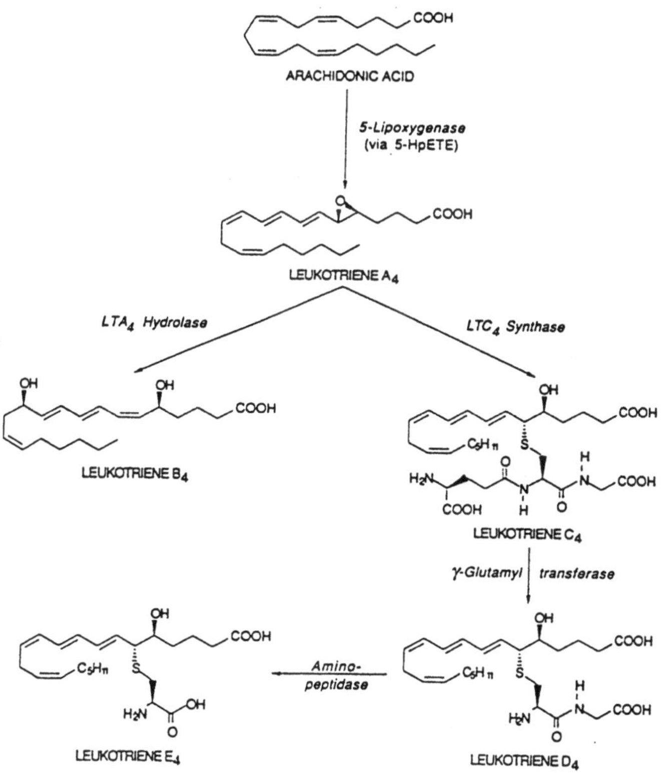

Fig. 6 Pathways of leukotriene biosynthesis. The primary product of leukotriene biosynthesis is leukotriene A_4 generated by the lipoxygenase and dehydrase activities of the 5-lipoxygenase. Leukotriene A_4 is converted by a LTA_4 hydrolase into leukotriene B_4 or by the LTC_4 synthase-catalyzed addition of glutathione into the cysteinyl leukotrienes C_4, D_4, and E_4.

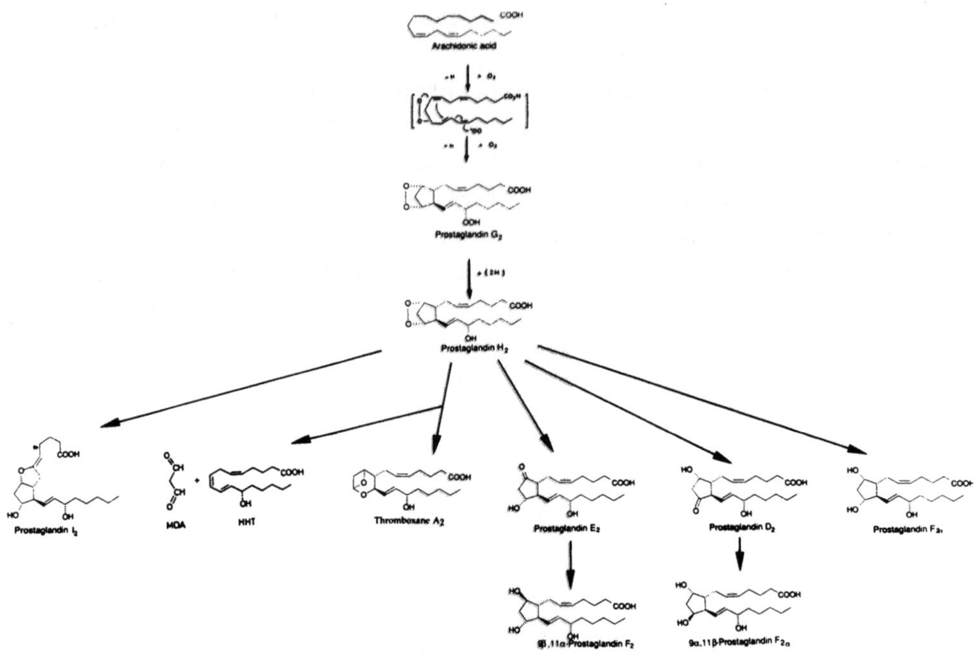

Fig. 7 Metabolism of arachidonic acid via the cyclooxygenase pathway. The bis-dioxygenase activity of the prostaglandin H synthase (PGHS) catalyzes the insertion of molecular oxygen at C11 of arachidonic acid concomitantly with a cyclization to an endoperoxide between C9 and C11. An introduction of a second molecule of oxygen at C15 and formation of a cyclopentane ring from C8 to 12 yields the 15S-hydroperoxy-9,11-endoperoxide prostaglandin G_2 (PGG_2). The hydroperoxidase activity of PGHS reduces the 15-hydroperoxy group of PGG_2 to produce PGH_2 which is transformed in a cell type specific manner into the different prostaglandins.

The biosynthesis of prostaglandins, mainly prostaglandin E_2 (PGE_2), $PGF_{2\alpha}$, and PGD_2, has been observed in skin preparations of mice, rats, guinea pigs, and humans (reviewed in Fürstenberger, 1990; Ruzicka, 1988; Ziboh, 1994). The main biosynthetic activity is localized in the epidermal layer. The initial transformation of arachidonic acid to PGH_2 is catalyzed by the bis-dioxygenase and hydroperoxidase activity of the prostaglandin H synthase (PGHS; Fig.7). Keratinocytes and epidermis in vivo express two forms of PGHS, i.e. a constitutive PGHS-1 and an inducible PGHS-2 (Müller-Decker et al., 1994a). Accordingly PGHS-1 was found to be constitutively expressed in normal epidermis, whereas PGHS-2 expression could not be detected. The induction of PGHS-2 mRNA and protein strictly correlated with the evolution of an inflammatory hyperplasia (hyperplastic transformation) either induced by phorbol ester or skin wounding (Scholz et al., 1994).

Interestingly, non-damaging hyperproliferative stimuli were unable to induce PGHS-2. Prostaglandin synthesis in keratinocytes has been shown to be enhanced by a wide variety of stimuli, including proinflammatory cytokines such as IL-1α, growth factors such as TGF-α, phorbol esters and UV-light (see above). Since all these stimuli also increase arachidonic acid release, it was suggested that prostaglandin biosynthesis is controlled at the level of phospholipase activation rather than by the activity of the PGHS isozymes. However, most of these stimuli concomitantly induce PGHS-2 expression in keratinocytes in an immediate-early manner. For example, the proinflammatory cytokine IL-1α has been found to activate and induce the expression of $cPLA_2$ and $sPLA_2$ concomitantly with the expression of PGHS-2 in mouse and human keratinocytes (unpublished results).

To summarize, these data show that keratinocytes express a broad spectrum of arachidonic acid-metabolizing enzymes and that keratinocyte activation in response to endogenous proinflammatory mediators and exogenous stimuli is accompanied by an upregulation at the transcriptional, translational or posttranslational level of key enzymes of arachidonic acid metabolism and generation of a keratinocyte-specific spectrum of eicosanoids (Fig. 8). However, arachidonic acid metabolism critically depends on the availability of appropriate concentrations of free arachidonic acid controlled primarily by the activity of PLA_2, the pace-making enzyme of eicosanoid biosynthesis.

Functions of eicosanoids

As already mentioned, keratinocyte activation is thought to be a critical event for the initiation of hyperplastic transformation of skin upon external damage. Keratinocyte activation, on the other hand, is consistently accompanied by an immediate activation of arachidonic acid release and eicosanoid biosynthesis (Fig. 8) indicating that eicosanoids are involved in the initiation phase of this highly coordinated tissue response.

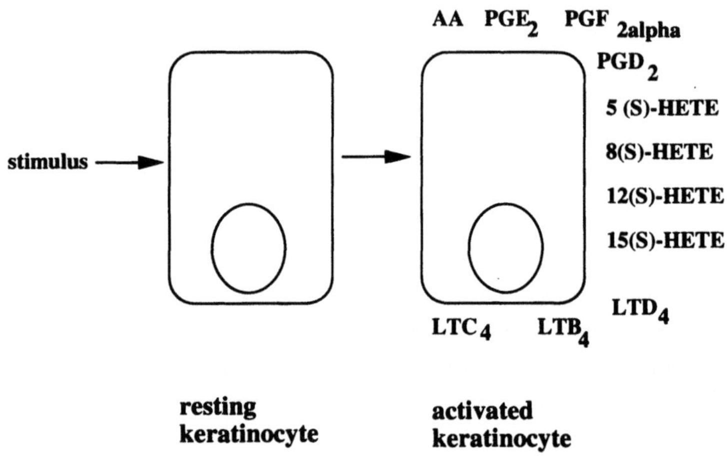

Fig. 8 Arachidonic acid-derived metabolites produced by activated keratinocytes.

Individual eicosanoids have indeed been shown to be involved in many steps of the acute inflammatory process, frequently by synergistic interaction with other members of the eicosanoid family or structurally unrelated proinflammatory mediators such as histamine, kinins, neuropeptids, cytokines (Greaves, 1988; Cunningham, 1990). These effects include the acute changes in cutaneous microvasculature observed upon intradermal injection of PGE_2 and PGD_2 (Fig. 9). Both prostaglandins induced a long-lasting erythema in consequence of cutaneous vasodilation. In addition, PGE_2 potentiated the bradykinin-, histamine or PAF-induced vasopermeability changes provoking edema formation (Bisgaard, 1990). A rapid increase of PGE_2 has indeed been observed in inflammatory skin reactions induced by chemical irritants and irradiation (see above).

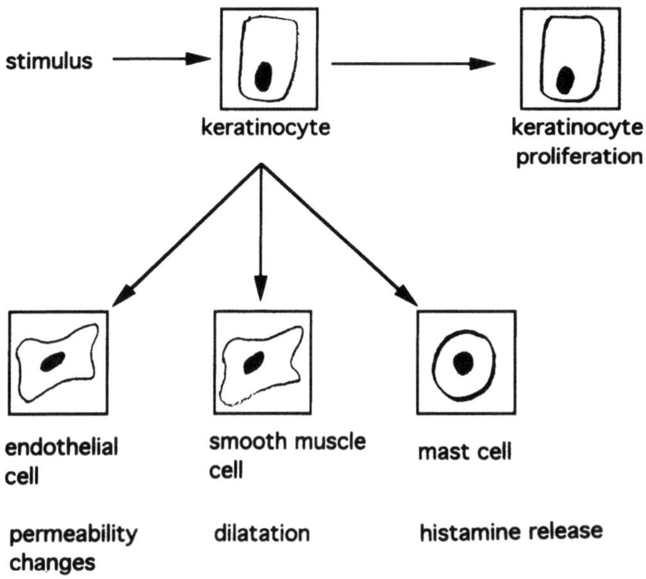

Fig. 9 Effects of PGE_2 on cells involved in hyperplastic transformation. Activation of keratinocytes is accompanied by an immediate release of PGE_2 which may be involved in the initiation of this response.

While a definite direct growth-promoting effect of PGE_2 could not consistently be demonstrated, a correlation between PGE_2 production and serum-stimulated keratinocyte growth is suggested by the fact that the PGHS inhibitor indomethacin concomitantly inhibited both effects in primary human keratinocytes and that addition of PGE_2 reversed the growth-inhibitory effect of the nonsteroidal-antiinflammatory agent (Pentland and Needleman, 1986). A similar observation was made in wounded or phorbol ester-treated mouse epidermis in vivo (reviewed in Fürstenberger and Marks, 1990). The induction of epidermal hyperproliferation and transient epidermal hyperplasia critically depended on an early burst of PGE_2 synthesis in that its inhibition by indomethacin prevented epidermal growth stimulation, an effect which was reversed by simultaneous treatment with PGE_2. Thus, while being inactive as a mitogen for resting keratinocytes, PGE_2 is thought to modulate the proliferative potential of activated keratinocytes by modulating the synthesis or activity of growth factors and cytokines released or the expression of the cognate receptors. This may also explain the growth-stimulating activity of an intradermal injection of PGE_2 into human skin (Eaglestein & Weinstein, 1975). On the other hand, topically applied PGE_2 was found to reverse hyperproliferation and hyperkeratosis in the skin of essential-fatty-acid-deficient mice (reviewed in Ziboh & Chapkin, 1988). This indicates that with a permanent hyperplastic state of epidermis, growth-inhibitory effects of PGE_2 may predominate. They could be due to a PGE_2 receptor-mediated increase of intracellular cAMP leading to a down-regulation of keratinocyte proliferation (Marks, 1983).

The cysteinyl leukotrienes are known to affect the microvascular tone and permeability although species-dependent variations are observed. In human skin LTC_4 and LTD_4 have been shown to increase microvascular permeability and to be potent vasodilators (Soter et al., 1983). Other lipoxygenase-derived arachidonic acid derivatives, in particular LTB_4, are kown to exhibit potent chemotactic activity towards inflammatory cells invading the tissue upon inflammation (Fig. 9; Ford-Hutchinson, 1990). Moreover, individual HETEs and LTB_4 stimulate or potentiate activation of the immigrated inflammatory cells to degranulate and release lytic enzyme, to generate and release active oxygen species, and to stimulate PGHS- and lipoxygenase-catalyzed oxygenation of arachidonic acid providing additional lipid mediators to the inflammatory milieu (Spector et al., 1988). However, eicosanoids do not only act as proinflammatory mediators as observed in the acute phase of inflammation, but they also exhibit anti-inflammatory effects in later phases of the inflammatory response (Zurier, 1990). A direct growth-promoting activity could be ascribed to LTB_4 which stimulated keratinocyte proliferation in human skin in vivo and in keratinocytes while LTC_4 and LTD_4 evoked only slight effects in keratinocyte cultures. Conflicting results have been reported with regard to the mitogenic activity of 12S-HETE even though high concentrations of enantiomeric 12-HETE have been detected in hyperproliferative diseased epidermis (reviewed in Ruzicka, 1988; Fürstenberger, 1990).

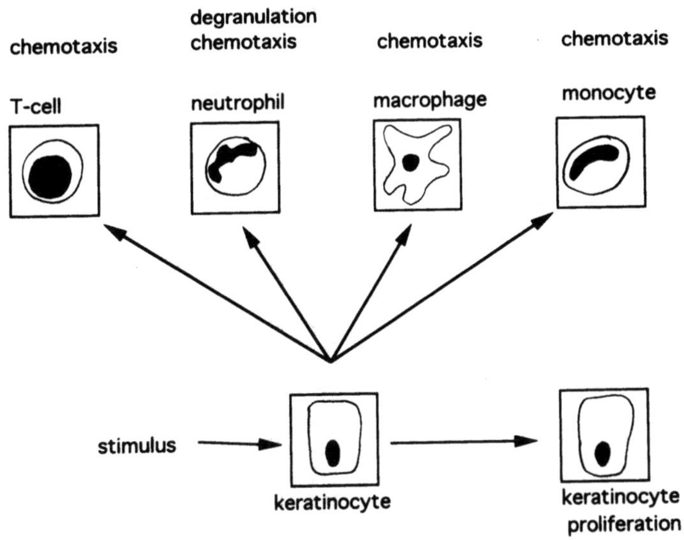

Fig. 10 Effects of LTB_4 on cells involved in hyperplastic transformation. Upon activation keratinocytes have been shown to produce LTB_4 on their own (Janßen-Timmen et al. 1994) or through transcellular metabolism of leukocyte-derived LTA_4 (Iversen et al., 1994) indicating that LTB_4 may be involved in the initiation of hyperplastic transformation of skin.

In summary, these data support the view, that eicosanoids represent a network of functionally important intercellular signal molecules which are responsible for the integration of all skin cells in a coordinated tissue reaction in response to external toxicity or damage. In addition, as an immediate response to keratinocyte activation stimulation of arachidonic acid release and prostaglandin biosynthesis are involved in the initiation of the tissue response.

The cytokine network in hyperplastic transformation

Cytokines represent another complex network of intercellular signaling molecules involved in hyperplastic transformation of skin. In concert with eicosanoids these factors have been shown to act as proinflammatory mediators but also to regulate proliferation and differentiation of various cell types involved in hyperplastic transformation of skin (Bull & Dowd, 1992). They are transiently produced and released from cells and exert their biological activities via specific cell-surface receptors (Dower et al., 1990). Cytokines include the interleukins (IL), the hematopoetic colony-stimulating factors (CSF), tumor necrosis factors α and β (TNF-α and -β), the interferons (IFN) and a still growing number of factors which are involved in immune and inflammatory reactions (Burke et al., 1993). Originally thought to be restricted to immune cells, cytokine production has been observed in many other cells (Kelso, 1989) and keratinocytes have been identified as a potential source of these low-molecular-weight-(glyco)proteins (Mc Kay & Leigh, 1991; Bos & Kapsenberg, 1993; Fig. 2). There is increasing evidence that ILs and TNFs may play a critical role in the activation of keratinocytes and the initiation of inflammation and epidermal hyperplasia (hyperplastic transformation) in skin exposed to toxic agents (Kupper, 1990a).

Keratinocyte-derived cytokines

Among the different cytokines released from activated keratinocytes (Mc Kay & Leigh, 1991 and Fig. 2) IL-1 is thought to play a key role for two reasons. It is immediately released in its active form upon injury (Kupper, 1990b; McKenzie & Sauder, 1990;) and it initiates an epidermal cytokine cascade (Luger & Schwarz, 1990).

Two different forms of this cytokine, namely IL-1α and IL-1β, which are encoded by two distinct genes (March et al., 1985; Di Giovine & Duff, 1990), have been identified, with IL-1α being the major species released by activated keratinocytes (Hauser et al, 1986; Didierjean et al., 1991). IL-1α is synthesized as a 31-kDa precursor molecule which is released from keratinocytes, either unprocessed or processed, as a 17 kDa species. Both show biological activity (Moslay et al., 1987), while the corresponding IL-1β precursor released from activated macrophages has to be proteolytically cleaved in order to be biologically active (Black et al., 1988; Molineaux et al., 1993). Resting keratinocytes are known to produce IL-1α. In normal epidermis, this cytokine is assumed to be associated with the membrane (Stevenson et al., 1993) or stored in the intercellular space between differentiated keratinocytes and eliminated by normal desquamation. Upon injury of the epidermis IL-1α is thought to be immediately released from these stores. In addition, keratinocytes can be activated to synthesize significant amounts of IL-1α in an autocrine manner by IL-1α, by a variety of growth factors, lipopolysaccharide, irritant agents such as phorbol ester tumor promoters and the contact allergen urushiol (Luger & Schwarz, 1990), as well as by thermal (Hannum et al, 1990) and physical injury including UVB-irradiation (Kupper et al., 1987; Nozacki et al., 1991; Griswold et al., 1991). Keratinocytes also express the 80 kDa IL-1 receptor. 100 - 4000 binding sites are found in resting cells (Sauder, 1989, Dower et al., 1990), the number of which can be strongly upregulated upon exposure of the keratinocytes to phorbol esters, cytokines, and UVB-irradiation (Parker et al., 1988; Blanton et al, 1989; Dower et al, 1990). Moreover, an increase of the extracellular Ca^{2+} concentration has been shown to upregulate the expression of the receptor without affecting IL-1α expression indicating a differential regulation of IL-1α and IL-1 receptors in keratinocytes (Blanton et al., 1989). According to Kupper (Kupper, 1990b) an activated keratinocyte is a phenotype with increased expression of IL-1 receptors and/or production of IL-1α responding to IL-1α in an autocrine or paracrine manner. Another cytokine that is constitutively expressed in keratinocytes is the tumor necrosis factor-α (TNF-α; Barker et al., 1991). Thus both IL-1α and TNF-α are considered as keratinocyte derived primary cytokines.

As already mentioned Il-1α is known to induce its own biosynthesis. Moreover, various genes coding for additional cytokines have been found to be strongly induced leading to the generation of IL-6, IL-8, IL-10, granulocyte/macrophage colony-stimulating factor (GMCSF), TNF-α and others (Nickoloff et al., 1990; Barker et al., 1991; Fig. 11). Moreover, it could be demonstrated that expression of some of these additional cytokines has also been found to be induced by exogenous stimuli, e.g. IL-6 by irritant phorbol ester tumor promoters and other mitogens, lipopolysaccharide and physical stress such as UVB irradiation (Kupper et al., 1989; Kirnbauer et al., 1991). These so-called secondary cytokines are produced only upon activation of keratinocytes e.g. by the primary or initiating cytokines IL-1α or TNF-α. In summary, keratinocytes are an important source of cytokine production with IL-1α acting as a shield to monitor any damage or toxicity to skin by triggering a cytokine cascade generating a network of multifunctional and multitargeted proteins.

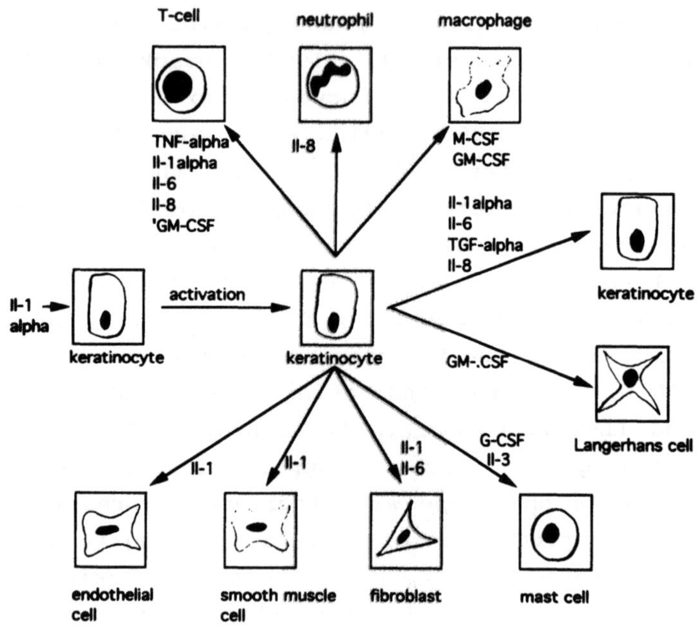

Fig. 11 Cytokines produced by activated keratinocytes and their target cells in hyperplastic transformation of skin. TNF, tumor necrosis factor; GM-CSF, granulocyte/macrophage colony stimulating factor; G-CSF, granulocyte stimulating factor; M-CSF, macrophage colony stimulating factor; TGF, transforming growth factor.

In normal epidermis or in resting keratinocytes the biological effects of the constitutively produced and sequestered IL-1α are suppressed by the presence of an endogenous interleukin-1 receptor antagonist (IL-1ra) in epidermis (Haskill et al., 1991; Gruaz-Chatellard et al., 1991) and in keratinocyte cultures (Gruaz-Chatellard et al., 1991; Kutsch et al., 1993). Keratinocytes have been found to store an unglycosylated IL-1ra intracellularly (Haskill et al., 1991) which may be released together with IL-1α upon appropriate stimuli or after cell damages. In addition, monocytes are known to constitutively produce and secrete a glycosylated form of IL-1ra (Haskill et al., 1991). Both IL-1ra isoforms when tested in a bioassay or receptor binding assay exhibited similar competitive inhibitory activities (Arend, 1991; Dinarello & Thompson, 1991; Haskill et al., 1991;). IL-1ra production in keratinocytes has been found to increase with terminal differentiation providing an excess of IL-1ra over IL-1α in the suprabasal compartment of epidermis (Gruaz-Chatellard et al., 1991; Bigler et al., 1992). Since IL-1α binds with higher affinity to the 80 kd receptor IL-1ra levels have to exceed those of IL-1α to maintain a protective effect in the normal epithelium.

The differential regulation of IL-1 and IL-1ra by cytokines like TNF-α and IL-4 (Kutsch et al., 1993) has been shown to alter the agonist/antagonist ratio (Vannier et al., 1992, Dinarello, 1993). In vivo this seems to be important for the triggering effect of keratinocytes at the onset of pathological disorders but also during healing or recovery processes.

Functions of cytokines

In vivo and in vitro studies have underlined the importance of IL-1α in the initiation of acute inflammation and epidermal hyperplasia in skin. Intradermal injection of IL-1α induces a dose-dependent erythema which is associated with an infiltrate of monocytes and neutrophils. IL-1α has been detected in suction blister fluids and skin chamber exudates from skin upon UV-irradiation, mechanical trauma and toxic insults (Bull & Dowd, 1992). In mouse skin in vivo inflammatory cell infiltration, edema and epidermal hyperplasia induced by the phorbol ester TPA are inhibited by subcutaneous injection of an anti-IL-1α antibody indicating IL-1α to be involved in hyperplastic transformation (Lee et al., 1994). In addition to activation of the cytokine cascade as mentioned above, IL-1α-exposed keratinocytes are stimulated to express intercellular adhesion molecules which serve as binding sites for monocytes, T-cells and neutrophils invading the epithelium (Nickoloff et al., 1990) . Moreover, IL-1α has been found to upregulate the epidermal arachidonic acid cascade (Pentland & Mahoney, 1990) by the coordinated induction of $cPLA_2$-activity and de novo synthesis of PGHS-2 protein (unpublished data) and to stimulate proliferation of cycling (Birchall et al., 1987), but not growth-arrested, keratinocytes (Ristow, 1990). Systemic effects of IL-1 include the activation of endothelial cells, macrophages, neutrophils, T-cells, fibroblasts, vascular smooth muscle cells, and others (Sauder 1989; Fig. 12). Depending on the target cell cytokine release, eicosanoid biosynthesis, receptor upregulation, adhesion molecule synthesis, chemotaxis, extracellular matrix production, and/or cell replication have been found to be stimulated by IL-1. Thus in the early phase of

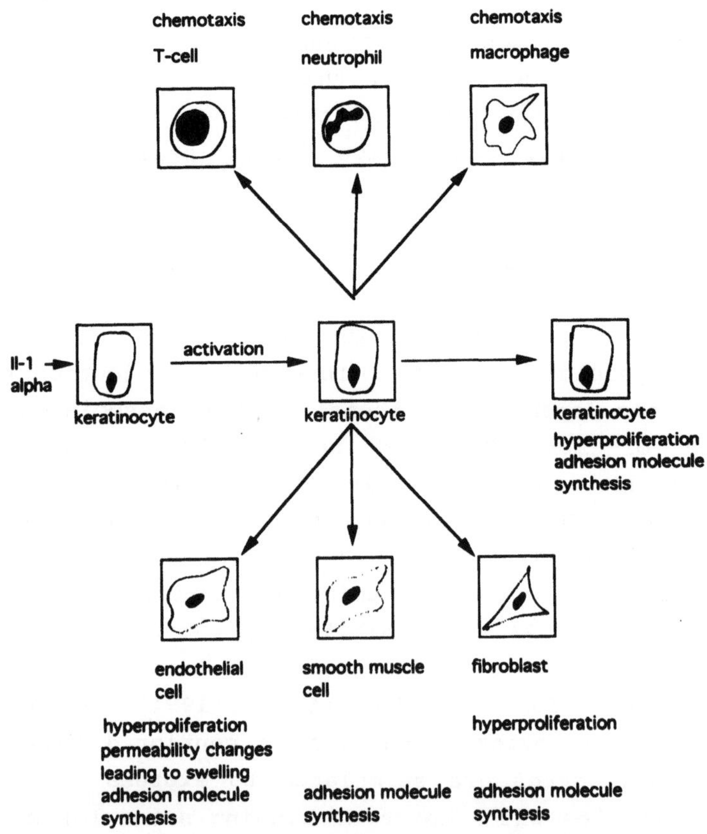

Fig. 12 Effect of IL-1α on cells involved in hyperplastic transformation of skin

inflammation IL-1 augments the concerted expression of intercellular adhesion molecules and cytokine and eicosanoid release in endothelial cells. Simultaneously, neutrophils and macrophages are chemoattracted by IL-1 and as a consequence the activated inflammatory cells adhere to the blood vessels. Upon further stimulation with chemotactic and secondary cytokines they migrate into the skin and feed back on the keratinocyte, thereby amplifying the response.

Although the effects of IL-1α have not been investigated in more detail in skin, some general principles delineated from extensive research on cytokine biology may hold true also for IL-1 in this tissue. Even though IL-1α acts as a primary cytokine, it does in part not act alone but synergistic with other cytokines or lipid mediators. Moreover, other cytokines can mediate similar biological effects as IL-1. For instance, overlapping abilities have been ascribed to IL-1 and TNF-α, IL-1 and IL-6 or IL-1 and IL-8. One and the same IL-1 effect, e. g. stimulation of proliferation, is induced in several target cells e. g. in fibroblasts, endothelial cells and growing keratinocytes. The outcome of the individual response critically depends on the IL-1-receptor number on target cells and on the ratio of IL-1 to IL-1 receptor antagonist within skin compartments (see above). Like every cytokine IL-1 has multiple biological effects on one or different target cell types (Sauder, 1989).

<u>Development of an in vitro skin irritancy test</u>

The extraordinary signaling capacity of keratinocytes indicates a key role of this cell type in the induction of an inflammatory hyperplasia, and signal release by keratinocytes is considered to provide a suitable in vitro parameter for monitoring this tissue response. Within the cutaneous networks of eicosanoids and cytokines arachidonic acid and IL-1α can be regarded as primary or initiating signals which are present in resting keratinocytes phospholipid-bound or sequestered inside the cell. Upon injury an immediate increase in the intracellular concentration of free arachidonic acid gives rise to the biosynthesis of eicosanoids which may be regarded as

TABLE 1

Time-course of the release of proinflammatory mediators from HPKII cells treated with chemicals of graded in vivo irritancy

compound[a]	**[^{14}C]-arachidonic acid**[b]		**IL-1α-Release**[c]	
	(M)	time-course	(M)	time-course
acetone	1.4	late[d]	2.7	early
ethanol	2.0	early	3.4	early
glycerol	2.7	late	5.6	delayed
$NiSO_4$	10^{-1}	late	5.0×10^{-3}	n.e.
benzoic acid	10^{-2}	late	10^{-2}	---
triethanolamine	10^{-1}	delayed	10^{-1}	delayed
cyclohexanol	10^{-1}	delayed	5.0×10^{-2}	late
phenol	2.5×10^{-2}	early	10^{-1}	---
$SnCl_2$	10^{-2}	delayed	10^{-3}	late
$ZnCl_2$	10^{-2}	delayed	10^{-4}	delayed
acrylamide	10^{-2}	early	10^{-1}	late
Tween 80	5.0×10^{-4}	delayed	10^{-4}	delayed
sodium dodecylsulfate	10^{-4}	delayed	10^{-4}	delayed
benzalkonium chloride	2.0×10^{-5}	early	5.0×10^{-5}	early
A23187	5.0×10^{-7}	early	10^{-5}	delayed

a, HPKII keratinocytes were treated with a single concentration of each compound as indicated for 1-, 4-, 12-, or 24 hr.
b, Media were collected for eicosanoid extraction. Lipids were analyzed by thin layer chromatography. Quantitation was done by radiodensitometry.
c, Media were collected for IL-1α-quantitation by enzyme immuno assay.
d, Mediator concentrations were found to be elevated over basal values derived from vehicle-treated cells: early, i.e. after 1 hr of incubation; delayed, i.e. after 4-12 hr of incubation; or late, i.e., after 24 hr of incubation. A minus means that IL-1α concentrations were not found to be elevated. n.e., IL-1α concentrations were not evaluable since quench effects were intolerably high at concentrations $\geq$ 5 mM $NiSO_4$.

TABLE 2

Potencies of chemicals of graded in vivo irritancy to induce proinflammatory mediator release from HPKII cells

Time-course of induction[a]	$1-[^{14}C]$-AA-release compound[b]	SC50 (M)[c]	IL-1α-release compound	ED10 (M)[d]
			ACE	>3.1 x 10^{0}
	ETOH	1.4 x 10^{0}	ETOH	2.4 x 10^{0}
EARLY	Phenol	2.6 x 10^{-2}	Phenol	(-)
	ACA	2.5 x 10^{-3}		
	BKCl	2.3-3.1 x 10^{-6}	BKCl	2.8 x 10^{-5}
	A23187	2.5 x 10^{-6}		
			GLOH	3.0 x 10^{0}
	TEA	4.0 x 10^{-2}	TEA	>1.0 x 10^{-1}
DELAYED	CHOH	2.8-3.2 x 10^{-2}		
	$SnCl_2$	3.0-3.3 x 10^{-2}		
	$ZnCl_2$	2.6 x 10^{-2}	$ZnCl_2$	9.5 x 10^{-5}
	Tween80	1.9 x 10^{-4}	Tween80	8.6 x 10^{-5}
	SDS	3.2-7.0 x 10^{-5}	SDS	4.5 x 10^{-5}
			A23187	4.4 x 10^{-5}
	ACE	>1.4 x 10^{0}		
	GLOH	6.8 x 10^{-1}		
	$NiSO_4$	1.4 x 10^{-1}	$NiSO_4$	n.e.
LATE	BA	1.0 x 10^{-2}	BA	>1.0 x 10^{-1}
			ACA	1.0 x 10^{-1}
			CHOH	5.3 x 10^{-2}
			$SnCl_2$	3.1 x 10^{-3}

a, derived from TABLE 1.
b, ETOH = ethanol; ACA = acrylamide; BKCl = benzalkonium chloride; A23187 = Ca^{2+}-ionophore A23187; TEA = triethanolamine; CHOH = cyclohexanol; SDS = sodiumdodecyl sulfate; ACE = acetone; GLOH = glycerol; BA = benzoic acid.
c, half-maximum stimulatory concentrations (SC50) estimated for a 12-hr treatment.
d, cells were treated with increasing concentrations of each compound for 20 hr. ED10 means a concentration which stimulates IL-1α concentrations 10-fold over values derived from vehicle-treated cells. >, no 10-fold elevated concentration of cytokine was found; n.e. IL-1α concentrations were not evaluable since quench effects were intolerably high at concentrations ≥5 mM $NiSO_4$.

secondary signals. The instantaneous release of IL-1α, most probably due to cell destruction, induces a cascade of secondary cytokines and expression of cell adhesion molecules. Based upon these observations we set out to develop an in vitro irritation assay using human keratinocytes and keratinocyte-derived proinflammatory key mediators, i.e. eicosanoids and IL-1α, and cell viability as endpoints (Müller-Decker et al., 1992, 1994b).

Fifteen structurally unrelated, pharmacologically relevant compounds were tested with respect to the dose-response characteristics (Table 1) and time-courses (Table 2) of arachidonic acid and IL-1α release. On the basis of the time-course of mediator release the chemicals were classified as inducers of an immediate early, delayed, or late response in keratinocytes. Inducers of an early release of arachidonic acid did not necessarily induce an early IL-1α release and vice versa. Stimulation of arachidonic acid release generally occurred at lower concentrations of the test compounds than IL-1α release or at similar concentrations with more rapid kinetics, suggesting arachidonic acid release to be the more sensitive endpoint. However, inducers of an early release are not necessarily more potent than those eliciting a delayed or late response, thus indicating that a detailed kinetic analysis of the parameters is of critical importance for determining the potencies of the individual compounds. Evaluation of the half-maximum stimulatory concentrations for arachidonic acid release and the tenfold-stimulatory concentrations for IL-1α release allowed ranking according to increasing potencies. Although the assay of arachidonic acid release (SC50 values) was found to be more sensitive when compared with the ED10 data for IL-1α-release for 11 of 15 compounds, combined measuring of these endpoints led to a better grading of the irritant potencies of the chemicals. An example is provided by $SnCl_2$, $ZnCl_2$, and acrylamide (ACA) which exhibited similar potencies in the arachidonic acid assay but could clearly be discriminated by the IL-1α assay.

A strict correlation between arachidonic acid release and impairment of cell viability was not found. A comparison of SC50 values with IC50 values showed that the cell viability assay was more sensitive for 7 (acetone, ethanol, $NiSO_4$, phenol, $ZnCl_2$, $SnCl_2$, A23187), equally sensitive for 3 (Tween 80, sodium dodecyl sulfate, benzalkonium chloride), or less sensitive for 5 (glycerol, benzoic acid, triethanolamine, cyclohexanol, acrylamide) of the 15 compounds. In some cases the extracellular levels of arachidonic acid continued to increase when impairment of cell viability had already reached about 90 %. Whether this is due to a paracrine stimulation of surviving cells by dying cells, a process which may take place in vivo during acute inflammation, is not known. While damaging of cells by test chemicals was apparently not necessary for the release of arachidonic acid, it seemed to be a prerequisite for that of IL-1α since 13 out of 15 test chemicals [except Tween 80 and triethanolamine (TEA)] evoked the release of IL-1α at similar or higher concentrations than those eliciting the 50 % impairment of cell viability. Whether liberated IL-1α in the extracellular fluid is a mixture of the mature and the premature form or de novo synthesized IL-1α, in particular in the case of delayed and late inducers, is not clear at present.

In conclusion, the in vitro evaluation of arachidonic acid and IL-1α release and cell viability clearly indicates that human keratinocytes in vitro respond to structurally unrelated chemicals of graded irritant potential (according to the Draize rabbit scores) with a graded response (Müller-Decker et al., 1994b). A sufficient in vivo/in vitro correlation has been observed for the surfactants benzalkonium chloride, sodium dodecyl sulfate, and Tween 80 for which adequate human in vivo data are available (Müller-Decker et al., 1992).

Any in vitro test developed as an alternative to the animal test has to be validated before use in toxicological laboratories and for acceptance by regulatory authorities. The process by which reliability and the relevance of a test are established has been defined as validation (Balls et al., 1990; Balls & Clothier 1991; Goldberg et al., 1993). In the past, in vitro data were correlated in many studies with Draize data neglecting the predominantly qualitative nature of the data obtained under non-standardized test conditions and remarkable interspecies differences. Therefore, the rabbit Draize data available for the set of test chemicals used in this study are of rather limited value for the purpose of in vivo/in vitro correlations. This situation calls for a controlled clinical study aiming at a qualitative and quantitative evaluation of the symptoms of skin inflammation including the determination of the eicosanoid- and IL-1α content of suction blister fluids from exposed and unexposed human skin. Upon adequate validation the in vitro skin irritancy test pursued here is proposed to be used as supplement, adjunct or alternative to animal tests in the hazard assessment of chemicals.

References

Andersen S, Sigursen W, Laegreid A, Volden G, Johansen B (1994) Elevated expression of human non-pancreatic phospholipase A_2 in psoriatic tissue. Inflammation 18:1-12

Arend WP (1991) Interleukin 1 receptor antagonist. A new member of the interleukin 1 family. J Clin Invest 88:1445-1451

Axelrod J (1990) Receptor-mediated activation of phospholipase A_2 and arachidonic acid release in signal transduction. Biochem Soc Trans 18:503-507

Balls M, Blaauboer B, Brusick D, Frazier J, Lamb D, Pemberton M, Reinhardt C, Roberfroid M, Rosenkranz H, Schmid B, Spielmann H, Stammati A-L & Walnum E (1990) Report and recommendations of the CAAT/ERGATT Workshop on the validation of toxicity test procedures. ATLA 18:313-337

Balls M & Clothier RH (1991) comments on the scientific validation and regulatory acceptance of in vitro toxicity tests. Toxic in Vitro 5:535-538

Barker JNWN, Mitra RS, Griffiths CEM, Dixit VM, Nickoloff BJ (1991) Keratinocytes as initiators of inflammation. The Lancet 337:211-214

Bigler CF, Norris DA, Weston WL and Arend WP (1992) Interleukin-1 receptor antagonist production by human keratinocytes. J Invest Dermatol 98:38-44

Birchall N, Kupper T & Mc Guire J (1987) Recombinant interleukin-1α is mitogenic for human keratinocytes. J Cell Biol 105-109

Bisgaard H (1990) Effects of eicosanoids on microcirculation in the skin. In: Eicosanoids and the skin, Ruzicka T, ed., CRC Press, Boca Raton, pp. 157-167

Black RA, Kronheim SR, Cantrell M, Deeley MC, March CJ, Prickett KS, Wignall J, Conlon PJ, Hopp TP & Mochizuki DY (1988) Generation of biologically active interleukin 1β by proteolytic cleavage of the inactive precursor. J Biol Chem 262:2941-2944

Blanton B, Kupper TS, McDougall J & Dower S (1989) Regulation of interleukin 1 and its receptor on human keratinocytes. Proc Natl Acad Sci USA 86:1273-1277

Bos JD & Kapsenberg ML (1993) The skin immune system: progress in cutaneous biology. Immunol Today 14:75-78

Bull HA & Dowd PM (1992) Prostaglandin synthetase, interleukin-1 and inflammation in the skin. Prostagl Leuko and Essential Fatty Acids 46:167-173
Burgoyne RD & Morgan A (1990) The control of free arachidonic acid levels. Trends Biochem Sci 15:365-366
Burke F, Naylor MS, Davies B and Balkwill F (1993) The cytokine wall chart. Immunol Today 14:165-170
Burral BA, Cheung M Chiu A, Goetzl EJ (1988) Enzymatic properties of the 15-lipoxygenase of human cultured keratinocytes. J Invest Dermatol 91:294-297
Chang WC, Ning CC, Lin MT, Huang JD (1992) Epidermal growth factor enhances a microsomal 12-lipoxygenase activity in A431 cells. J Biol Chem 267:3657-3666
Chen X, Kurre U, Jenkins NA, Copeland NG, Funk CD (1994) cDNA cloning, expression, mutagenesis of C-terminal isoleucine, genomic structure, and chromosomal localizations of murine 12-lipoxygenases. J Biol Chem 269:13979-13987
Clark JD, Lin LL, Kriz RW, Ramesha CS, Sultzman LA, Lin YL, Milona N, Knopf JL (1991) Novel arachidonic acid-selective cytosolic PLA_2 contains a Ca^{2+}-dependent translocation domain with homology to PKC and GAP. Cell 65:1043-1051
Cunningham F (1990) Lipid mediators in inflammatory skin disorders. J Lipid Med 2:61-74
Didierjean L, Groves RW & Saurat JH (1991) Interleukin 1α in normal skin. J Invest Dermatol 96:294-295
Di Giovine FS & Duff GW (1990) Interleukin 1: the first interleukin. Immunol Today 11:13-20
Dinarello CA & Thompson RC (1991) Blocking IL-1: Interleukin-1 receptor antagonist in vivo and in vitro. Immunol Today 12:404-410.
Dinarello CA (1993) Modalities for reducing interleukin 1 activity in disease. Immunol Today 14:260-264
Dower SK, Smith CA & Park LS (1990) Human cytokine receptors. J Clin Immunol 10:289-299
Eaglestein WH & Weinstein GD (1975) Prostaglandin and DNA synthesis in human skin. Possible relationship to ultraviolett light effect. J Invest Dermatol 64:386-389
Fitzpatrick FA & Murphy RC (1989) Cytochrome P450-dependent metabolism of arachidonic acid: formation and biological actions of epoxygenase-derived eicosanoids. Pharmacol Rev 40:229-241
Ford-Hutchinson AW (1990) Leukotriene B_4 in inflammation. Crit Rev Immunol 10:1-12
Fürstenberger G (1990) Role of eicosanoids in mammalian skin epidermis. Cell Biol Rev 24:1-111
Fürstenberger G & Marks F (1990) The role of eicosanoids in normal, hyperplastic and neoplastic skin (Ruzicka T, ed.) CRC Press, Boca Raton, pp. 107-124
Fürstenberger G, Hagedorn H, Jacobi T, Besemfelder E, Stephan M, Lehmann WD, Marks F (1991) Characterization of an 8-lipoxygenase activity induced by the phorbol ester tumor promoter 12-O-tetradecanoylphorbol-13-acetate in mouse skin in vivo. J Biol Chem 266:15738-15745
Fürstenberger G, Hess M, Kast R, Marks F (1994a) Expression of two $cPLA_2$ isoforms in mouse epidermis in vivo. In: Eicosanoids and other Bioactive Lipids in Cancer, Inflammation and Radiation Injury (Development in Oncology, Vol 71; Nigam S et al., eds.) Kluwer, Boston, in press

Fürstenberger G, Csuk-Glänzer IB, Marks F, Keppler D (1994b) Phorbolester-induced leukotriene biosynthesis and tumor promotion in mouse epidermis. Carcinogenesis, in press

Fuse I, Iwanaga T, Tai HH (1989) Phorbol ester, 1,2-diacyglycerol, and collagen induce inhibition of arachidonic incorporation in human platelets. J Biol Chem 264:3890-3895

Glaser KB, Mobilio D, Chang JY, Senko N (1993) Phospholipase A_2 enzymes: regulation and inhibition. Trends Pharmacol Sci 14:92-98

Goldberg AM, Frazier JM, Brusick D, Dickens MS, Flint O, Gettings SD, Hill RN, Lipnick RL, Renskers KJ, Bradlaw JA, Scala RA, Veronesi B, Green S, Wilcox NL & Curren RD (1993) Report of the Validation and Technology Transfer committee of the John Hopkins Center for Alternatives to Animal Testing. Xenobiot 23:563-572

Greaves MW (1988) Inflammation and mediators. Br J Dermatol 119:419-426

Green FA (1989) Generation and metabolism of lipoxygenase products in normal and membrane-damaged cultured human keratinocytes. J Invest Dermatol 83:486-491

Griswold DE, Connor JR, Dalton BJ, Lee JC, Simon P, Hillegaas L, Sieg DJ & Hanna N (1991) Activation of Il-1 gene in UV-irradiated mouse skin: association with inflammatory sequelae and pharmacologic intervention. J Invest Dermatol 97:1019-1023

Gruaz-Chatellard D, Baumberger C, Saurat JH & Dayer JM (1991) Interleukin-1 receptor antagonist in human epidermis and cultured keratinocytes. FEBS Letters 294:137-140

Hammarström S, Lingren JA, Marcelo C, Duell EA, Anderson TF, Voorhees JJ (1979) Arachidonic acid transformations in normal and psoriatic skin. J Invest Dermatol 73:180-183

Hannum CH, Wilcox CJ, Arend WP, Joslin FG, Dripps DJ, Heimdal PL, Armes LG, Sommer A, Eisenberg SP & Thompson RC (1990) Interleukin-1 receptor antagonist activity of a human interleukin-1 inhibitor Nature 343:336-340

Haskill S, Martin G, VanLe L, Morris J, Peace A, Bigler CF, Jaffe GJ, Hammerberg C, Sporn SA, Fong S, Arend WP & Ralph P (1991) cDNA cloning of an intracellular form of the human interleukin-1 receptor antagonist associated with epithelium. Proc Natl Acad Sci USA 88:3681-3685

Hauser C, Saurat JH, Schmitt A, Jaunin F & Dayer JM (1986) Interleukin is present in normal human epidermis. J Immunol 136:3317-3323

Holtzman MJ, Turk J, Pentland AP (1989) A regiospecific monooxygenase with novel stereopreference is the major pathway for arachidonic acid oxgenation in isolated epidermal cells. J Clin Invest 84:1446-1453

Hussain H, Shornick LP, Shannon VR Wilson JD Funck CD, Pentland AP, Holtzman MJ (1994) Epidermis contains platelet-type 12-lipoxygenase that is overexpressed in germinal layer keratinocytes in psoriasis. Am J Physiol 266:C243-C253

Iversen L, Fogh K, Ziboh VA, Kristensen P, Schmedes A, Kragballe K (1993) Leukotriene B_4 formation during human neutrophil keratinocyte interactions: evidence for transformation of leukotriene A_4 by putative keratinocyte leukotriene A_4 hydrolase. J Invest Dermatol 100:293-298

Iversen L, Kristensen P, Gron B, Ziboh VA, Kragballe K (1994) Human epidermis transforms exogenous leukotriene A_4 into

peptide leukotrienes: possible role in transcellular metabolism. Arch Dermatol Res 286:261-267
Janßen-Timmen U, Vickers P, Beilecke U, Lehmann WD, Stark HJ, Fusenig NE, Rosenbach T, Goerig M, Radmark O, Samuellson B, Habenicht AJR (1994) 5-lipoxygenase expression in cultured keratinocytes. Adv Prost Thromb Leukotr Res 22, in press
Kang-Rotondo CH, Miller CC, Morrison AR, Pentland AP (1993) Enhanced keratinocyte prostaglandin synthesis after UV injury is due to increased phospholipase A_2 activity. Am J Physiol 264:C396-C401
Kast R, Fürstenberger G, Marks F (1991) Activation of a keratinocyte phospholipase A_2 by bradykinin and 4ß-phorbol 12-myristate 13-acetate. Eur J Biochem 202:941-950
Kast R, Fürstenberger G, Marks F (1993a) Activation of a cytosolic phospholipase A_2 by transforming growth factor α in HEL-30 keratinocytes. J Biol Chem 268:16795-16802
Kast R, Fürstenberger G, Marks F (1993b) Phorbol ester TPA- and bradykinin-induced arachidonic acid release from keratinocytes is catalyzed by a cytosolic phospholipase A_2 ($cPLA_2$). J Invest Dermatol 101: 567-572
Kelso A (1989) Cytokines: structure, function and synthesis. Current opinion in Immunol 2:215-225
Kirnbauer R, Köck A, Neuner P, Förster E, Krutmann J, Urbanski A, Schauer E, Ansel JC, Schwarz T, Luger TA (1991) Regulation of epidermal cell interleukin-6 production by UV light and corticosteroids. J Invest Dermatol 96:484-489
Kishino J, Ohara O, Nomura K, Kramer RM, Arita H (1994) Pancreatic-type phospholipase A_2 induces group II phospholipase A_2 expression and prostaglandin biosynthesis in rat mesangial cells. J Biol Chem 269:5092-5098
Kramer R (1993) Structure, function and regulation of mammalian phospholipase A_2. Adv Sec Mes Phosphoprot Res 28:81-89
Kudo J, Murakami M, Hara S, Inoue K (1993) Mammalian non-pancreatic phospholipase A_2. Biochem Biophys Acta 117:217-231
Kupper TS, Chua AO, Flood P, Mc Guire J & Gubler U (1987) Interleukin 1 gene expression in cultured human keratinocytes is augmented by ultraviolet irradiation. J Clin Invest 80:430-436
Kupper TS, Min K, Sehgal PB, Mizutani H, Birchall N, Ray A, May LT (1989) Production of IL-6 by keratinocytes: implications for epidermal inflammation and immunity. Ann NY Acad Sci 557:454-465
Kupper TS (1990a) Immune and inflammatory processes in cutaneous tissues. Mechanisms and speculations. J Clin Invest 86:1783-1789
Kupper TS (1990b) The activated keratinocyte: a model for inducible cytokine production by non-bone marrow-derived cells in cutaneous inflammatory and immune responses. J Invest Dermatol 95:146S-150S
Kutsch CL, Norris DA & Arend WP (1993) Tumor necrosis factor a induces interleukin-1α and interleukin-1 receptor antagonist production by cultured human keratinocytes. J Invest Dermatol 101:79-85
Lambeau G, Ancian P, Barhanin J, Lazdunski M (1994) Cloning and expression of a membrane receptor for secretory phospholipase A_2. J Biol Chem 269:1575-1578

Lee SW, Morhenn VB, Ilnicka M, Eugui EM & Allison AC (1991) Autocrine stimulation of interleukin-1α and transforming growth factor α production in human keratinocytes and its antagonism by glucocorticoids J Invest Dermatol 96:688-691
Lee WY, Butler AP, Locniskar MF, Fischer SM (1994) Signal transduction pathway(s) in phorbol esters and autocrine induction of Interleukin-1α mRNA in murine keratinocytes. J Biol Chem 269:17971.17980
Lehmann WD, Stephan M, Fürstenberger G (1993) Profiling assay for lipoxygenase of linoleic and arachidonic acid by gas chromatography-mass spectrometry. Analyt Biochem 204:158-170
Lin LL, Lin AY, Knopf JL (1992) Cytosolic phospholipase A_2 is coupled to hormonally regulated release of arachidonic acid. Proc Nat Acad Sci USA 89:6147-6151
Lin LL, Wartman M, Lin AY, Knopf JL, Seth A, Davis R (1993) $cPLA_2$ is phosphorylated and activated by MAP kinase. Cell 72:269-278
Löschke M, Krieg P, Lehmann WD, Marks F, Fürstenberger G (1994) Purification and characterizatioh of the epidermal 8(S)-lipoxygenase. In: Eicosanoids and other Bioactive Lipids in Cancer, Inflammation and Radiation Injury (Development in Oncology, Vol 71; Nigam S et al., eds.) Kluwer, Boston, in press
Luger TA & Schwarz T (1990) Evidence for an epidermal cytokine network. J Invest Dermatol 95: 100S-104S
March CJ, Moslay B, Larsen A, Cerretti DP, Braedt G, Price V, Gillis S, Henney CS, Kronheim SR, Grabstein K, Conlon PJ, Hopp TP and Cosman D (1985) Cloning, sequence and expression of two distinct human interleukin-1 complementary DNAs. Nature 315:641-647
Marks F (1983) Prostaglandins, cyclic nucleotides, and the effect of phorbol ester tumor promoters on mouse skin in vivo. Carcinogenesis 4:1465-1470
Marks F (1990) Hyperplastic transformation: the response of skin to irritation and injury. In Skin pharmacology and toxicology (Galli CL, Marinovich DM, Hensby CN, eds.) Plenum Press, New York, pp 121-145
Marks F & Fürstenberger G (1993) Proliferative responses of the skin to external stimuli. Environ Health Perspect 101:95-102
Mc Kay IA & Leigh IM (1991) epidermal cytokines and their roles in cutaneous wound healing. Brit J dermatol 124:513-518
Mc Kenzie RS & Sauder DN (1990) The role of keratinocyte cytokines in inflammation and immunity. J Invest Dermatol 95:105S-107S
Milstone LM & Edelson RL (1988) Endocrine, metabolic, and immunologic functions of keratinocytes. Ann NY Acad Sci 548
Molineaux SM, Casano FJ, Rolando AM, Peterson EP, Limjuco G, Chin J, Griffin PR, Calaycay JR, Ding GFJ, Yamin TT, Palyha OC, Luell S, Fletscher D, Miller DK, Howard AD, Thornberry NA, Kostura MJ (1993) Interleukin 1β (IL-1β) processing in murine macrophages requires a structurally conserved homologue of human IL-1β converting enzyme. PNAS 90:1809-1813
Moslay B, Urdal DL, Prickett KS, Cosman D, Conlon SG & Dower SK (1987) The interleukin-1 receptor binds the human interleukin-1α precursor but not the interleukin-1β precursor. J Biol Chem 262:2941-2944.

Müller-Decker K, Fürstenberger G & Marks, F (1992) Development of an In Vitro alternative assay to the Draize skin irritancy test using human keratinocyte-derived proinflammatory key mediators and cell viability as test parameters. In Vitro Toxicol 5:191-209

Müller-Decker K, Scholz K, Marks F, Fürstenberger G (1994a) Differential expression of prostaglandin H synthase isozymes during multistage carcinogenesis in mouse skin. Mol Carcinogenesis, in press

Müller-Decker K, Fürstenberger G, Marks, F (1994b) Keratinocyte-derived proinflammatory key mediators and cell viability as in Vitro parameters of irritancy: a possible alternative to the Draize skin irritation test. Toxicol Appl Pharmacol 126:in press

Needleman P, Turk J, Jakschik BA, Morrison AR, Lefkowith JB (1986) Arachidonic acid metabolism. Ann Rev Biochem 55:69-102

Nickoloff BJ, CEM Griffiths, JNWN Barker (1990) The role of adhesion molecules, chemotactic factors, and cytokines in inflammatory and neoplastic skin disease - 1990 update. J Invest Dermatol 94:151S-157S

Nozaki S, Abrams JS, Pearce MK & Sauder DN (1991) Augmentation of granulocyte/macrophage colony stimulating factor expression by ultraviolet irradiation is mediated by interleukin 1 in pam 212 keratinocytes. J Invest Dermatol 96:888-897

Pace-Asciak CR & Asotra S (1989) Biosynthesis, catabolism, and biological properties of HPETEs, hydroperoxide derivatives of arachidonic acid. Free Rad Biol Med 7:409-433

Parker KP, Sauder DN & Killian PL (1988) Presence of Il-1 receptors on keratinocytes

Pentland AP & Needleman P (1986) Modulation of keratinocyte proliferation by endogenous prostaglandin synthesis. J Clin Invest 77:246-251

Pentland AP & Mahoney BS (1990) Keratinocyte prostaglandin synthesis is enhanced by Il-1. J Invest Dermatol 1:43-46

Ristow HJ (1990) Interleukin-1 does not stimulate DNA synthesis of cultured human keratinocytes growth-arrested in growth-factor-depleted medium. J Invest Dermatol 96:688-691

Ruzicka T (1988) The physiology and pathophysiology of eicosanoids in the skin. Eicosanoids 1:59-72

Sauder DN (1989) Interleukin 1. Arch Dermatol 125: 679-682

Scholz K, Fürstenberger G, Müller-Decker K, Marks F (1994) Differential expression of prostaglandin H synthase in normal and phorbolester-treated mouse epidermis in vivo, in press

Spector AA, Gordon JA, Moore SA (1988) Hydroxyeicosatetraenoic acids. Progr Lipid Res 27:271-283

Soter NA, Lewis RA, Corey EJ, Austen KF (1983) Local effects of synthetic leuktrienes (LTC_4, LTE_4, LTB_4) in human skin. J Invest Dermatol 80:115-119

Stevenson FT, Bursten SL, Fanton C, Locksley RM and Lovett DH (1993) The 31-kda precursor of interleukin 1α is myristoylated on specific lysines within the 16-kda N-terminal piece. Proc Natl Acad Sci USA 90:7245-7249

Takahashi Y, Reddy GR, Ueda N, Yamamoto S, Arase S (1993) Arachidonate 12-lipoxygenase of platelet-type in human epidermal cells. J Biol Chem 268:16443-16448

Vannier E, Miller LC & Dinarello CA (1992) Coordinated

antiinflammatory effects of interleukin 4: interleukin 4 suppresses interleukin 1 production but up-regulates gene expression and synthesis of interleukin 1 receptor antagonist. Proc Natl Acad Sci USA 89:4076-4080

Yamamoto S (1992) Mammalian lipoxygenases: molecular structure and functions. Biochim Biophys Acta 1128:117-131

Ziboh VA & Chapkin RS (1988) Metabolism and function of skin lipids. Progr Lipid Res 27:81-105

Ziboh VA (1994) Essential Fatty Acids/Eicosanoid Biosynthesis in the Skin: Biological Significance. J Soc Exp Biol Med 205: 1-11

Zurier RB (1990) Role of prostaglandins E in inflammation and immune responses. Adv Prost Thromb Leukotr Res 21:947-953

The Network of Intracellular Signal Processing as a Main Site of Cellular Responses in Toxicity

Friedrich Marks
German Cancer Research Center
Research Program Tumor Cell Regulation
D-69120 Heidelberg
Germany

The "brain of the cell"

The ability to communicate is one of the characteristic properties of cells and may actually be considered as the fundamental condition of life. Communication takes place through the exchange of signals. Biological signals are symbols, i.e. they have a distinct meaning. To respond adequately the receiver has to both recognize and decipher a signal. For this purpose prior information is required, which may either have been acquired or is genetically fixed.

Signal transduction requires a physical medium. However, the significance of a signal is by no means encoded in its structure. There is a rather arbitrary connection between form and meaning resulting in an enormous flexibility of communicative systems. This principle prevailing in human language holds equally true for intercellular communication. A prominent example is provided by the stress hormone adrenalin with its numerous physiological effects on almost all organs of the body. Depending on the target tissue the hormonal signal has quite different meanings, because differentiation takes place exclusively in the receptor cells. Another example in proof of the irrelevance of the signal structure is cyclic AMP, which has acquired a variety of different meanings in the course of evolution. Thus it serves as a gene-regulatory emergency signal in bacteria, as a cell attractant and maturation-inducing hormone in the slime-mold Dictyostelium discoideum and as an intracellular second messenger with a wide variety of functions in the cells of our body. It is always the receiver who coordinates the signal and its specific meaning. Thus, cellular communication has both a syntactic and a semantic aspect.

NATO ASI Series, Vol. H 93
Modulation of Cellular Responses in Toxicity
Edited by C. L. Galli, A. M. Goldberg, M. Marinovich

To provide a symbol, e.g. a word or a picture, with a distinct meaning we need our brain. During the processing of sensory input signals rather diffuse patterns of excitation are observed in the brain which involve innumerable interconnections between millions of cells and do not allow a precise cellular localization of single events (Black, 1991). Nevertheless, the result is generally a rather exact allocation of meaning which in turn is the precondition for a proper response.

Signal-processing is not restricted to neurones but is a characteristic property of every single cell. In the following I shall use the metapher of "the brain in the cell" in order to emphasize the close relationship between signal processing in neuronal and molecular networks. Occasionally it has been stated that the genome resembles a "brain" on the subcellular level. This is certainly not true: the genome is nothing but a memory store for primary protein structures. A brain, however, is much more, namely a device for the interpretation of symbols. Thus, the cellular brain would at least include both the genome and the network of chemical interactions required for signal processing. Only by an interaction of these two entities the meaning of a symbol can be deciphered.

A term such as "meaning" has, of course, a certain teleological after-taste which may be inacceptable for many a scientist. However, if we restrict ourselves solely to describing structures and molecular interactions in physico-chemical terms we will certainly fail to cope with the complexity of living systems, a situation similar to that a behavioristic hard-liner is confronted with in psychology. It must be emphasized, however, that this does not imply anything like a "ghost in the molecule", but that we are talking about the phenomenon of communicative interactions between biomolecules and cell structures resulting in the emergence of properties which can neither be explained on the basis of molecular structures alone nor be reduced to structural parameters without losing exactly what proves to be their biological significance. Each biomolecule gets its meaning only out of its "semantic milieu", i.e. from the living organism, just as a word does from the framework of language.

Carcinogenesis: an illustrative example of the interaction of toxic events with cellular signal processing

The great majority of toxic events occurs on the level of inter- and intracellular signaling. This holds true in particular for almost all microbial, plant and animal toxins, but also for a wide variety of man-made poisons. Cancer provides an especially impressive example of a disorder which is due to a toxic damage of cellular signal processing.

Cancer may be defined as the result of a population explosion on the cellular level. Tumors are indeed characterized by an imbalance between the birth rate and the decay rate of cells. Such a defect may be due to an overproduction of "young" tissue cells or to a delay in the maturation and death rate of their "old" counterparts. Therefore, the corresponding disturbances have to be looked for in fundamental cellular processes such as mitosis and programmed cell death (apoptosis, terminal differentiation, senescence). These processes are genetically regulated, checked against one another, and stay normally under a strict control by signals of the organism. During carcinogenesis this control becomes disturbed.

Cancer is a drastic example of the principle "minimal cause - maximal effect". The disease starts with a gene mutation being a single molecular event in a single cell. At the end we are confronted with the ruin of an individual consisting of some 100 trillions of cells. The disease develops slowly depending on many different factors. Thus, for the development of the mutated stem cell into a malignant tumor additional gene mutations are required and the different sectors of local and systemic defence reactions have to be overridden (Fearon & Vogelstein, 1990). Normally, this takes decades.

Many of those genes which are involved in carcinogenesis have been recently identified. Depending on whether they are deleted or overactivated by the oncogenic mutation one speaks of suppressor - or oncogenes. Physiologically, these genes play important roles in cellular life. In particular, they control the cell's ability to receive and process intercellular signals and to proceed through the cell cycle in an orderly fashion.

Representatives of genotoxic carcinogens are summarized in Fig.1. These mutagenic agents are joined by a large number of so-called non-genotoxic carcinogens, i.e. agents which do not exhibit mutagenic potential in the commonly used tests (Perera, 1991). Many of the non-genotoxic carcinogens are strong irritants and potent stimulators of cellular proliferation (Grasso et al., 1991). These findings have given rise to a renaissance of Virchow's concept of cancer being the result of a chronic hyperproliferative and inflammatory process. It is, however, widely accepted, that cancer starts with a gene mutation by a - frequently unidentified - mutagen, and that non-genotoxic carcinogens either facilitate this genotoxic event (cocarcinogenesis) or accelerate the development of the tumor from the mutated cell (tumor promotion).

Several non-genotoxic carcinogens such as, for instance, dioxin and the so-called peroxisome proliferators interact specifically with cellular receptors which serve the processing of endogenous signals of still unknown nature (Grasso et al., 1991; Green, 1992; Poellinger et al., 1992). Also endogenous signals themselves may promote tumor development. Well-known examples are provided by sexual hormones as well as by growth factors when overexpressed due to genetic manipulations. Prominent representatives of the non-genotoxic carcinogens are the tumor promoters which have been well defined on the basis of animal experiments (Fig.1). Tumor promoters and genotoxic carcinogens induce cancer in an extremely synergistic manner. Combined treatment with these agents does not only result in an extraordinary acceleration of tumorigenesis but also allows to divide the process of tumor-development into several well-defined stages (Fig.1).

The classical animal model of multistage carcinogenesis is provided by chemically induced skin cancer in mice (Marks, 1989, in press). In this model the phorbol esters and related toxins have been identified as the most potent tumor promoters. By using these agents as experimental tools a rather deep insight into the molecular events of tumor development was obtained. It turned out that phorbol esters and related tumor promoters evoke a tissue reaction which down to the molecular details resembles a wound response (Marks & Fürstenberger, 1990; Marks et al., 1992). Indeed, also mechanical wounding

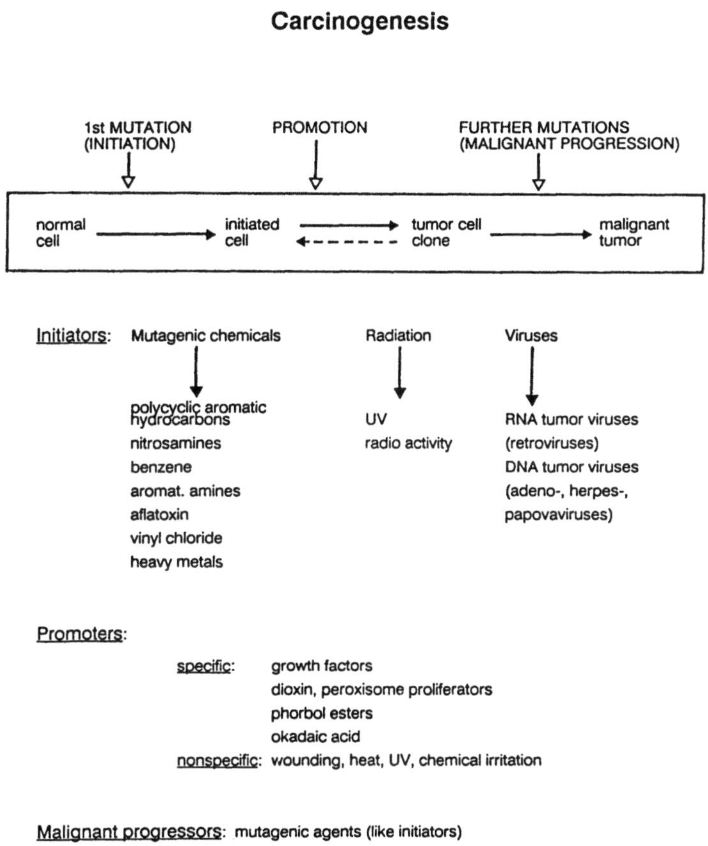

Fig.1: Carcinogenesis.
Cancer development is a multistage process which traditionally is subdivided into the stages of initiation, promotion and malignant progression. While initiation and progression correlate with irreversible genotoxic events (somatic gene mutations etc.) promotion seems to result mainly from reversible disturbances at the epigenetic level. Tumor promoters are therefore classified as a subgroup of the so-called non-genotoxic carcinogens. Some tumor promoters specifically interact with components of intracellular signal processing ("tumor promoter-receptors") whereas others unspecifically induce tissue damage and irritation.

has been found to provide a powerful tumor promoting stimulus in initiated tissue. However, the effect of phorbol esters is due to a massive interaction with pathways of intracelluar signal processing rather than to tissue destruction. Actually, these tumor-promoters mimic the action of diacylglycerol, an intracellular second messenger, which is generated upon interaction of cells with a wide variety of intercellular signals.

Three conclusions may be drawn from these observations:
1.) Virchow's concept of a causal relationship between chronic irritation and cancer may be correct at least as far as tumor promotion is concerned. Provided that many tumors remain latent in the absence of promotion, cancer may indeed be looked upon as the result of "permanently repeated and disturbed tissue regeneration" (Orth, 1911), and a tumor may be understood as a wound that does not heal (Dvorak, 1986) or - vice versa - a wound as a tumor that heals itself (Haddow, 1972).
2.) The process of tumor development can be extremely promoted by disturbing the mechanisms of cellular signal processing. In the case of wounding it is a question of those signals by which the different cell types involved in the healing process communicate. Such wound hormones or cytokines have in fact been shown to exhibit a tumor-promoting effect when repeatedly applied or excessively released from gene-manipulated cells.
3.) All the genes which are critically involved in oncogenesis code for signal-transducing proteins, i.e. are important for the social behavior of cells, in particular for the adaptation of cellular population dynamics to the demands of the organism.

To arrive at a better understanding of cellular responses in toxicity and in particular in carcinogenesis, it is essential to know how signal processing in cells works. Although we are just at the beginning of understanding these relationships some basic principles have become apparent already indicating a high degree of similarity between intercellular (i.e. neuronal or endocrine) and intracellular (i.e. molecular) signal processing. On both levels we see a complex network pattern which operates in a nonlinear manner due to a high degree of feedback interactions as well as multiplicity and redundancy of the processing units. This provides the systems with plasticity, i.e. enables them to adapt and to learn.

Both systems decipher the meaning of signals by adjusting them to the information they have previously acquired, be it the memory of the brain, or the genome and other molecular memory stores in a single cell. The nervous and the endocrine system establish long-distance communication along nerve fibers and blood vessels. On the subcellular level such connections result from interactions between molecules and substructures (Fig.2).

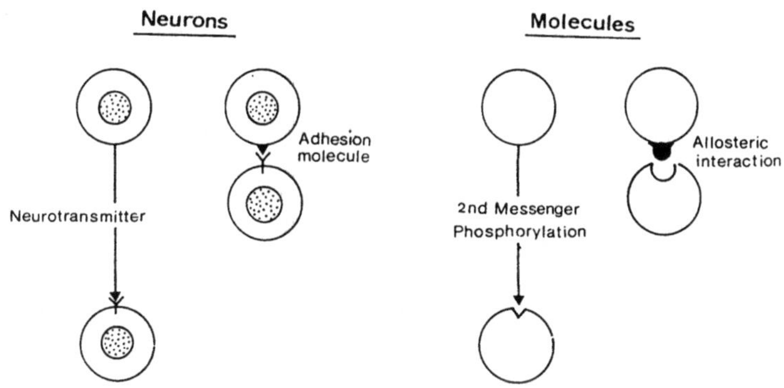

Fig.2: Signal transduction between cells and between molecules. Cells -as for example neurons- communicate either via diffusible signal molecules (such as neurotransmitters, hormones, cytokines etc.) or by means of direct contact via adhesion molecules. The same principles of communication hold true for the protein molecules involved in intracellular signal processing. Like cells they may communicate via either diffusible signal molecules (second messengers) and chemical reactions (in particular protein phosphorylation) or by direct contact (allosteric interaction).

How cells receive and decipher signals

A cell responds to intercellular signals by means of receptors, i.e. protein molecules which interact non-covalently with the signal molecule due to structural complementarity (lock-key principle). Like any signal-transducing molecule a receptor consist of at least two parts, a receiver module for discrimination of the input signal and a transmitter module for the emission of an output signal. The latter consists generally of an allosteric conformational change which is "recognized" by other signal-transducing proteins located downstream in the signaling cascade. Many intercellular signal molecules such as proteins, peptides, amino acids, amines and nucleotides/nucleosides cannot penetrate the lipid barrier of the plasma membrane and interact with receptors at the cell surface (Fig.3). Other signal molecules such as thyroid hormones, steroids and retinoids are able to enter the cell finding their receptors in the cytoplasm or in the nucleus (Barritt, 1992).

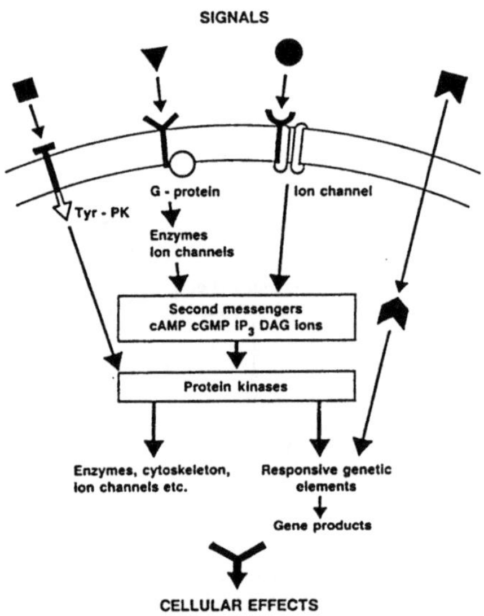

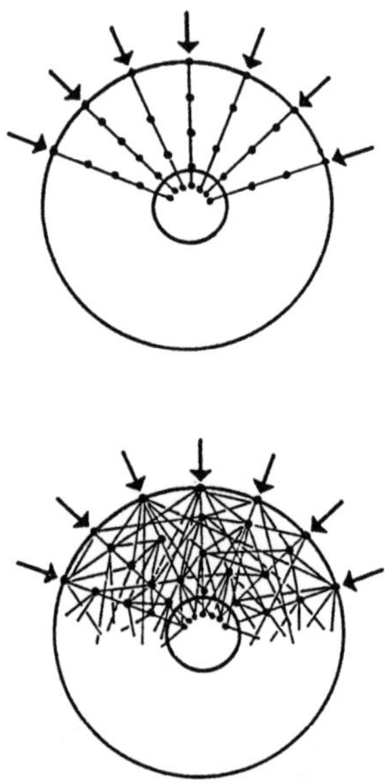

Fig.3: Standard pathways of outside-inside signaling in cells. Depending on their chemical structures signals interact with receptor proteins localized either at the cell surface or in the cytoplasm. The signal-activated receptors are able to contact effector molecules such as (from the left to the right) tyrosine-specific protein kinases (Tyr-PK), G-proteins, ion channels or gene-regulatory DNA sequences. In certain cases receptor and effector may be localized on one and the same protein molecule (see also Fig.19). Effector molecules modulate the input signal in amplitude and frequency and translate it into the cell's signaling language. The latter makes use of a variety of second messengers and covalent protein modifications in particular phosphorylation. It must be emphasized that the left diagram represents an extremely simplified picture. The variability of cellular signal cascades is actually much greater. Moreover, there are no linear pathways but a complex pattern of feedback interactions and cross-talking (see the right diagram). Signal reception, thus, results in diffuse excitation patterns rather than in a precisely defined sequence of chemical interactions.

The effector molecules which are controlled by activated receptors include enzymes, G-proteins, ion channels and regulatory genomic sequences. Effector molecules modulate the signals and translate them into an intracellular "language", which - just as extracellular signaling - makes use of chemical interactions by means of signal molecules (second messengers), chemical reactions or direct contacts between signal-transducing protein molecules (Fig.2). The variability of second messengers can not yet be estimated. Well-known representatives are the cyclic nucleotides (cAMP, cGMP), diacylclycerol (DAG), the inositol phosphates and Ca^{2+} ions. These intracellular signal molecules are produced by corresponding effector enzymes such as adenylate cyclase, guanylate cyclase and phospholipase C, or passed into the cell through Ca^{2+} channels. Second messengers control the function of other downstream effector proteins including protein kinases, ion channels, components of the cytoskeleton etc.

<u>The system of G-proteins: a microcomputer</u>

For a large number of receptors the immediate downstream-effectors are G-proteins. G-proteins are guanine nucleotide-binding proteins with an intrinsic GTPase activity (Boege et al., 1991; Lindner & Gilman, 1992; Hepler & Gilman, 1992). Their main function is signal transduction and modulation. For this purpose G-proteins get in contact with other effector proteins, in particular with ion channels and enzymes which catalyze second messenger formation such as adenylate cyclase and phospholipase C beta. These interactions require the activation of the G-protein by another signal, which may consist in an interaction with an activated receptor molecule (Fig.3). As shown in Fig.4 G-protein activation involves an exchange of bound GDP by GTP. Receptor-coupled G-proteins are heterotrimeric molecules which upon activation dissociate into two subunits α and β/γ. Both subunits have been shown to influence different pathways of signal transduction. The active state of a G-protein is only short-lived since the bound GTP is rapidly hydrolyzed by the intrinsic GTPase activity. Both the activating GDP/GTP exchange reaction and the inactivating GTP

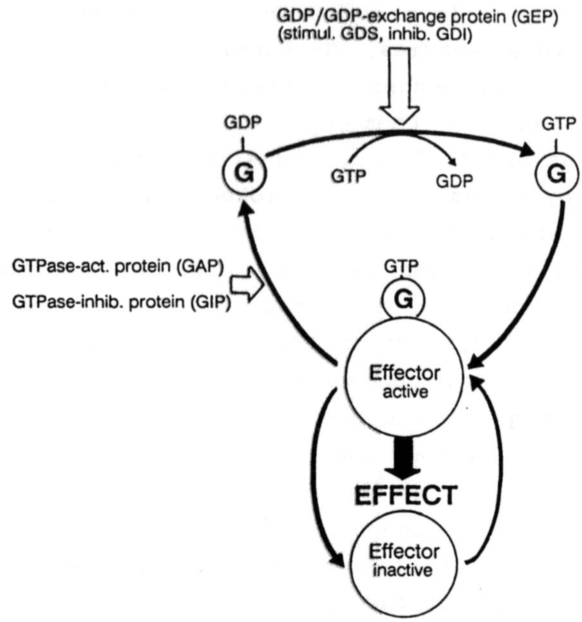

<u>Fig.4</u>: The activation cycle of G-proteins.
G-proteins are molecular switches which are in the OFF position when binding GDP and in the ON position when binding GTP.
Thus, activation means exchange of GDP by GTP and inactivation degradation of GTP to GDP plus inorganic phosphate.
The latter reaction is catalyzed by an intrinsic GTPase activity. Both activation and inactivation are regulated by other proteins such as GDP-GTP-exchanging proteins (GEP) and GTPase-activating proteins (GAP). An example of GEP's is provided by ligand-receptor complexes.

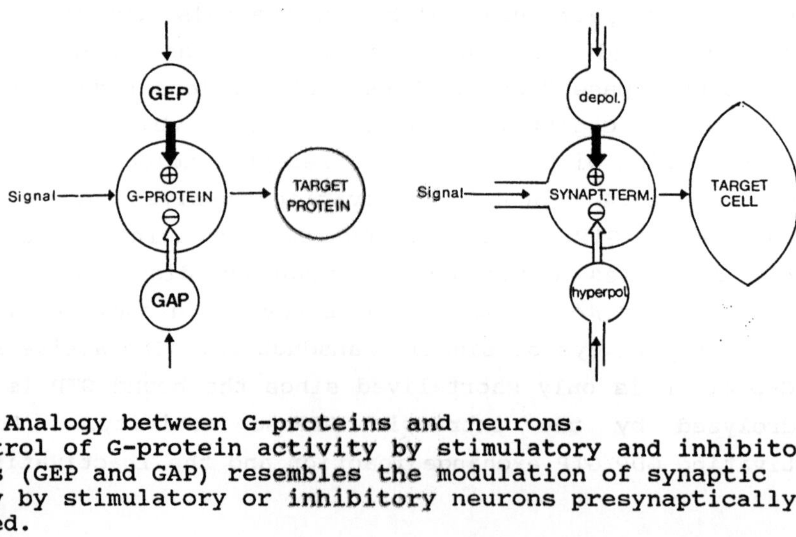

<u>Fig.5</u>: Analogy between G-proteins and neurons.
The control of G-protein activity by stimulatory and inhibitory proteins (GEP and GAP) resembles the modulation of synaptic activity by stimulatory or inhibitory neurons presynaptically connected.

hydrolysis are under the control of accessory proteins (Fig.4) which are components of other signaling pathways. This system provides a striking analogy with the presynaptic modulation of neurons (Fig.5).

The enormous variability of G-proteins and their interactions with other signal-processing elements of the cell can not yet be assessed at all. Beside being receptor-controlled effector molecules G-proteins also control mRNA translation and microtubuli association (Bourne et al., 1991). Moreover, the Ras superfamily of so-called small (or monomeric) G-proteins (Takai et al., 1992) represents a large group of regulatory proteins with key functions in the control of cellular vesicle transport, organization of the cytoskeleton and transduction of mitogenic signals (see Fig.21).

Together with the associated activator, inhibitor and effector molecules G-proteins form a signal-processing network, which transforms, integrates and modulates input signals in their amplitude and frequency. Since the interaction between different G-proteins provide biochemical "AND" and "BUT NOT" logic gates, the G-protein network has been looked upon as a cellular microcomputer (Ross, 1992). Moreover, due to their intrinsic GTPase activity G-proteins are exponential timers, which upon activation become inactivated at characteristic rates. This may help to transform the digital mode of signal processing at the molecular level into an analog behavior of the cell.

The receptors connected to G-proteins form a large family consisting of several hundred members (Dohlman et al., 1987; Findlay & Eliopoulos, 1990). They interact with numerous hormones and neurotransmitters, but also with environmental signals such as light (rhodopsin, see Hargrave & McDowell, 1990), odorants (Lancer & Ben-Aric, 1993) and taste stimulants. G-protein-coupled receptors exhibit a common structural motif, i.e. seven transmembrane domains (Fig.6). Like most other signal-transducing proteins these receptors are subject to a sophisticated feedback control of their activity which resembles an analogous situation in sensory nerve systems (Fig.6).

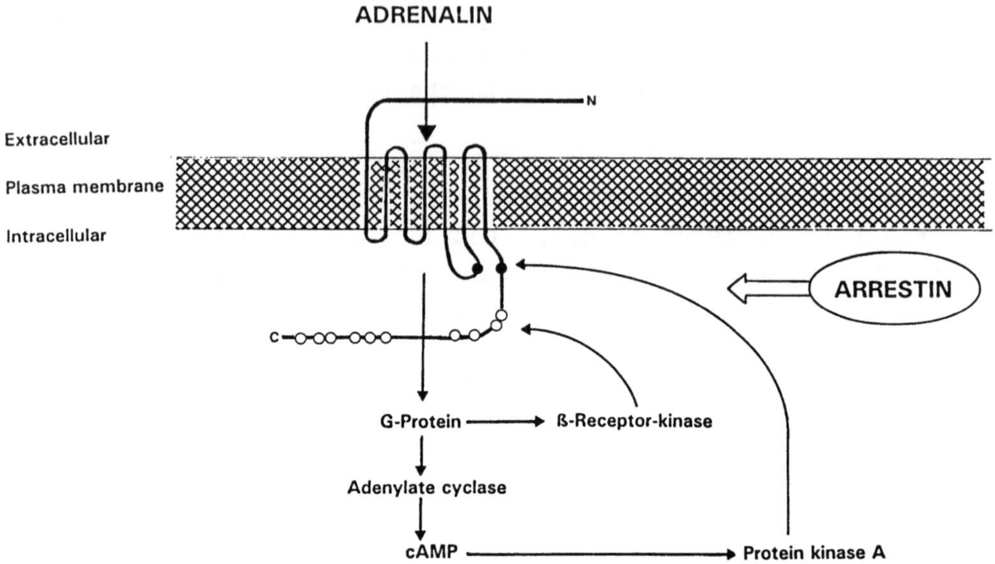

Fig.6: Feedback control of a G-protein-coupled receptor. G-protein-coupled receptors such as the ß-adrenergic receptor are characterized by 7 transmembrane domains and a series of phosphorylation sites on their cytoplasmic domain. For the ß-adrenergic receptor a phosphorylation of these sites has been shown to facilitate an interaction of the cytoplasmic domain with an inhibitory protein (arrestin) resulting in receptor inactivation. These phosphorylations are catalyzed by two protein kinases localized downstream in the signaling cascade, i.e. cAMP-activated protein kinase A and a specific ß-adrenergic receptor kinase. Both enzymes are directly or indirectly under the control of the receptor-coupled G-protein which upon activation dissociates into the two subunits α and β/γ. While α activates the adenylate cyclase, thus leading to cAMP production and an activation of protein kinase A, β/γ stimulates the ß-receptor kinase. Note the analogy between this molecular circuit and neuronal circuits such as found, for instance, in sensory nerves. For details see Lefkowitz, 1993; Wilson & Applebury, 1993.

G-proteins are important targets of toxic effects. Thus, bacterial toxins such as cholera, pertussis and diptheria toxin catalyze the ADP ribosylation of certain G-proteins which results - via an inhibition of the GTPase activity - in permanent stimulation. Mutations of the ras genes, also leading to an impaired GTPase activity of the corresponding G-proteins, are frequent events in neoplastic transformation and have been found in at least 30 % of human tumors (see below). Other oncogenes related to G-proteins are gsp and gip2, whereas vav encodes an activating protein (GEP, see Fig.4) for Ras.

Protein phosphorylation: the binary standard code of cellular signal processing

In the course of evolution the reversible phosphorylation of proteins has developed into the most efficient and versatile signal of intermolecular communication. It is found in the simplest prokaryotes and the most sophisticated brain neurons alike. In fact there is almost no cellular protein, which does not at least potentially provide a target of phosphorylation/dephosphorylation thus undergoing functional modulation. This holds true in particular for all components of the cellular signal processing machinery. The reactions are catalyzed by protein kinases and phosphatases with ATP or GTP (or phosphorylated proteins in prokaryotes) as phosphate donors. The number of identified protein kinases is still growing exponentially (Fig.7, see Hunter, 1991). Certainly, protein kinases represent the most multifarious enzyme family known.

Protein kinases are communicative molecules which receive and transduce signals. These properties are reflected by the molecular structure consisting both of receiver (input) and transmitter (output) modules (see, for instance, Fig.13). Protein kinases have been divided into two subgroups depending on whether they catalyze the O-phosphorylation of tyrosine or serine/threonine residues in substrate proteins. Recently, also kinases with dual specificity have been found. Furthermore, prokaryotes contain kinases which catalyze the phosphorylation of histidine and aspartic acid residues. Whether

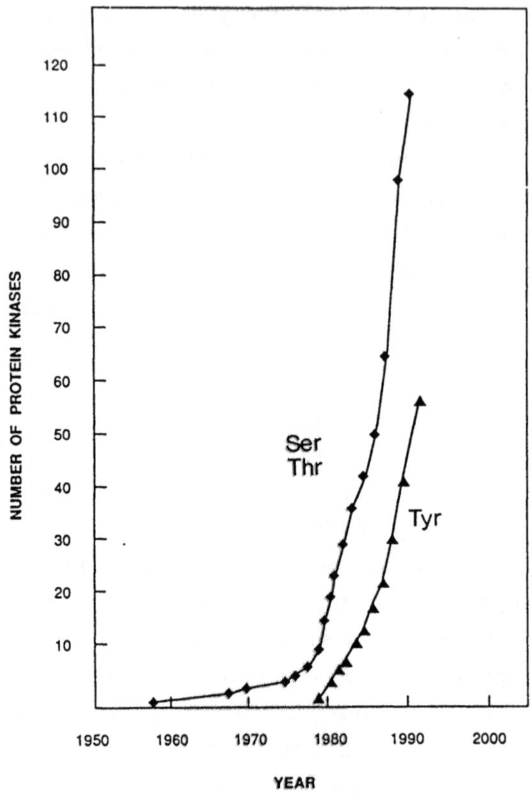

Fig.7: The discovery of protein kinases.
The diagram shows the exponential rate at which protein kinases (Ser/Thr- and Tyr-specific enzymes) were discovered up to 1990 (from Hunter, 1991).

or not a special amino acid residue is phosphorylated by a distinct kinase depends on its immediate environment in the polypeptide chain, i.e. a particular phosphorylation consensus sequence. This guarantees a certain degree of enzyme specificity and an enormous variability of interactions despite the monotony of the output signal, i.e. phosphorylation (Pearson & Kemp, 1991).

The activity of protein kinases is modulated by an extraordinarily high variety of input signals, in particular by receptors and other regulatory proteins, first and second messengers, nucleic acids and covalent modifications such as phosphorylation (Fig.8, see Hunter, 1991). By this means of "signal cross-talking" the cell is able to form complex signal processing networks on demand. Like G-protein activation/ inactivation the system of protein phosphorylation/dephosphorylation may be understood as a binary code of a biochemical microcomputer.

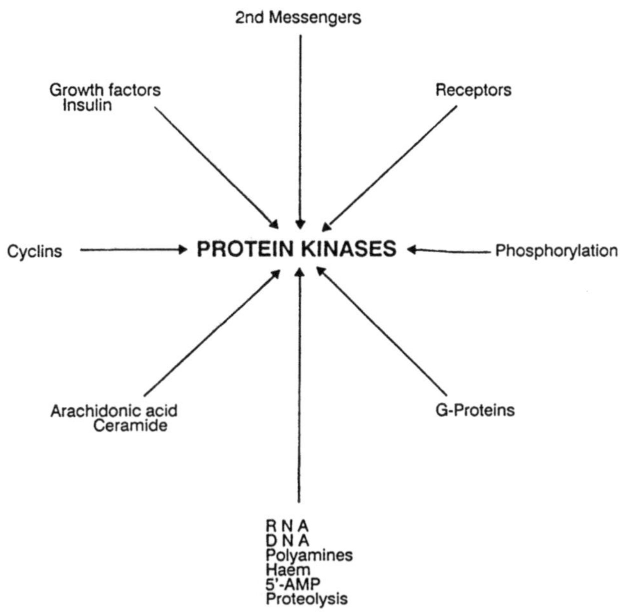

Fig.8: Protein kinases as receivers of various input signals. The diagram shows factors which have been found to regulate the enzymatic activity of different classes of protein kinases. Most of these factors are themselves under the control of the "cellular brain". As far as the variability of regulatory interactions is concerned protein kinases outdo by far any other enzyme family.

The majority of proteins has multiple phosphorylation sites, which may be recognized by different kinases. Their phosphorylation may result in a graded and hierarchic modulation of the proteins' function (Fig.9,10). It has been speculated that by this and other means the machinery of signal processing is enabled to acquire an analog behavior despite the digital characteristics of the underlying circuit elements (Parkinson, 1993). Again, one is faced with a striking analogy

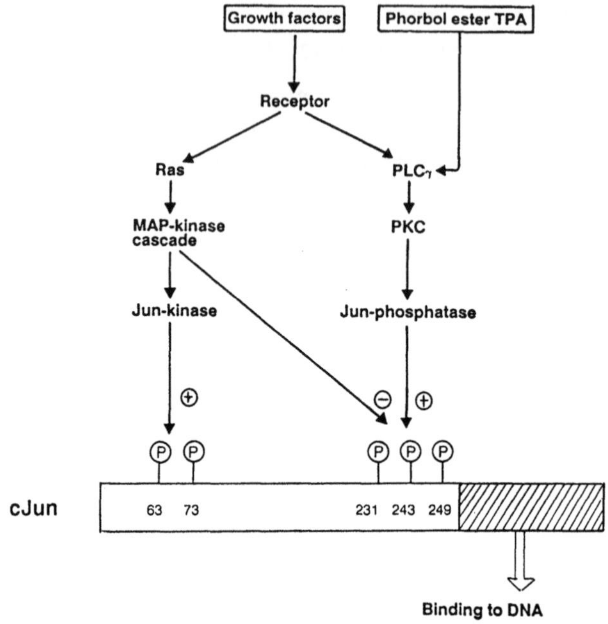

Fig.9: Control of the transcription factor c-Jun by multiple phosphorylation.
c-Jun exhibits 5 phosphorylation sites which regulate its gene-regulatory activity in a graded manner. For full activation of c-Jun the sites 63 and 73 have to be phosphorylated by a Jun-kinase whereas the inhibitory sites 231, 243 and 249 must be dephosphorylated by a Jun-phosphatase. Both enzymes are integrated in different signaling cascades which are activated by growth factor receptors and - in the case of Jun-phosphatase - a protein-kinase-C-dependent pathway. For this reason the gene-regulatory DNA sequences (AP-1 sites) which are controlled by c-Jun (in combination with c-Fos) have been called (phorbol ester) TPA-responsive elements. See Karin & Smeal (1992).

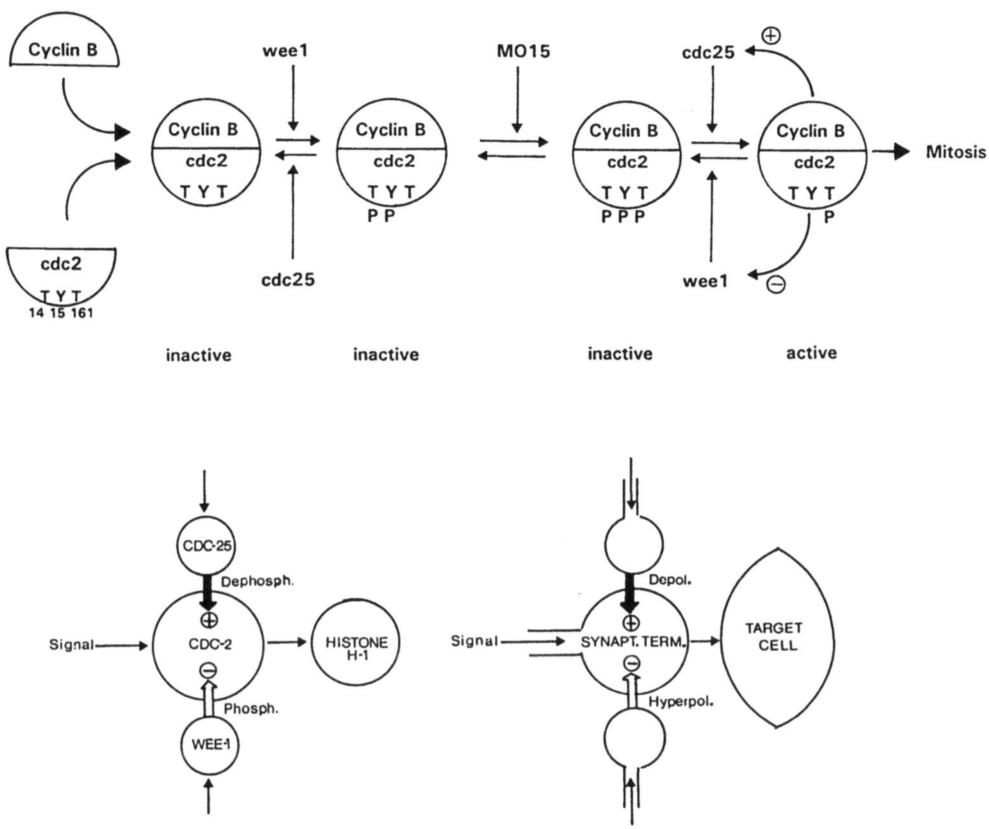

<u>Fig.10</u>: Control of a signal-transducing protein by regulatory proteins and multiple phosphorylation.
An illustrative example of a complex regulation of protein kinase activity is provided by the kinase cdc2 ($p34^{cdc2}$) which controls the entry of the cell into mitosis (Murray & Kirschner, 1991). The activation of cdc2 requires an interaction with the regulatory protein cyclin B as well as combined phosphorylation/dephosphorylation at Tyr 15 and Thr 14 and 161 (upper diagram). These reactions are catalyzed by protein kinases (wee1, MO15) and phosphatases (cdc25) which themselves are under the control of the active cyclin B/cdc2 complex (Nurse, 1990; Nigg, 1993). The positive feedback thus resulting helps to guarantee a precise timing of cdc2-activation at the G_2-M-transition of the cell cycle (Murray, 1993). As in the case of G-protein activation, a striking analogy between the biphasic control of cdc2 activity and presynaptic modulation of neuronal firing exists (lower diagram). It should be emphasized that beside cdc2 many other protein kinases are controlled by combined phosphorylation/dephosphorylation.

between the mode of signal processing in a single cell as compared with the nervous system. Actually, proteins with multiple phosphorylation sites may be understood as molecular counterparts of neurons with multiple synaptic contacts (Fig.11).

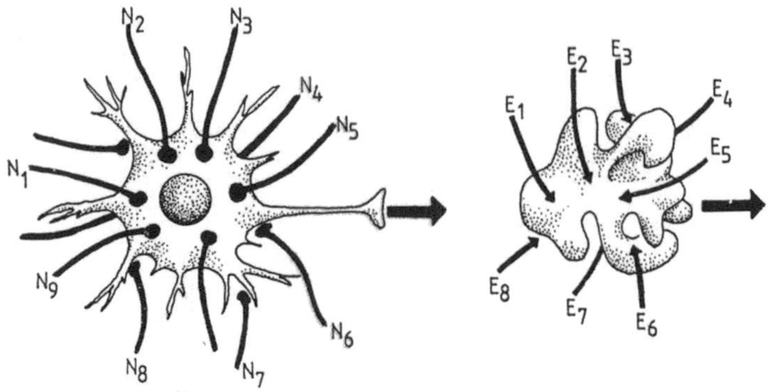

Fig.11: Multiple protein phosphorylation (right) as an analogy to multiple synaptic contacts on neurones (left). In both cases the multiplicity of interactions may provide a means for a stepwise and hierarchically structured control of activity, i.e. of the output signal at the axonal terminal of the neurone or at the catalytic center of the enzyme (strong arrows).

Protein kinases and phosphatases are targets of prominent effects in toxicity. Thus, the majority of proto-oncogenes code for protein kinases, which become deregulated upon gene mutation (see below). A number of highly toxic agents have been shown to interact specifically with the enzymes of the protein kinase C family. These agents include the bryostatins and the phorbol esters and related toxins, which induce severe inflammatory and hyperproliferative tissue reactions. This group of drugs also includes the most potent tumor promoters known (Fig.12, see Gschwendt et al., 1991).

The protein kinase C family is divided into 3 subfamilies (Fig.13, see Baser, 1993), i.e. the cPKC's ("classical"), the nPKC's ("novel") and the aPKC's ("atypical"). The 12 members of the PKC family which have been identified thus far (Dekker & Parker, 1994) provide an illustrative example of the multiplicity and apparent redundancy of molecular signal processors.

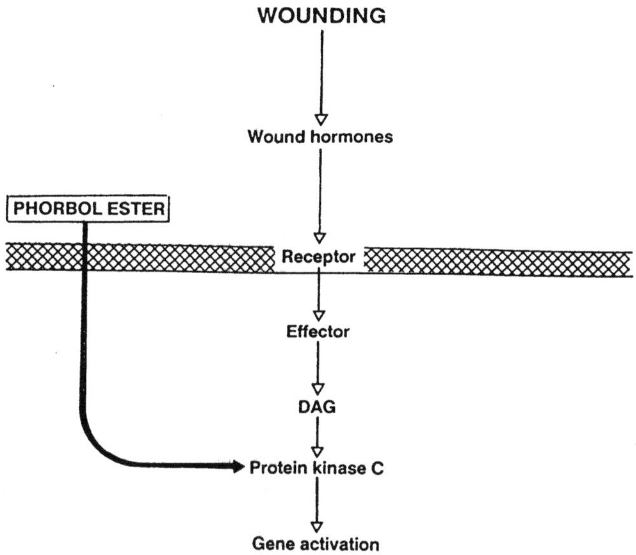

Fig.12: The mechanism of action of phorbol-ester-type tumor promoters.
These plant toxins mimic the stimulatory effect of the second messenger DAG on protein kinase C and thus make a short-cut in intracellular signal transduction pathways. Since the in vivo effects of phorbol esters widely resemble a wound response it is postulated that those signal transduction pathways are normally under the control of "wound hormones" (Marks et al., 1992). Recently, other "phorbol ester receptors" have been discovered in cells, including regulatory proteins (GAP's and GEP's) for certain small G proteins, which could be also involved in mitogenesis and tumor promotion (Dickmann et al., 1991; Gulbins et al., 1994).

However, an amazing degree of microheterogeneity and microspecificity is revealed upon closer examination of these enzymes (Dekker & Parker, 1994; Stabel & Parker, 1991). All PKC's require interaction with membrane phospholipids to become active. Those species which are localized in the cytoplasm undergo translocation to the membrane when receiving an activating signal. The latter is the second messenger diacylglycerol (DAG) in the case of cPKC's and nPKC's, whereas the input signal for aPKC's has not yet been unequivocally identified. cPKC's additionally require Ca^{2+} ions for full enzymatic activity. Thus, the enzymes of the three PKC subfamilies are under the control of different signaling pathways (Fig.14).

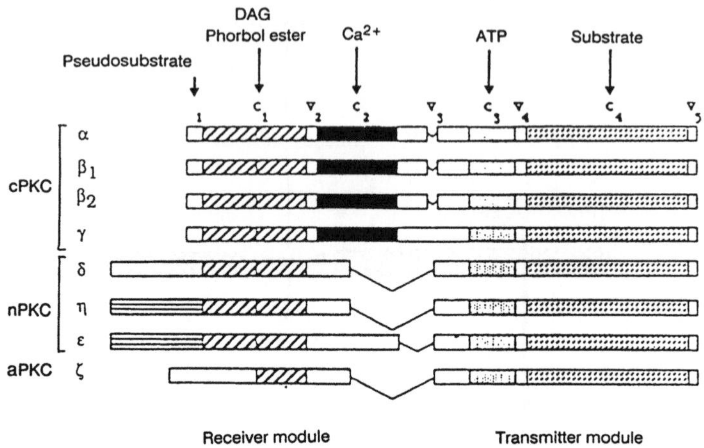

Fig.13: The protein kinase C family.
The scheme shows the modular structure of protein kinase C consisting of a receiver module (regulatory domain, left) and a transmitter module (catalytic domain, right). Each module is built up by constant (c) and variable (v) regions. The region C_1 consisting of two zinc finger structures interacts with the second messenger DAG (or with phorbol esters) while C_2 may represent a Ca^{2+}-binding site.
The PKC isoforms exhibiting both C_1 and C_2 form the subfamily of the Ca^{2+}-dependent cPKC's (c= classical), whereas PKC-isoforms without C_2 region are called Ca^{2+}-independent nPKC's (n = novel). The isoform ζ exhibiting a truncated C_1 region (only 1 zinc finger) is both Ca^{2+}- and DAG/phorbol ester-independent (aPKC, a = atypical). In the resting state enzyme activity is blocked by an interaction of the substrate binding site with a pseudosubstrate sequence.

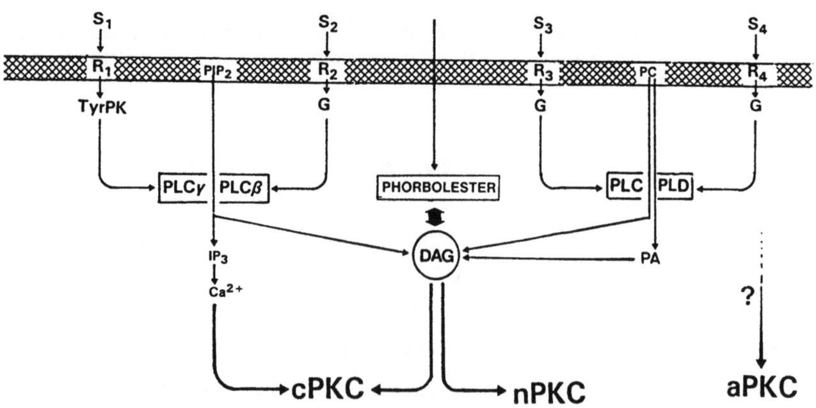

Fig.14: Different signal transducing pathways leading to PKC activation.
While the Ca^{2+}-dependent cPKC's are activated along the PIP_2 cascade, i.e. by intercellular signals which stimulate phospholipases γ or ß (left), the Ca^{2+}-independent cPKC's might also be activated by pathways involving a receptor-mediated stimulation of other phospholipase C types or of phospholipase D (right).
The mechanism of aPKC activation is still not clear. Recently, phosphatidylinositol 3,4,5-trisphosphate has been found to stimulate aPKC (Nakanishi et al., 1993).

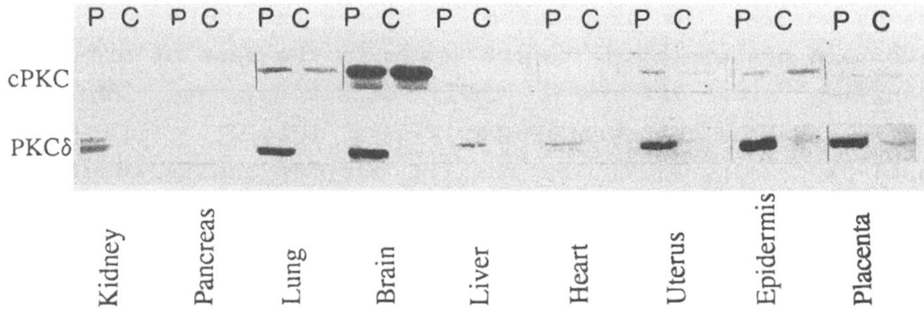

<u>Fig.15</u>: Tissue-specific expression of different PKC subtypes. The photograph shows the distribution of cPKC (mixture of α,β, γ) as compared with nPKC (type delta) in different mouse tissues (immunoblotting, P = particulate cell fraction, C = cytosolic fraction). From Leibersperger et al., 1991.

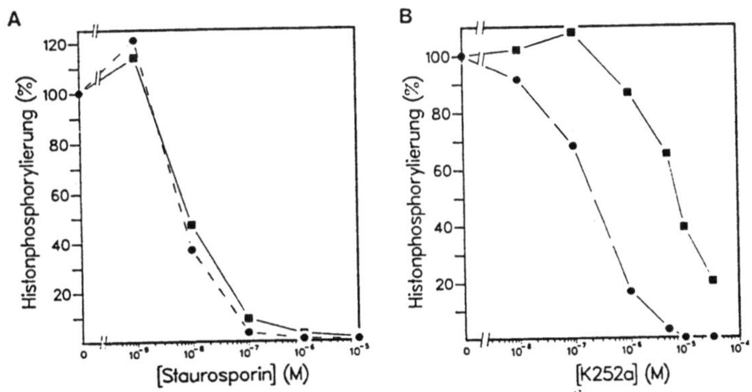

<u>Fig.16</u>: Differential effects of the protein kinase inhibitors staurosporine and K252a on purified PKC subtypes.
Broken line, cPKC (mixture of α,β,γ); solid line, nPKC (type delta).
For details see Gschwendt et al., 1989.

PKC isoenzymes are tissue-specifically expressed (Fig.15). Some of them such as PKCα or PKCδ are found in many different tissues whereas the expression of others is strictly confined to certain tissues, e.g. to the brain in the case of PKCγ. PKC isoenzymes also differ in their sensitivity to inhibitors (Fig.16) and in substrate specificity (Fig.17). Like many signal-processing molecules the PKC enzymes easily adapt to a signaling overload. When tissues or cells are treated with a strong PKC activator such as the phorbol ester TPA the enzymes of the C- and n-subfamilies (but not of the a-subfamily) become reversibly down-regulated, however at individual rates (Fig. 18). Since, in contrast to DAG, these agents are metabolically rather stable, they induce - prior to down regulation - an unphysiological overactivation of PKC. Considering the central role of PKC in intracelluar signal processing, it is easily conceivable that such an overactivation followed by down regulation may lead to profound disturbances in cellular physiology which are manifested as the severe toxic symptoms mentioned above. These symptoms include an impairment of fundamental cellular processes such as proliferation, secretion and movement.

The tumor-promoting effect of PKC activators (such as the phorbol esters) depends critically on the mitogenic effect of those agents (Marks et al., 1992). Indeed, PKC is directly involved in the activation of early-response genes of mitogenesis such as c-jun (see Fig.9 and Rahmsdorf & Herrlich, 1990) and appears, in addition, to interact with several other elements of the mitogenic signal transduction cascade exhibiting both positive and negative feedback effects. Moreover, PKC has been shown to activate, probably via MAP-kinase, phospholipase A_2, thus inducing the release of eicosanoids, i.e. physiological mediators of mitogenic and inflammatory effects (see G. Fürstenberger, this volume). PKC's also help to control the formation and the release of other proinflammatory mediators such as the cytokines. These combined actions easily explain the strong irritant effects of PKC-activating toxic agents. It must not be overlooked, however, that cells contain in addition to PKC other target proteins for phorbol esters, the activation of which may also participate in the toxic effects of those agents (see Fig.12).

Toxic reactions similar to those evoked by PKC activators are induced by inhibitors of protein phosphatases. The most prominent example for such a phosphatase inhibitor is provided by okadaic acid, a tumor-promoting irritant (Fujiki, 1992).

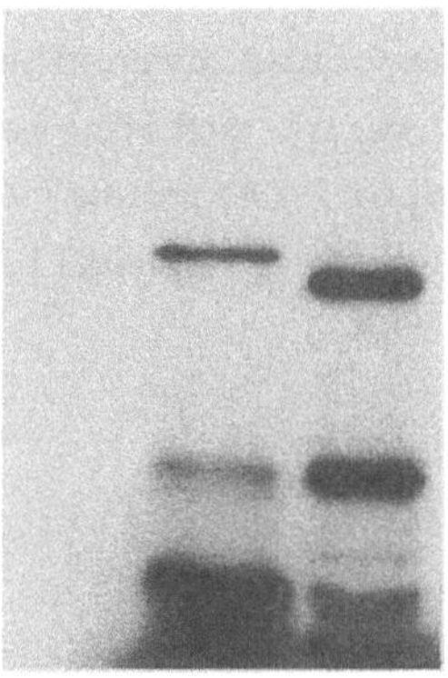

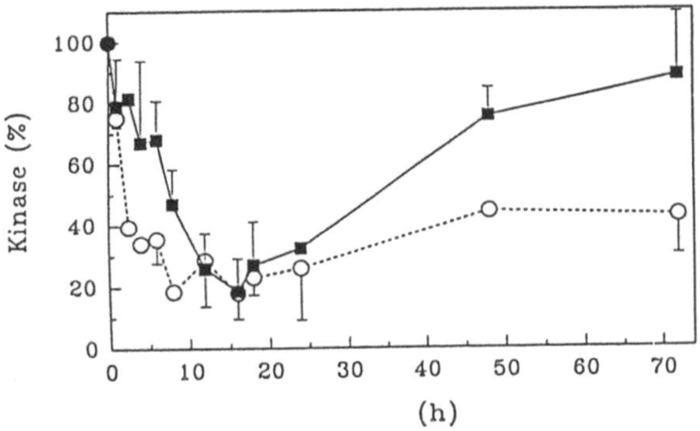

Fig.17: Substrate selectivity of different PKC subtypes. Mouse epidermis homogenate cleared of endogenous PKC by chromatography on Q-sepharose was incubated with $\gamma[^{32}P]$ ATP and either cPKC (α,β,γ) or nPKC (type δ). The photograph shows an autoradiogram obtained upon separation of the proteins by SDS-polyacrylamide-gel electrophoresis. Note the preferential phosphorylation of p37 (annexin-1) by cPKC, and of p50 (ribosomal elongation factor, eEF-1α) by nPKC. The upper bands represent the autophosphorylation of the enzymes.
(K. Kielbassa, M. Gschwendt, F. Marks, unpublished results)

Fig.18: Down-regulation of cPKC (dotted line) and nPKC, type δ (solid line) upon treatment of mouse skin with phorbol ester TPA (20 nmoles, at zero time).
Evaluation by immunoblotting. For details see Gschwendt et al., 1992; Leibersperger et al., 1991.

Tyrosine phosphorylation

Certain membrane receptors display an intrinsic protein kinase activity at their cytoplasmic transmitter domain, which is activated upon interaction of the extracellular receiver module with an input signal. Such a mechanism of signal transduction has been found for the insulin receptor and a series of growth factor receptors, i.e. for EGF, PDGF, FGF, IGF, CSF-1 and others (Fig.19, see Schlessinger & Ullrich,

1992). In all these cases the kinase activity is tyrosine-specific. Receptor activation results both in the autophosphorylation of the receptor protein and phosphorylation of other substrate proteins. Tyrosine phosphorylation also provides a primary output signal for other receptors which do not contain an intrinsic kinase activity but interact upon activation with separate cytoplasmic or membrane-bound tyrosine kinases (Mustelin & Burn, 1993). This mechanism of action is characteristic for the T-cell antigen receptor (Abraham & Karnitz, 1992; Izquierdo & Cantrell, 1992; Burns & Ashwell, 1993) and a series of cytokine receptors (Tan, 1993).

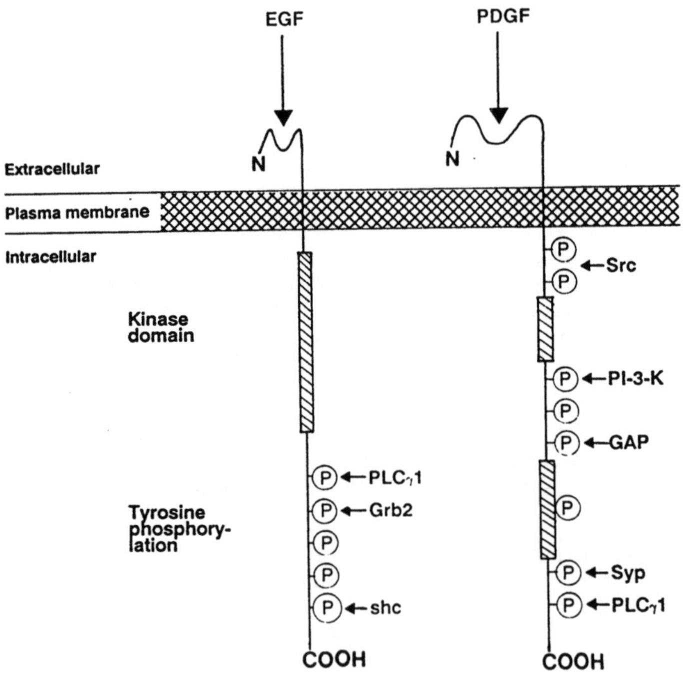

Fig.19: The modular structure of receptor tyrosine kinases. These transmembrane receptors bind signal molecules such as EGF or PDGF at their extracellular receiver domains. This reversible interaction results -via receptor dimerization- in an activation of the intrinsic tyrosine kinase activity on the intracellular transmitter module and in multiple autophosphorylation. The phosphotyrosine residues (P) thus generated specifically interact with a series of SH_2 proteins (see Fig. 20) some of which may become also phosphorylated on tyrosine. For details see Schlessinger & Ullrich, 1992.

Tyrosine phosphorylation provides both a signaling reaction and a means for establishing direct contacts between protein molecules (Pawson & Schlessinger, 1993).

The phosphotyrosine residue is specifically recognized by complementary receptor sites on proteins resulting in a reversible interaction according to the lock-key principle. These receptor sites are called SH_2-domains (src-homologous domains, referring to the src protein where they were orginally discovered). Many signal-transducing proteins carry one or more SH_2-domains which enable them to communicate with tyrosine-phosphorylated proteins (Fig.20). Frequently SH_2 domains and tyrosine-phosphorylation sites are found on one and the same molecule. Since SH_2-domains discriminate between different phosphotyrosine residues on the basis of adjacent structural parameters, this interaction acquires specificity and variability, thus fulfilling ideal conditions for the reversible construction of signal-transducing complexes and networks on demand (Fig.19,20).

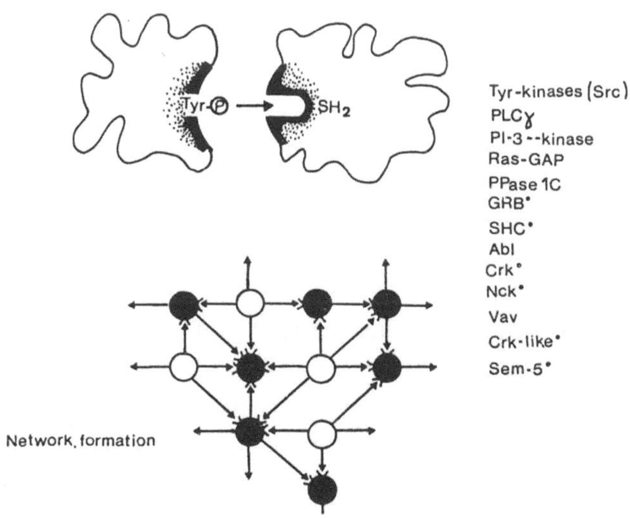

<u>Fig.20</u>: Direct contact between communicative protein molecules through the SH_2 domain.
The SH_2 domain - a structural motif of approximately 100 amino acids found in several proteins (see list at the right side) - "recognizes" phosphotyrosine residues in their specific microenvironment on polypeptide chains. By means of this interaction signal-transducing complexes and networks of proteins are formed reversibly and upon demand.

Besides SH_2-domains and phosphotyrosine residues proteins do express other contact sites, such as for instance SH_3-domains (Pawson & Schlessinger, 1993) and PH-domains (Musacchio et al., 1993), which probably bring about reversible interactions with regulatory proteins, cytoskeletal structures and G-proteins, as well as motifs which allow a specific binding to regulatory sites on nucleic acids. DNA binding of gene-regulatory proteins (transcription factors) is again frequently controlled by protein phosphorylation/dephosphorylation (for an example see Fig.9, see Jackson, 1992; Hunter & Karin, 1992) as well as by a direct interaction with intercellular signal molecules such as the thyroid and steroid hormones and retinoic acid (Fig.3, see Reichel & Jacob, 1993).

Cancer: a cellular "psychosis"

As already mentioned all (proto)-oncogenes which have been identified code for proteins involved in cellular signal processing (Cantley et al., 1991; Wynford-Thomas, 1991). These include growth factors, receptors, G-proteins, protein kinases and kinase substrates such as transcription factors (McMahon & Monroe, 1992; Gutman & Wasylyk, 1991). Frequently, oncogenic mutations are found among the tyrosine kinases, i.e. proteins which play a fundamental role in the transduction of mitogenic signals and the establishment of signal-processing networks. Another target protein of carcinogenesis holding a key position in mitogenic signal transduction is the G-protein Ras. Ras seems to focus mitogenic input signals and to transduce them to the genome (Fig.21, see Marshall, 1993). The proto oncogene vav has been identified as a Ras-activating factor (GEP, see Fig.4) which like protein kinase C is activated by phorbol esters tumor promoters (Gulbins et al., 1994).

Apart from kinases transcription factors are the most abundant targets of oncogenic mutations. They are encoded by both proto oncogenes and suppressor genes. While proto oncogene-derived transcription factors appear to be "ultimate targets" of mitogenic signal cascades, suppressor gene products may control the entry into and the passage through the cell cycle as well as the termination of cell proliferation by either reversible growth arrest or programmed cell death

(Marshall, 1991; Levine et al., 1991; Cobrinik et al., 1992). The latter may occur either by terminal differentiation or apoptosis. Cell cycle passage and reversible or irreversible growth arrest are under strict control of intercellular signals and the signal processing machinery of the cell.

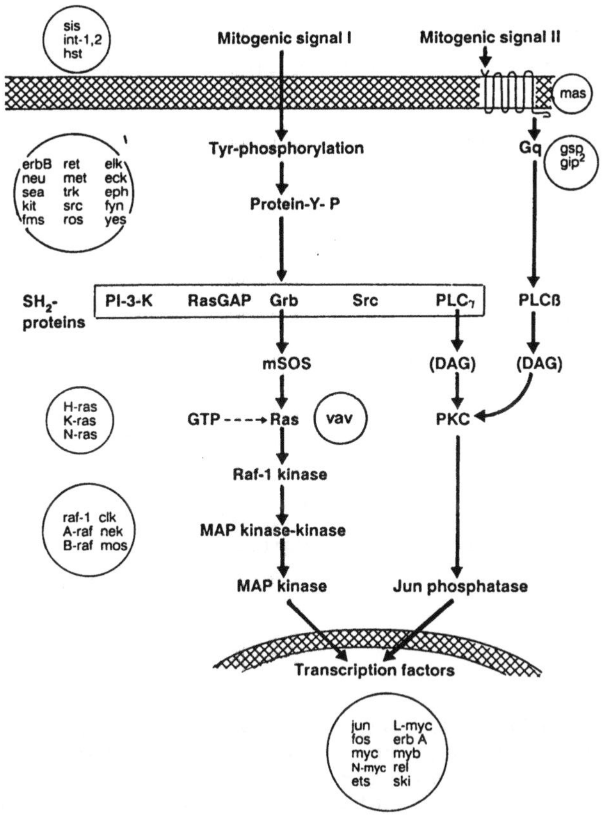

<u>Fig.21</u>: Cascade for the intracellular processing of mitogenic signals as major sites of oncogenic mutation.
The scheme demonstrates the synergistic effect of two mitogenic signals, one of which interacts with a receptor tyrosine kinase the other with a receptor G-protein complex. The great majority of proto oncogenes (encircled) have been shown to code for proteins which may be found at distinct places in these cascades. The signaling cascades leading into the nucleus are shown in a highly simplified manner. In reality numerous cross-talkings between these and other signaling pathways occur (see also Fig.3). For more details see Marshall, 1993.

The genotoxic effects of carcinogens thus result altogether in defects of the molecular interactions of signal processing. In the early stages of tumor development such disturbances may be balanced by the plasticity of the "cellular brain" and persist in a state of latency which is characteristic for the initiated state (see Fig.1). Non-genotoxic carcinogens such as the tumor promoters may overcome this latency by a massive disruption of cellular signal transduction resulting in an overstimulation of mitogenic cascades. As a consequence the cell loses its ability to interprete communication signals correctly according to their specific "meaning". The responses thus resulting are no longer in harmony with the organism. They include uncontrolled, invasive and metastasizing growth and uncoordinated functions which will finally kill the host.

If we figuratively spoke of the cellular machinery of signal-processing as the "brain of the cell", cancer may be regarded as a "psychotic" disorder of the cell. It should be emphasized, however, that this is a metapher, since in the general understanding the term "psychosis" refers to "consciousness". The latter, however, seems to be a property emerging only at a state of complexity which by far surpasses that of the signal-processing system of a single cell.

References:

Abraham RT, Karnitz LM, Secrist JP & Leibson PJ (1992) Signal transduction through the T-cell antigen receptor. Trends in Biochem Sci 17:434-438

Barritt GJ (1992) Communication within animal cells. Oxford Univ Press, Oxford etc.

Baser A (1993) The potential of protein kinase C as a target for anticancer treatment. Pharmac Ther 59:257-280

Black ID (1991) Information in the brain. MIT Press Cambridge (Mass.)

Boege F, Neumann E & Helmreich EJM (1991) Structural heterogeneity of membrane receptors and GTP-binding proteins and its functional consequences for signal transduction. Europ J Biochem 199:1-15

Bourne HR, Sanders DA & McCormick F (1991) The GTPase superfamily. Nature 348:125-132 and 349:117-127

Burns SM & Ashwell JD (1993) ZAPping the T-cell receptor. Current Biology 3:97-99

Cantley LC, Auger KR, Carpenter C, Duckworth B, Graziani A, Kapeller R & Soltoff S (1991) Oncogenes and signal transduction. Cell 64:281-302

Cobrinik A, Dowdy SF, Hinds PH, Mittnacht S & Weinberg RA (1992) The retinoblastoma protein and the regulation of cell cylcing. Trends in Biochem Sci 17:312-315

Dekker LV, Parker PJ (1994) Protein kinase C - a question of specificity. TIBS 19:73-77

Diekmann D, Brill S, Garrett MD, Totty N, Hsuan J, Monfries C, Hall C, Lim L & Hall A (1991) Brc encodes a GTPase-activating protein for p21rac. Nature 351:400-402

Dohlman HG, Caron MG & Lefkowitz R (1987) A family of receptors coupled to guanine nucleotide regulatory proteins. Biochemistry 26:2657-2664

Dvorak HF (1986) Tumors: wounds that do not heal. N Engl J Med 315:1650-1659

Fearon ER & Vogelstein B (1990) A genetic model for colorectal tumorigenesis. Cell 61:759-767

Findlay J & Eliopoulos E (1990) Three-dimensional modelling of G-proteins linked receptors. Trends in Pharmacol Sci 11:492-499

Fujiki H (1992) Is the inhibitor of protein phosphatase 1 and 2A activities a general mechanism of tumor promotion in human cancer development ? Molec Carcinogenesis 5:91-94

Grasso P, Sharratt M & Cohen AJ (1991) Role of persistent, non-genotoxic tissue damage in rodent cancer and relevance to humans. Annu Rev Pharmacol Toxicol 31:253-287

Green S (1992) Nuclear receptors and chemical carcinogenesis. Trends in Pharmacol Sci 13:251-255

Gschwendt M, Leibersperger H & Marks F (1989) Differentiative action of U252a on protein kinase C and a calcium-unresponsive, phorbol ester/phospholipd-activated protein kinase. Biochem Biophys Res Commun 164:974-982

Gschwendt M, Kittstein W & Marks F (1991) Protein kinase C activation by phorbol esters: do cysteine-rich regions and pseudosubstrate motifs play a role ? TIBS 16:167-169

Gschwendt M, Kittstein W, Lindner D & Marks F (1992) Differential inhibition by staurosporine of phorbol ester, bryostatin and okadaic acid effects on mouse skin. Cancer Lett 66:139-146

Gulbins E, Coggershall KM, Wittinghofer A, Burn P, Katzar S & Altman A (1994) Dual regulation of the Ras-exchange protein vav by diglycerides and tyrosines. J Cell Biochem Suppl 18B:257

Gutman A & Wasylyk B (1991) Nuclear targets for transcription regulation by oncogenes. Trends in Genetics 7:49-54

Haddow A (1972) Molecular repair, wound healing and carcinogenesis: tumour production a possible overhealing ? Adv Cancer Res 16:181-234

Hargave PA & McDowell JH (1992) Rhodopsin and phototransduction: a model system for G-protein linked receptors. FASEB J 6:2322-2331

Hepler JR & Gilman AG (1992) G-Proteins. Trends in Biochem Sci 17:383-327

Hunter T (1991) Protein kinase classification. Methods in Enzymol 200:3-37

Hunter T & Karin M (1992) The regulation of transcription by phosphorylation. Cell 70:375-387

Izquierdo M & Cantrell DA (1992) T-cell activation. Trends in Cell Biol 2:268-271

Jackson SP (1992) Regulating transcription factor activation by phosphorylation. Trends in Cell Biol 2:104-108

Karin M & Smeal T (1992) Control of transcription factors by signal transduction pathways: the beginning of the end. Trends in Biochem Sci 17:418-422
Lancer D & Ben-Aric N (1993) Olfactory receptors. Current Biology 3:668-674
Lefkowitz RJ (1993) G-protein-coupled receptor kinases. Cell 74:409-412
Leibersperger H, Gschwendt M, Gernold M, Marks F (1991) Immunological demonstration of a calcium-unresponsive protein kinase C of the δ-type in different species and murine tissues. J Biol Chem 266:14778-14784
Levine AJ, Momand J & Finlay CA (1991) The p53 tumor suppressor gene. Nature 351:453-456
Lindner ME & Gilman AG (1992) G-proteins. Sientific American 267:36-43
McMahon SB & Monroe JG (1992) Role of primary responses genes in generating cellular responses to growth factors. FASEB J 6:2707-2715
Marks F (1989) Skin cancer (excluding melanomas). In: Handbook of Exp Pharmacol, vol 87/II, Pharmacology of the Skin II, Greaves MW, Schuster S, eds, Springer-Verlag, Berlin etc. p 165-194
Marks F (1989) Chemical carcinogenesis - the multistage approach. Interdisc Sci Rev 14:233-240
Marks F & Fürstenberger G (1994) Tumor promotion in the skin. In: Chemical Induction of Cancer, Vol IV, Arcos JC, ed, Academic Press
Marks F & Fürstenberger G (1990) The conversion stage of skin carcinogenesis. Carcinogenesis 11:2085-2092
Marks F, Fürstenberger G & Gschwendt M (1992) Skin tumour promotion and the wound response: two sides of a coin. In: The Environmental Threat of the Skin (Marks R, Plewig G, eds), p 297-307, Dunitz London.
Marshall MS (1993) The effector interactions of $p21^{ras}$. Trends in Biochem Sci 18:250-254
Marshall CJ (1991) Tumor suppressor genes. Cell 64:313-326
Murray AW & Kirschner MW (1991) What controls the cell cycle. Scientific American 264:34-41
Murray AW (1993) Turning on mitosis. Current Biology 3:291-293
Musacchio A, Gibson T, Rice P, Thompson J & Saraste M (1993) The PH domain: a common piece in the structural patchwork of signalling proteins. Trends in Biochem Sci 18:343-348
Mustelin T & Burn P (1993) Regulation of src family tyrosine kinases in lymphocytes. Trends in Biochem Sci 18:215-220
Nakanishi H, Brewerk A & Exton JH (1993) Activation of the ς isozyme of protein kinase C by phosphatidyl inositol 3,4,5-trisphosphate. J Biol Chem 268:13-16
Nigg EA (1993) Cellular substrates of $p34^{cdc2}$ and its companion cylcin-dependent kinases. Trends in Cell Biol 3:296-301
Nurse P (1990) Universal control mechanism regulating onset of M-phase. Nature 344:503-507
Orth J (1911) Präcarcinomatöse Krankheiten und künstliche Krebse. Z Krebsforsch 10:42
Parkinson JS (1993) Signal transduction schemes of bacteria. Cell 73:856-871
Pawson T & Schlessinger J (1993) SH2 and SH3 domains. Current Biology 3:434-442

Pearson RB & Kemp BE (1991) Protein kinase phosphorylation site sequences and consensus specificity motifs. Methods in Enzymol 200:62-81

Perera FP (1991) Perspectives on the risk assessment for nongenotoxic carcinogens and tumor promoters. Environ Health Perspect 94:231-235

Poellinger L, Göttlicher M & Gustafsson JA (1992) The dioxin and peroxisome proliferator-activated receptors: nuclear receptors in search of endogenous ligands. Trends in Pharmacol Sci 13:241-245

Rahmsdof HJ & Herrlich P (1990) Regulation of gene expression by tumor promoters. Pharmac Ther 48:157-188

Reichel RR & Jacob ST (1993) Control of gene expression by lipophilic hormones. FASEB J 7:427-436

Ross EM (1992) Twists and turns on G-protein signalling pathways. Current Bioloy 2:517-519

Schlessinger J & Ullrich A (1992) Growth factor signalling by receptor tyrosine kinases. Neuron 9:383-391

Stabel S & Parker PJ (1991) Protein kinase C. Pharmac Ther 51:71-95

Takai Y, Kaibuchi K, Kikuchi A & Kawata M (1992) Small GTP-binding proteins. Int Rev Cytol 133:187-227

Tan YH (1993) Yin and Yang of phosphorylation in cytokine signaling. Science 262:376-377

Wilson CJ & Applebury ML (1993) Arresting G-protein coupled receptor activity. Current Biology 3:683-686

Wynford-Thomas D (1991) Oncogenes and anti-oncogenes; the molecular basis of tumour behaviour. J Pathol 165:187-201

Actin Involvement in Cell Toxicity

Marina Marinovich, Barbara Viviani, Corrado Lodovico Galli
Lab. of Toxicology, Inst. of Pharmacological Sciences
University of Milan
via Balzaretti, 9, 20133 Milano
Italy.

The cytoskeleton is an extensive, highly organised, three-dimensional network in the cytoplasm of cells. The cytoskeletal elements, termed microtubules (MT), intermediate filaments (IF) and microfilaments (MF), are responsible for the maintenance of cellular architecture, shape and internal organisation, and for regulating cell migration and motility (Stossel, 1988; Cooper, 1991). Furthermore, the cytoskeleton, via association with the cell membrane, appears to function in intracellular communication and cellular responses to membrane events.

MF are essentially composed of the protein actin, wich has a molecular weight of 43000. Actin is present in monomer, globular form (G-actin); in order to exert its physiological activity, G-actin must be polymerised into filaments 6 nm wide (F-actin, MF).

The filament grows at the two extremities, the pointed and the barbed ends (70 +/2- monomers each second and 20+/1-, respectively) (Pollard and Craig, 1982).

The polymerisation of actin can be divided into four steps:
- activation (binding of divalent ions and conformational changes of monomers)
- nucleation : formation of trimers
- elongation: the bi-directional growth of oligomers
- annealing : the end-to-end joining of two filaments

The actin monomer can bind either ATP or ADP; however ATP is more tightly bound. Actin-bound ATP is hydrolysed shortly after monomer incorporation into the filament, but hydrolysis is not kinetically coupled to polymer assembly; actin filaments will tend to disassemble rapidly at low ATP levels (Pollard and Cooper, 1986).

NATO ASI Series, Vol. H 93
Modulation of Cellular Responses in Toxicity
Edited by C. L. Galli, A. M. Goldberg, M. Marinovich

Even if MF contain essentially actin other proteins are associated to the MF with different functions (Pollard and Craig, 1982b).

The assembly G-->F-actin is very rapid. It is caused by receptor-dependent and independent stimuli and it mediates several important cell activities (Caterina and Devreotes, 1991; Meyer and Howard, 1987). There is, for instance, a close correlation of the dose-response curves for actin polymerisation and the mean rate of locomotion of neutrophils(Howard and Meyer, 1984; Meyer and Howard, 1987).

Reorganisation of cytoskeletal protein assemblies following stimulation of the plasma membrane by a variety of ligands appears to be a common event. Both actin polymerisation and depolymerisation can result from the ligand:receptor interaction. This leads to changes in the gross morphology and ultrastructural organisation of the cells linked to MF formation, and are often followed by a generalised stimulation of MF-based motility.

The dynamic characteristics of actin are fundamental for all eukaryotic cells and of great interest from a toxicological point of view. Because of the complex functional role of actin, which varies with cell type, degree of differentiation (Kwiatkowski, 1988; Sham et al. 1991), and cell cycle (Bravo et al, 1982; Dornish et al., 1987), its disruption or alteration may result in a variety of cellular changes (Marinovich, 1991).

Actin organisation in the cell

Although many details of the mechanism of actin polymerisation are still to be determined, a general picture is emerging correlating polyphosphoinositides (PPI) turnover with cytoskeletal remodelling (Lassing and Lindberg, 1985). Phosphatidylinositol 4,5-bisphosphate (PIP_2) can bind to profilin (an actin binding protein normally associated to actin), dissociate the profilactin complex, and thus render actin available for polymerisation. This suggests that the phosphatidylinositol (PI) cycle, which plays important roles in cellular regulation, might also control MF-based motility.

However, there are studies showing that in vivo PIP_2 concentrations decrease dramatically shortly after chemotactic stimulation when actin polymerisation is increased (Carson et al 1986; Hall et al., 1989) and also that chemotactic peptide stimulation can raise the cellular content of F-actin despite the absence of PIP_2 hydrolysis (Bengtsson et al 1988). PIP_2 could potentiate the uncovering of barbed filament ends blocked by gelsolin or other filament-capping proteins (Luna and Hitt, 1992) or control actin assembly through one or more second messengers such as diacylglycerol (DG). DG may potentiate one of the two rate-limiting processes associated with actin assembly (activation or nucleation). The formation of new actin nucleation sites at the membrane is triggered by DG (Shariff and Luna, 1992), and membrane permeant DG induces actin polymerisation and cytoskeletal changes in lymphocytes, neutrophils and leukaemia cells (Zimmermann et al, 1988; Apgar et al. 1991).

Omann et al (1987) and Särndahl et al (1989) have proposed that chemoattractant-receptor interaction stimulates phospholipase activity presumably via G-proteins. This hypothesis has been confirmed by Bengtsson et al (1990) on neutrophils using direct activators of G-proteins as γ-S-GTP or AlF_4. The assumption of G-protein as a direct trigger of actin polymerisation could fit with the role of PIP_2 if G proteins participate in the regulation of any of the enzymes responsible for PIP_2 formation.

More recently it has been reported that Rho and ras proteins (belonging to the superfamily of GTP-binding proteins) are required specifically for the formation of focal adhesions and stress fibres of actin but not for membrane ruffles in fibroblasts (Ridley and Hall, 1992; Ridley et al, 1992). Moreover, prenylated rho proteins have been shown to play a critical role in regulating the state of actin MF (Fenton et al, 1992)

In RBC ghost the phospholipase C (particulate form) responsible for hormone receptor- and G-protein-regulated polyphosphoinositide hydrolysis is located on the cytoskeleton (Vaziri and Downes, 1992). This contrasts with the common conceptions that PLC associates directly (through undescribed mechanisms) with the plasma membrane or that cytoplasm PLC translocates to the plasma membrane to associate with activated G-proteins. The association of a population of PLC with a component of the cytoskeleton could provide a locally high

concentration and appropriate orientation of the effector enzyme, allowing more efficient communication with the G-protein.
More recently, a role for the products of the metabolism of arachidonic acid in the actin organisation has been demonstrated. In a number of cell types, epidermal growth factor (EGF) evokes dramatic morphological changes, cortical actin polymerisation and stress fibre breakdown by means of arachidonic acid metabolites. In particular, lipoxygenase metabolites (leukotrienes) appear to induce actin polymerisation, and cyclooxygenase products (prostaglandins) seem to cause depolymerisation of cytoplasm actin filaments (Peppelenbosch et al 1993)

Xenobiotics acting on cytoskeleton

It appears evident that:
- MF are an important cellular structure from a quantitative (actin represents about 10% of cell protein) and a qualitative point of view;
- actin polymerisation plays a pivotal role in several cell responses to stimuli, and it is controlled by basic cell systems (PPI cycle, AA metabolism, etc.);
- MF have close links with other structures and signal transduction systems in the cell.

It is arguable that the cytoskeleton can constitute an easy target for toxic compounds, and that the resulting injury to this structure may have dramatic consequences for the cell.

A number of substances that interact selectively with cytoskeleton components have been identified (Scapigliati et al, 1988) The majority of the data concerns neurotoxic agents and the MT system. It is only recently that more interest has been shown in other cytoskeletal components.
The toxic agents that have been shown to have an effect on MF belong to various categories including drugs, toxins, metals and pesticides. In Table 1-3 some examples are reported of compounds recognised to express their toxicity by interacting with actin MF.

Of particular interest from a toxicological point of view is the interaction between toxins and cytoskeleton, particularly actin (Table 1).

Toxin	**Effect**
Tetanus t.	Aberrant distribution of actin filaments in Sertoli cells (Eisel et al., 1993)
Escherichia coli t.	Increase of actin polymerization (Fiorentini et al, 1988)
Clostridium botulinum C2 t.	ADP-ribosylation of G-actin (Aktories et al, 1986)
Clostridium botulinum neurotoxin	Actin cleavage *in vitro* (Das Gupta and Tepp 1993)
Clostridium botulinum exoenzyme C3	ADP-ribosylation of Rho proteins with loss of actin stress fibers (Just et al. 1992; Chardin et al, 1989)
Clostridium perfr. iota t.	ADP-ribosylation of G-actin (Schering et al., 1988)
Clostridium spiroforme t.	ADP-ribosylation of G-actin (Popoff et al. 1989)
Clostridium difficile iota and B t.	Rearrangement of actin and α-actinin filaments, and of actin, vinculin and talin, respectively (Fiorentini et al.,1989; Ottlinger and Lin, 1988)
Clostridium difficile ADP-ribosyltransferase	ADP-ribosylation of actin (Popoff et al., 1988)
Clostridium baratii t. *Clostridium butyricum* t. (A-G neurotoxins)	Actin cleavage *in vitro* (Das Gupta and Tepp, 1993)
Microcystis aeruginosa (microcystin-LR hepatotoxin)	Microfilament reorganization in rat hepatocytes (Eriksson et al. 1989)
Phalloidin	Stabilization of F-actin Increasing rate of actin polymerization (Dancker et al., 1975)
Snake venom cardiotoxin	Induction of actin polymerization (Chen and Chu, 1988)

Table 1.Toxins and cytoskeleton

Toxins, in fact, affecting directly actin, provide a specific tool for evaluating the role of catin in cell functions.
Botulinum C2 toxin ADP-ribosylates actin in cells in a concentration-dependent manner. Substantially ADP-ribosylated actin behaves like a capping protein at the barbed ends of actin filaments (Aktories et al, 1986). The ADP-ribosylation of cellular actin causes depolymerisation of cytoskeleton associated F-actin to monomeric G-actin and blocks the ability of actin to polymerise.

Both capping of actin filaments and trapping of monomeric actin in its G-form induce the destruction of the MF network, thereby causing the cytopathic effects of *botulinum* C2 toxin. Recently it has been shown that besides *botulinum* C2 toxin, various other bacterial toxins modify actin probably acting through this mechanism (Aktories et al, 1987; Aktories and Wegner, 1989). *Botulinum* C2 toxin-induced ADP-ribosylation is highly specific for β/γ actin (non-muscle) in its G-form (Aktories et al,1986), whereas *Clostridium perfringens* iota toxin is able to ADP-ribosylate all the actin isoforms (Schering et al., 1988; Aktories et al, 1990).
Phalloidin, a mushroom toxin, has also become an important tool in studying MF (Dancker et al., 1975; Frimmer, 1987)
Metals are an other class of toxic agents active on the cytoskeleton (Table 2).
Methyl mercury (MeHg) disrupts microtubules but effects on other cytoskeleton components require further investigation (Brown et al., 1988; Vogel et al 1985). At 0.5 μM MeHg the number of MT was reduced in PtK2 cells (Sager, 1988). Only secondary effects on vimentin or actin filaments were observed at concentrations of MeHg which caused extensive microtubule disassembly.
Also in the case of lead and its organic salts the main effect is on tubulin (Zimmermann et al, 1985,1987).
Cadmium is able to alter the organisation of actin filaments in Madin-Darby canine kidney cells (Dìaz-Barriga et al, 1989), and so is zinc. This latter also decreases cell attachment (Mills et al, 1992).

Other toxic compounds having the cytoskeleton as target are the solvent hexanedione (Moretto et al, 1991; Durham, 1988), the pesticides benomyl (Davidse and Flach, 1977; Marinovich et al., 1994), paraquat

(Cappelletti and Maci, 1993; Cappelletti et al, 1994) and the organic derivatives of tin, mainly tributyl- and triphenyltin. These compounds depolymerise F-actin in different types of cells (Marinovich et al.,

METAL	FORM	EFFECT
Aluminum	Al^{3+}	promotion of tubulin assembly (MacDonald et al 1987)
Cadmium	Cd^{2+}	disruption of actin filaments (Prozialeck et al 1991, Diaz-Barriga et al, 1989)
Cromium	Cr^{6+}	depolymerization of MT and redistribution of MF and loss of MF at high concentrations (Li et al, 1992)
Lead	$(CH_3)_3PbCl$ $(C_2H_5)_3PbCl$	inhibition of microtubule assembly (Faulstich et al 1984, Zimmerman et al 1985, Roderer and Doenges 1983); induction of perinuclear coil formation of neurofilaments in mouse neuroblastoma cells (Zimmermann et al 1987)
Mercury	CH_3HgCl	disruption of microtubules (Brown et al 1988, Vogel et al 1985, Sager 1988)
Zinc	Zn^{2+}	depolymerization of microtubules; increase in F-actin (Mills et al, 1992)

Table 2. Metals and cytoskeleton

1990; Chow and Orrenius, 1994) and they have also been shown to inhibit the actin polymerisation of neutrophils both in vitro and in vivo (Marinovich et al, 1990; Colosio et al. 1991).

LITHIUM	promotion of the assembly of tubulin and actin (Pan and Ware, 1988; Colombo et al, 1991)
GLUCOCORTICOIDS	loss of actin stress fibers (Hughes-Fulford et al, 1992)
ADRIAMYCIN (doxorubicin)	MF clusters in glomerular epithelial associated cells with deformation of cell shape and detachment of the the cell membrane from the GBM decrease of actin polymerization in cell-free system (Ito et al, 1986) alteration of G-actin assembly and G/F-actin ratio; MF pattern alterations in non-muscle cell (Colombo et al, 1990)
RETINOIDS	formation of actin stress-fibers (Rao, 1985)
cis-PLATIN	alteration of microtubule disassembly (Boekelheide et al, 1992)
DOXYCYCLIN	reduction of F-actin (Gabler et al, 1994)

Table 3. Cytoskeleton affecting drugs

Molecular mechanisms of actin toxicity

The most relevant and known mechanisms whereby xenobiotics interfere with microfilament structure or their assembly are the following:

1 ATP depletion

2 cross-linking of actin (alteration of -SH groups)

3 denaturation of actin-binding protein (ABP) by Ca^{2+}-activated proteases

4 ADP-ribosylation of actin

5 interaction with trigger signal for polymerisation (disturbance

of the membrane "active" components, G-proteins)
6 phosphorylation of ABP
7 interference with prenylation of proteins
8 actin carbamylation
9 physical modification of cellular compartment

Mechanisms listed in points 1-3 are typical of oxidative damage. In fact, one of the earliest signs of quinone and oxidative stress-induced cytotoxicity in hepatocytes is the loss of microvilli and the appearance of multiple surface protrusions called blebs, due to disturbances of the cytoskeleton organisation. During the metabolism of menadione a net increase in the amount of cytoskeleton associated proteins and a depletion of cytoskeletal protein thiols occurs (Bellomo and Mirabelli, 1987; Mirabelli et al.,1989). Monomeric actin has four thiol groups, three of which are naturally masked in the presence of physiological concentrations of ATP. Since the metabolism of menadione in hepatocytes causes ATP depletion, the unmasking of sulfhydryl groups in actin due to ATP depletion may increase their susceptibility to oxidative damage.

Confirming this hypothesis, addition of dithiothreitol to menadione-treated cells was found to protects the cells both from the appearance of surface blebs and the occurrence of alterations in cytoskeletal protein composition. These alterations were associated with oxidative modifications of actin, including ß-mercaptoethanol-sensitive cross-linking of actin to form dimers, trimers and high molecular weight aggregates which also contained other cytoskeletal proteins, i.e. α-actinin and actin-binding-protein (Mirabelli et al, 1989). The calcium ion also plays a critical role in the maintenance of the structure and function of the cytoskeleton. In physiological conditions Ca^{2+} controls the MF disassembly phenomenon more than actin polymerisation. Mirabelli et al. (1989) have demonstrated that the multiple effects of oxidative stress on the cytoskeleton are mediated by oxidative as well as by Ca^{2+}-dependent mechanisms. The calcium ion can exert its effect either directly, by promoting the dissociation of polypeptides from the cytoskeleton, or indirectly, through the activation of a leupeptin-inhibitable protease that specifically degrades the ABP.

An oxidative mechanism has also been invoked to explain the toxic effect on the cytoskeleton caused by the well-known pesticide, paraquat (Cappelletti and Maci, 1993; Cappelletti et al. 1994)
As already discussed (see Table 2) ADP-ribosylation of actin is the predominant mechanism by which toxins affect the cytoskeleton. ADP-ribosylation of actin inhibits actin polymerisation and G-actin associated ATP-ase activity. The only information about the molecular site of interaction is that C2 and iota toxin ADP-ribosylate actin in arginine 177. So far there is no evidence that the functions of actin are endogenously regulated by ADP-ribosylation. Nevertheless it is fascinating to speculate that also the actin-ADP-ribosylating toxins mimic endogenous eukaryotic ADP-ribosyltransferase.
Moreover, cholera toxin and pertussis toxin ADP-ribosylate $G_{s\alpha}$ and $G_{i\alpha}$s, respectively (Casey and Gilman, 1988), the GTP-binding regulatory components of the adenylate cyclase system. This fact may be important in the manifestation of cytoskeletal damage, considering the pivotal role of G-proteins in triggering actin polymerisation (Bengtsson et al, 1990).

The activation of protein kinase C and protein kinase A has been demonstrated to promote assembly (Apgar, 1991; Downey et al, 1992, 1991; Lamb et al, 1988). Phosphorylation of the myosin light chain can promote actin assembly, possibly by altering the conformation of G-actin to one that favors polymerisation (Eto et al, 1991; Lamb et al, 1988). The exposure of human neutrophils to okadaic acid (a marine dinoflagellates toxin) and calyculin A, inhibitors of phosphatases, induces assembly and reorganisation of actin (Downey et al, 1993) with consequent alteration of cell shape and motility. In isolated hepatocytes, the hepatotoxin and phosphatase inhibitor microcystin-LR induces blebbing and alterations of the actin cytoskeleton, correlating with inhibition of protein phosphatases (Eriksson et al, 1989).

Posttranslational modification by covalent attachment of polyisoprene intermediates to a carboxyterminal CAAX-box (Schmidt et al, 1984) is required for the biological function of proteins such as p21ras, the supergene family of ras-related proteins, nuclear lamins, and subunits of heterotrimeric G-proteins (Farnsworth et al.,1989; Hancock et al 1989;Casey et al, 1989; Yamane et al. 1991; Yoshida et al 1991). Cells grown in the presence of lovastatin, which inhibits hydroxymethylglutaryl-coenzyme A (HMG-CoA) reductase and prevents synthesis of intermediates required for protein prenylation, develop a round, refractile morphology due to a selective loss of actin cables (Fenton et al., 1992). Mevalonate (the product of the HMG-CoA reductase reaction) reverses the lovastatin-induced morphological change by inducing a rapid reorganisation of actin cables with coincident reversion to flat morphology. These data indicate that prenylated protein(s) play a critical role in regulating the state of intracellular actin, and that the mechanism of lovastatin's effect upon cellular architecture thus appears to be a depletion of the prenyl-pyrophosphate(s) required for the prenyltransferase reaction(s). Probably by this mechanism lovastatin strongly inhibits the cytotoxicity of modeccin in Vero and BER-40 cells and partially inhibits the cytotoxicity of Ricin, *Pseudomonas* toxin, and diphtheria toxin (Oda and Wu 1994). Prenylated small G-proteins such as rab proteins (Plutner et al 1991; Goud et al 1990) could be involved in vesicle trafficking of protein toxins, especially modeccin.

Recently, Kuckel et al (1993) provided evidence that non-enzymatic carbamylation of the lysine residues prevents the polymerisation of actin and that carbamylated actin inhibits the polymerisation of nascent, unmodified actin. This mechanism could be at the basis of the loss of ordered filament structure and shape of the lens fibre cell, predisposing it to cataract development after exposure to methylisocyanate (Harding and Rixon,1985). In addition, the effect of benomyl, a fungicide, on tubulin (Davidse and Flach, 1977) and actin (Marinovich et al, 1994) could be due to its ability to carbamylate cell protein, including cytoskeletal elements, through a metabolite.

References

Aktories K, Bärmann M, Chhatwal G, Presek P (1987) New class of microbial toxins ADP-ribosylates actin. Trends Pharmacol. Sci. 8:158-160.

Aktories K, Bärmann M, Ohishi I, Tsuyama S, Jakobs KH, Habermann E (1986) Botulinum C2 toxin ADP-ribosylates actin. Nature 322:390-392.

Aktories K, Geipel U, Wille M, Just I (1990) Characterization of the ADP-ribosylation of actin by *Clostridium botulinum* C2 toxin and *Clostridium perfringens* iota toxin. J. Physiol. Paris 84:262-266.

Aktories K, Wegner (1989) ADP-ribosylation of actin by clostridial toxins. J. Cell Biol. 109:1385-1387.

Apgar JR (1991) Regulation of the antigen-induced F-actin response in rat basophilic leukemia cells by protein kinase C. J. Cell Biol.112: 1157-1163.

Bellomo G, Mirabelli F (1987) Oxidative stress injury studied in isolated intact cells. Mol. Toxicol. 1:281-293.

Bengtsson T, Rundquist I, Stendahl O, Wymann MP, Andersson T (1988) Increased breakdown of phosphatidylinositol 4,5-bisphosphate is not an initiating factor for actin assembly in human neutrophils. J. Biol. Chem. 263: 17385-17389.

Bengtsson T, Särndahl E, Stendahl O, Andersson (1990) Involvement of GTP-binding proteins in actin polymerization in human neutrophils. Proc. Natl. Acad. Sci. USA 87:2921-2925.

Boekelheide K, Arcila ME, Eveleth J (1992) cis-Diamminedichloro-platinum (II) (Cisplatin) alters microtubule assembly dinamics. Toxicol. Appl. Pharmacol. 116:146-151.

Bravo R, Small JV, Fey SJ, Larsen PM Celis JE (1982) Architecture and polypeptides composition of HeLa cytoskeletons, modification of cytoarchitectural polypeptides during mitosis.J. Mol. Biol. 1 54:121-143.

Brown DL, Reuhl KR, Bormann S, Little JE (1988) Effects of methyl mercury on the microtubule system of mouse lymphocytes. Toxicol. Appl. Pharmacol. 94:66-75.

Cappelletti G, Incani C, Maci R (1994) Paraquat induces irreversible actin cytoskeleton disruption in cultured human lung cells. Cell Biol. Toxicol. 10:1-9.

Cappelletti G, Maci R (1993) Actin filaments disassembly: a novel step in the genesis of paraquat toxicity. Bull. Environ. Contam. Toxicol. 50:717-723.

Carson M, Weber A, Zigmond SH (1986) An actin-nucleating activity in polymorphonuclear leukocytes is modulated by chemotactic peptides. J. Cell Biol. 103:2707-

Casey PJ, Gilman AG (1988) G protein involvement in receptor-effector coupling. J. Biol. Chem. 263:2577-2580.

Casey PJ, Solski PA, Der CJ, Buss JE (1989) $p21^{ras}$ is modified by a farnesyl isoprenoid. Proc. Natl. Acad. Sci. USA 86:8323-8327.

Caterina MJ, Devreotes PN (1991) Molecular insights into eukaryotic chemotaxis. FASEB J. 5:3078-3085.

Chardin P, Boquet P, Madaule P, Popoff MR, Rubin EJ, Gill DM (1989) The mammalian G-protein *rho* C is ADP-ribosylated by *Clostridium botulinum* toxin exoenzyme C3 and affects actin microfilaments in Vero cells. EMBO J. 8:1087-1092.

Chen Y-H, Chu S-T (1988) Snake venom cardiotoxin induces G-actin polymerization. Biochim. Biophys. Acta 966:266-268.
Chow SC, Orrenius S (1994) Rapid cytoskeleton modification in thymocytes induced by the immunotoxicant tributyltin. Toxicol. Appl. Pharmacol. 127:19-26.
Colombo R, Milzani A, Dalle Donne I (1991) Lithium increases actin polymerization rates by enhancing the nucleation step. J. Mol. Biol. 217:401-404.
Colombo R, Milzani A, Necco A, Vailati G (1990) Doxorubicin effects on contractile structures and molecules. Cytotechnology 3:9-19.
Colosio C, Tomasini M, Cairoli S, Foà V, Minoia M, Marinovich M, Galli CL (1991) Occupational triphenyltin acetate poisoning: a case report. Br. J. Ind. Med. 48:136-139.
Cooper JA (1991) The role of actin polymerization in cell motility. Annu. Rev. Physiol. 53:585-605.
Dancker P, Löw I, Hasselbach W, Wieland TH (1975) Interaction of actin with phalloidin: polymerization and stabilization of F-actin. Biochim. Biophys. Acta 400:407-414.
DasGupta BR, Tepp W (1993) Protease activity of botulinum neurotoxin type E and its light chain:cleavage of actin. Biochem. Biophys. Res. Comm.190:470-474.
Davidse LC, Flach W (1977) Differential binding of methyl benzimidazol-2-yl-carbamate to fungal tubulin as a mechanism of resistance to this antimitotic agent in mutant strains of *Aspergillus nidulans*. J. Cell Biol. 72:174-193.
Dìaz-Barriga F, Carrizales L, Yanez L,Hernàndez JM, Domìnguez Robles MC, Palmer E, Saborio JL (1989) Interaction of cadmium with actin microfilaments in vitro. Toxic. in Vitro 3:277-284.
Dornish JM, Randen I, Juul N, Pettersen EO (1987) Is accumulation of actin restricted to G1? Biochem. Soc. Transact. 15:856-857.
Downey GP, Chan CK, Grinstein S (1992) Phorbol ester-induced actin assembly in neutrophils: role of protein kinase C. J. Cell Biol. 116: 695-706.
Downey GP, Erzurum SC, Young SK, Schwab B, Elson EL, Worthen GS (1991) Biophysical properties and microfilament assembly in neutrophils modulation by ciclic AMP. J. Cell Biol. 114:1179-1190.
Downey GP, Takai A, Zamel R, Grinstein S, Chan CK (1993) Okadaic acid-induced actin assembly in neutrophils: role of protein phosphatases. J. Cell. Physiol. 155:505-519.
Durham HD (1988) Aggregation of intermediate filaments by 2,5-hexanedione: comparison of effects on neurofilaments, GFAP-filaments and vimentin-filaments in dissociated cultures of mouse spinal cord-dorsal root ganglia. J. Neuropathol. Expt. Neurol. 47:432-442.
Eisel U, Reynolds K, Riddick M, Zimmer A, Niemann H, Zimmer A (1993) Tetanus toxin light chain expression in Sertoli cells of transgenic

mice causes alterations of the actin cytoskeleton and disrupts spermatogenesis. EMBO J. 12:3365-3372.

Eriksson JE, Paatero GIL, Meriluoto JAO, Codd GA, Kass GEN, Nicotera P, Orrenius S (1989) Rapid microfilament reorganization induced in isolated rat hepatocytes by microcystin-LR, a cyclic peptide toxin. Expt. Cell Res. 185:86-100.

Eto, MF, Morita F, Nishi N, Tokura S, Ito T, Takahashi K (1991) Actin polymerization promoted by a heptapeptide, an analogue of the actin-binding S site on myosin head. J. Biol. Chem. 266:18233-18236.

Farnsworth CC, Wolda SL, Gelb MH, Glomset JA (1989) Human lamin B contains a farnesylated cysteine residue. J. Biol. Chem. 264:20422-20429.

Faulstich H, Stournaras C, Doenges KH, Zimmermann H-P (1984) The molecular mechanism of interaction of Et_3Pb^+ with tubulin. FEBS Lett. 174:128-131.

Fenton RG, Kung H-F, Longo DL, Smith MR (1992) Regulation of intracellular actin polymerization by prenylated cellular proteins. J. Cell Biol. 117:347-356.

Fiorentini C, Arancia G, Caprioli A, Falbo V, Ruggeri FM, Donelli G (1988) Cytoskeletal changes induced in HEp-2 cells by the Cytotoxic Necrotizing Factor of *Escherichia coli*. Toxicon 26:1047-1056.

Fiorentini C, Arancia G, Paradisi S, Donelli G, Giuliano M, Piemonte F, Mastrantonio P (1989) Effects of *C. difficile* toxin A and B on cytoskeleton organization: a comparative morphological study. Toxicon 27:1209-1218.

Frimmer M (1987) What we have learned from phalloidin. Toxicol. Lett. 35:169-182.

Gabler WL, Smith J, Tsukuda N (1994) Doxycycline reduction of F-actin content in human neutrophils and fibroblasts. Inflammation 18:107-118.

Goud B, Zahraoui, Tavitian A, Saraste J (1990) Small GTP-binding proteins associated with Golgi cisternae. Nature 345:553-556.

Hall AI, Warren V, Dharmawardhane S, Condeelis J (1989) Identification of actin nucleation activity and polymerization inhibitor in ameboid cells: their regulation by chemotactic stimulation. J. Cell Biol. 109:2207-2213.

Hancock JF, Magee AI, Childs JE, Marshall CJ (1989) All ras proteins are polyisoprenylated but only some are palmitoylated. Cell 57:1167-1177.

Harding J, Rixon K (1985) Lens opacities induced in rat lenses by methyl isocyanate. Lancet ii:762.

Howard TH, Meyer WH (1984) Chemotactic peptide modulation of actin assembly and locomotion in neutrophils. J. Cell Biol. 98:1265-1271.

Hughes-Fulford M, Appel R, Kumegawa M, Schmidt J (1992) Effect of dexamethasone on proliferating osteoblasts: inhibition of prostaglandin E2 synthesis, DNA synthesis and alterations in actin cytoskeleton. Expt. Cell Res. 203, 150-156.

Ito K, Ger Y-C, Kawamura S (1986) Actin-filament alterations in glomerular epithelial cells of adriamycin-induced nephrotic rats. Acta Pathol. Jpn. 36:253-260.

Just I, Mohr C, Schallehn G, Menard L, Didsbury JR, Vandekerckhove J, vanDamme J, Aktories K (1992) Purification and characterization of an ADP-ribosyltransferase produced by Clostridium limosum. J. Biol. Chem. 267:10274-10280.

Kuckel CL, Lubit BW, Lambooy PK, Farnsworth PN (1993) Methylisocyanate and actin polymerization: the *in vitro* effects of carbamylation. Biochim. Biophys. Acta 1162:143-148.

Kwiatkowski DJ (1988) Predominant induction of gelsolin and actin-binding protein during myeloid differentiation. J. Biol. Chem. 263:13857-13862.

Lamb NJ, Fernandez A, Conti MA, Adelstein R, Glass D, Welch WJ, Feramisco JR (1988) Regulation of actin microfilament integrity in living nonmuscle cells by the cAMP-dependent protein kinase and the myosin light chain kinase. J. Cell. Biol. 106:1955-1971.

Lassing I, Lindberg U (1985) Specific interaction between phosphatidylinositol 4,5-biphosphate and profilactin. Nature (Lond.) 314-472-474.

Li W, Zhao Y, Chou I-N (1992) Cytoskeletal injury induced by hexavalent chromate. Toxic. *in vitro* 6:433-444.

Luna EJ, Hitt AL (1992) Cytoskeleton-plasma membrane interactions. Science 258:955-964.

MacDonald TL, Humphreys WG, Martin RB (1987) Promotion of tubulin assembly by aluminum ion *in vitro*. Science 236:183-186.

Marinovich M (1991) The role of actin in the transduction of toxic effect. Pharmacol. Res. 24:319-336.

Marinovich M, Guizzetti M, Galli CL (1994) Mixtures of benomyl, pirimiphos-methyl, dimethoate, diazinon and azinphos-methyl affect protein synthesis in HL-60 cells differently. Toxicology 94:173-185.

Marinovich M, Sanghvi A, Colli S, Tremoli E, Galli CL (1990) Cytoskeletal modifications induced by organotin compounds in human neutrophils. Toxicol. *in Vitro* 4:109-113.

Meyer WH, Howard TH (1987) Actin polymerization and its relationship to locomotion and chemokinetic response in maturing human promyelocytic leukemia cells. Blood 70:363-367.

Mills JW, Zhou J-H, Cardoza L, Ferm VH (1992) Zinc alters actin ilaments in Madin-Darby canine kidney cells. Toxicol. Appl. Pharmacol. 116:92-100.

Mirabelli F, Salis A, Vairetti M, Bellomo G, Thor H, Orrenius S (1989) Cytoskeletal alterations in human platelets exposed to oxidative stress are mediated by Ca^{2+}-dependent mechanisms. Arch. Biochem. Biophys. 270:478-488.

Moretto G, Monaco S, Passarin MG, Benedetti MD, Rizuto N (1991) Cytoskeletal changes induced by 2,5-hexanedione on developing humans neurons *in vitro*. Arch. Toxicol. 65:409-413.

Oda T, Wu HC (1994) Effect of lovastatin on the cytotoxicity of ricin, modeccin, *Pseudomonas* toxin, and diphtheria toxin in brefeldin A-sensitive and -resistant cell lines. Exptl. Cell Res. 212:329-337.

Omann GM, Allen RA, Bokoch GM, Painter RG, Traynor AE, Sklar LA (1987) Signal transduction and cytoskeletal activation in the neutrophil. Pharmacol. Rev. 67:285-322.

Ottlinger ME, Lin S (1988) *Clostridium difficile* toxin B induces reorganization of actin, vinculin, and talin in cultured cells. Expt. Cell Res. 174: 215-229.

Pan X-X, Ware BR (1988) Actin assembly by lithium ions. Biophys. J. 53:11-16.

Peppelenbosch MP, Tertoolen LGJ, Hage WJ, de Laat SW (1993) Epidermal growth factor-induced actin remodeling is regulated by 5-lipoxygenase and cyclooxygenase products. Cell 74, 565-575.

Plutner H, Cox ADn Pind S, Khosravi-Far R, Bourne J, Schwaninger R, Der CJ, Balch WE (1991) Rab 1b regulates vesicular transport between the endoplasmic reticulum and successive Golgi compartments. J. Cell Biol. 115:31-43.

Pollard TD, Cooper JA (1986) Actin and actin-binding proteins: a critical evaluation of the mechanisms and functions. Annu. Rev. Biochem. 55:987-1035.

Pollard TD, Craig SW (1982) Mechanism of actin polymerization. TIBS february:55-58.

Pollard TD, Craig SW (1982b) Actin-binding proteins. TIBS March:88-92.

Popoff MR, Milward FW, Bancillon B, Boquet P (1989) Purification of the *Clostridium spiroforme* binary toxin and activity of the toxin on HEp-2 cells. Infection and Immunity 57:2462-2469.

Popoff MR, Rubin EJ, Gill DM, Boquet P (1988) Actin-specific ADP-riboyltransferase produced by a *Clostridium difficile* strain. Infection and Immunity 56:2299-2306.

Prozialeck WC, Niewenhuis RJ (1991) Cadmium (Cd^{2+}) disrupts intercellular junctions and actin filaments in LLC-PK1 cells. Toxicol. Appl.Pharmacol. 107:81-97.

Rao KMK (1985) Phorbol esters and retinoids induce actin polymerization in human leukocytes. Cancer Lett. 28:253-262.

Ridley AJ, Hall A (1992) The small GTP-binding protein rho regulates the assembly of focal adhesions and actin stress fibers in response to growth factors. Cell 70:389-399.

Ridley AJ, Paterson HF, Johnston CL, Diekmann D, Hall A (1992) The small GTP-binding protein rac regulates growth factor-induced membrane ruffling. Cell 70:401-410.

Röderer G, Doenges KH (1983) Influence of trimethyl lead and inorganic lead on the *in vitro* assembly of microtubules from mammalian brain. Neurotoxicology 4:171-180.

Sager PR (1988) Selectivity of methy mercury effects on cytoskeleton and mitotic progression in cultured cells. Toxicol. Appl. Pharmacol. 94:473-486.

Särndahl E, Lindroth M, Bengtsson T, Fällman M, Gustavsson J, Stendahl O, Andersson T (1989) Association of ligand-receptor complexes with actin filaments in human neutrophils: a possible regulatory role for a G-protein. J. Cell Biol. 109:2791-2799.

Scapigliati G, Rappuoli R, Silvestri S, Pallini V (1988) Cytoskeletal alterations as parameter for assessment of toxicity. Xenobiotica 18:715-724.

Schering B, Bärmann M, Chhatwal GS, Geipel U, Aktories K (1988) ADP-ribosylation of skeletal muscle and non-muscle actin by *Clostridium perfringens* iota toxin. Eur.J.Biochem. 171:225-229.

Schmidt RA, Schneider CJ, Glomset JA (1984) Evidence for post-translational incorporation of a product of mevalonic acid into Swiss 3T3 cell proteins. J. Biol. Chem. 259:10175-10180.

Sham RL, Packman CH, Abboud CN, Lichtman MA (1991) Signal transduction and the regulation of actin conformation during myeloid maturation: studies in HL 60 cells. Blood 77:363-370.

Shariff A, Luna EJ (1992) Diacylglycerol-stimulated formation of actin nucleation sites at plasma membranes. Science 256:245-247.

Stossel TP (1988) The mechanical responses of white blood cells. Gallin JI, Goldstein IM, Snyderman R (eds) Inflammation: Basic principles and clinical correlates. Raven Press, Ltd., New York, pp. 325-342.

Vaziri C, Downes CP (1992) Association of a receptor and G-protein-regulated phospholipase C with the cytoskeleton. J. Biol. Chem. 267, 22973-22981.

Vogel DG, Margolis RL, Mottet NK (1985) The effects of methyl mercury binding to microtubules. Toxicol. Appl. Pharmacol. 80:473-486.

Yamane H, Farnsworth CC, Xie H, Evans T, Howald W, Gelb MH, Glomset A, Clarke S, Fung BK-K (1991) Membrane-binding of the small G protein G25K contains an S-(all-trans-geranylgeranyl)cysteine methyl ester at its carboxyl terminus. Proc. Natl. Acad. Sci. USA 88:286-290.

Yoshida Y, Kawata M, Katayama M, Horiuchi H, Kita Y, Yakai Y (1991) A geranylgeranyltransferase for rhoA p21 distinct from the farnesyltransferase for ras p21. Biochem. Biophys. Res. Commun. 175:720-728.

Zimmermann A, Gehr P, Keller HU (1988) Diacylglycerol-induced shape changes, movements and altered F-actin distribution in human neutrophils. J. Cell Sci. 90:657-666.

Zimmermann H-P, Doenges KH, Röderer G (1985) Interaction of triethyl lead chloride with microtubules *in vitro* and in mammalian cells. Expt Cell Res. 156:140-152.

Zimmermann H-P, Plagens U, Traub P (1987) Influence of triethyl lead on neurofilaments *in vivo* and *in vitro*. NeuroToxicology 8:569-578.

Cell Death by Apoptosis: Morphological, Genetic and Biochemical Features and Toxicological Implications

Giuseppina Palladini[1], *Filippo Taddei*[1] *and Giorgio Bellomo*[2]

[1]Department of Internal Medicine and Medical Therapeutics, First Medical Clinic, University of Pavia, Policlinico S.Matteo, I-27100 PAVIA and [2]Department of Medical Sciences, University of Torino, Via Solaroli 17, I-28100 NOVARA, Italy

Apoptosis is a type of cell death that plays an important role in early development and growth of normal adult tissue. It is regulated by physiological stimuli and is present in many species and tissues.Originally described as a "shrinkage necrosis" (Kerr, 1971) apoptosis is characterized by a marked reduction in cell volume and an increase in buoyant density (Ohyama et al., 1981 and Wyllie et al., 1982). Apoptotic bodies are characterized by their small size, dense cromatin, nuclear fragmentation, randomly assorted organelles in the cytoplasm, loss of characteristic membrane architecture and appareance of blebbing (Wyllie, 1980 et al.; Kyprianou et al., 1989). Significant alterations in membrane composition take place. The best-defined biochemical event in apoptosis involves nuclear DNA and the internucleosomal DNA fragmentation, probably caused by the activation of one or more endonucleases or DNA fragmenting enzymes (Wyllie, 1980b ; Arends et al., 1990). Apoptosis is associated with death of isolated cells, rather than contiguous patches or areas of tissue; there is no inflammatory infiltrate; nuclear shrinkage occurs relatively early, but changes to the organelles and loss of membrane integrity are relatively late; the dying cells are phagocytosed by neighbouring cells, rather than immigrant professional phagocyte, the DNA is rapidly broken down (Kerr et al., 1972 ; Wyllie et al., 1980).

Necrosis, on the other hand, affects many adjoining cells (Searl et al., 1982). It is characterized by cell swelling, with early loss of plasma-membrane integrity and major changes to the organelles, and the nucleus tends to swell. Necrosis is accompanied by an inflammatory infiltrate of phagocytic cells. DNA degradation, if it occurs, is a late event (Dive et al., 1992). Necrosis also differs morphologically from apoptosis, since, at least in

NATO ASI Series, Vol. H 93
Modulation of Cellular Responses in Toxicity
Edited by C. L. Galli, A. M. Goldberg, M. Marinovich

early stages, there is dissolution of organized cytoplasmatic structures, while the nucleus remains intact.

Kerr et al. (1972) observed that the ultrastructure of most mammalian cells undergoing Programmed Cell Death (PCD) changed similarly, a process they termed apoptosis. They then suggested that perhaps all PCD occurred via apoptosis. This supposition has become so widely accepted that many authors use these terms interchangeably. The term "programmed cell death" is commonly used synonymously with "apoptosis" but PCD is related to but non synonymous with apoptosis and both are clearly distinguishable from necrotic cell death. Apoptosis is a descriptive term, whereas PCD implies that decision to die was made cell-autonomously, indipendently of any other cells. PCD may be considered as a two-phase process. In the first phase, cell commits to a variety of stimuli. Cells in this stage may be considered to be in a "latent" phase of PCD. Ultimately, however, the cells enter the "active" phase of PCD in which they undergo a dramatic series of morphological and phyisiological changes that culminate in death. This active phase of cell death has been identified as a characteristic morphologic evolution of dying cells, termed apoptosis (Kerr et al., 1972). Alternatively apoptosis and PCD could be considered separate phenomena, PCD being the unique active cell suicide process. In this case apoptosis become an alternative cell death pathway to necrosis, not involved in ischemic or toxic injury (necrosis) but characteristic of cell aging and turnover in tissues.

Among the model systems employed to study PCD are: (i) the death of T-cells during negative selection in the mouse thymus (Jenkinson et al., 1989 ; Mercep et al., 1989) and (ii) the loss of intersegmental muscle (ISMs) of the moth *Manduca sexta* at the end of methamorphosis (Finlayson, 1956; Lockshin et al., 1965)

Schwartz et al. (1993) compared the patterns of cell death displayed by T-cells and the intersegmental muscle and found that they differed in terms of cell-surface morphology, nuclear ultrastructure, DNA fragmentation, and polyubiquitin gene expression. Unlike the T-cells, which are known to die via apoptosis, they found that the intersegmental muscle display few of the features that characterize apoptosis, such as membrane blebbing chromatin condensation and DNA fragmentation (Cohen, 1991). Additionally, although the death of both T-cells and the ISMs require de novo gene expression (Wyllie et al., 1984 and Schwartz et al., 1990) only the ISMs exhibited dramatic increases in the expression of the polyubiqutin gene. Two possible explanations for these results have been postulated. (i) Apoptosis could include a variety of different steps, and the T cells and ISMs may represent different ends of this spectrum. (ii) ISMs do not actually undergo apoptosis but, rather, die by a different molecular mechanism. Taken together, these data suggest that some cells may use a PCD mechanism that is distinct from apoptosis.

CHARACTERISTIC	PCD	APOPTOSIS	NECROSIS
Morphology	cell condensation, fragmentation	cell condensation, fragmentation	often cell swelling, lysis
Membrane barrier	persists	persists	early failure
Chromatin	condensed, electron-dense	margination	pycnosis
Protein synthesis	death blocked by actinomycin D, cycloheximide	death sometimes blocked by actinomycin D, cycloheximide	not affected by protein synthesis inhibitors
Paradigmatic cells or organs	interdigital tissue, intersegmental muscles, labial gland	thymocyte + glucocorticoids, postcastration prostate	liver + hepatotoxins
Alterations in DNA structure	no laddering of DNA	internucleosomal cleavage	diffuse DNA degradation
First manifestation	decreased protein synthesis	endonuclease activation	blebbing or swelling

Table I: Characteristics of Programmed Cell Death (PCD), Apoptosis and Necrosis

Genetic aspects of apoptosis

It is now widely accepted that apoptosis is a gene-directed process and can be seen as part of the repertoire available to the cell to respond to external (presence or absence of soluble signalling molecules, interaction with other cells, interaction with substrate) and internal (genotype, cell lineage, development stage, cell cycle stage, metabolic state, DNA damage) stimuli (Williams et al.,1990; Williams et al.,1992; Raff, 1992; Mapara et al, 1993) Until recently, most of the information available on the genetics of apoptosis is derived from studies on the nematode *Caenorhabditis elegans* (Ellis et al., 1991), but important information about the intracellular molecular signals involved in stimulation and suppression of apoptosis in mammalian cell is recently emerging.

The genetics of apoptosis can be over-semplified by postulating the existence of two categories of genes: (i) genes that encode for the death machinery such as the tumor suppressor gene *p53*, whose overexpression in a variety of cultured cell systems could induce apoptosis (Yonish-Rouach et al., 1991), the proto-oncogene *c-myc* in mammalian cells (Evan et al., 1992) and *ced-3* and *ced-4* genes in *Caenorhabditis elegans* (Hedgecock et al., 1983; Yuan et al., 1990). (ii) genes whose products prevent the expression of genes encoding for death machinery or inhibit the activity of pre-existing death machinery such as *bcl-2* and *ced-9* genes (Nunez et al., 1990; Borzillo et al., 1992).

The *p-53* tumor suppressor gene is the most widely mutated gene in human tumorigenesis (Hollstein et al., 1991; Levine et al., 1991). *p53* encodes a trascriptional activator (Fields et al, 1990) whose target may include genes that regulate genomic stability (Livingstone, 1992; Yin et al. 1992), cellular responses to DNA damage (Kastan, 1992) and cell-cycle progression (Michalovitz et al, 1990). Introduction of wild-type *p53* into cell-lines that have lost endogenous *p53* function can cause growth arrest (Mercer, 1990) or induce apoptosis (Yonish-Rouach et al.1991; Shaw et al., 1992). Wild-type *p53* induces apoptosis in a myeloid cell line and clear-cut results from *p53* gene knockout mice (Clarke et al., 1993) indicate that DNA damaging agents such as ionizing radiations require *p53* to induce apoptosis in thymocytes, while glucocorticoids do not. In thymocytes and probably in some other cell types, *p53* is an essential component of the pathway leading from DNA damage to apoptosis, although in other cells it can lead to growth arrest and DNA repair. Since *p53* has been implicated in controlling a checkpoint during the G1 phase of the cell cycle that may monitor the state of the DNA before entry into S phase (Kastan, 1992; Lane, 1992), the different cellular responses (apoptosis versus G1 arrest) may result from the activation of distinct target genes by *p53*. Alternatively, activation of the same target genes in the two cell types could have different consequences.The fact that elevated levels of *p53* can lead to the initiation of apoptosis is consistent with earlier studies that demonstrated a link between *p53* expression and cell death (Crabtree, 1989; Smith et al, 1989).

The *C-myc* oncogene has been implicated in the control of normal cell proliferation by many studies (Heikkila et al., 1987; Loke et al., 1988; Prochownik et al., 1988). In particular it is one of the immediate early growth response genes that are rapidly induced in quiescient cells upon mitogenic induction (Eilers et al., 1991), suggesting that it could plays a role in mediating the transition from quiescence to proliferation (Blackwood et al., 1991).

In tumor cells, elevated or deregulated expression of *c-myc* is so widespread to suggest a critical role in carcinogenesis (Fields et al.,1990b; Spencer et al,1991). Although it is unclear wether is deregulation or overexpression of *c-myc* that makes up the major determinant in *c-myc* oncogene activation, it is nonetheless evident that *c-myc* activation disrupts the growth regulation machinery of the cell. Since *c-myc* can induce both proliferation and apoptosis, the cellular decision between these two responses is determined by other signals such as the presence of growth factors or other survival stimuli. In fact its presence under conditions of growth arrest, such as growth factor deprivation, or its activation in cells already arrested, can induce apoptosis (Evan et al., 1992). Almost all tested tumor cells possess a deregulated *c-myc* gene and this may explain why they undergo apoptosis so readily in the presence of cytotoxic and growth-inhibiting drugs (Cotter et al, 1990; Lennon et al, 1990). Successful proliferation in normal cells requires the active suppression of programmed cell death, thereby providing an inbuilt failsafe to guard against uncontrolled proliferation and so allowing the same basic machinery to regulate the two necessarily linked processes, proliferation and apoptosis.

In conclusion the effects of *p53* and *c-myc* are cell type- and stimulus-specific and are not needed for all forms of apoptosis induction.

In the nematode *Caenorhabditis elegans* over 100 of the 1090 somatic cells formed during development die in a process resembling apoptosis (Hedgecock et al., 1983). In most cells the dying process is regulated by *ced-3* and *ced-4* genes: they probably act cell-autonomously to promote cell death. (Yuan et al., 1990). Apoptosis is not detectable in mutants lacking *ced-3* or *ced-4* and they have extra cells (Ellis et al., 1986). The action of the *ced-3* and *ced-4* genes is antagonized by expression of *ced-9* probably since cell death also fails to occur in activating mutants of *ced-9* (Hengartner et al. , 1992). The *ced-9* gene is needed in these cells to prevent *ced-3* and *ced-4* killing activity, and all the defects due to *ced-9* mutations are completely suppressed by mutations in *ced-3* or *ced-4* (Hengartner et al., 1992).

In mammals, the genes encoding for cell death machinery have not been identified and characterized in detail. The *bcl-2* gene encoding for an intracellular membrane protein, seems to exert a series of effects comparable to those of *ced-9*. In fact expression of *bcl-2* can prevent apoptosis when certain growth factors are withdrawn (Nunez et al, 1990; Borzillo et al., 1992) or when cell death is mediated by *c-myc* expression (Bissonette, 1992; Fanidi, 1992) . Important evidences (Alnemri et al., 1992) suggest that high levels of *bcl-2* enhance cell survival under conditions of repressed *c-myc* expression, probably by mobilisation of Ca^{2+} from mitochondria to cytoplasm. This in turn can activate protein kinase C and help in survival (McConkey et al., 1990). At least one mechanism apparently responsible for the high levels of *bcl-2* protein involves loss of trascriptional repression of the *bcl-2* gene by the tumor suppressor *p53*, since *p53* normally down-*regulates bcl-2* gene expression. However *bcl-2* expression does not protect target cells against DNA fragmentation and cell killing by cytotoxic T lymphocytes (Vaux et al., 1992), comparably to *ced-9* expression that does not protect against apoptosis of male-linker cell in *C. elegans* (Ellis et al., 1986), a type of cell death that does not require the presence of *ced-3* and *ced-4* (in fact it occurs even when *ced-3* and *ced-4* are mutated). If *bcl-2* acts like *ced-9*, it may function by inhibiting the putative mammalian homologues (probably the interleukin-1β converting enzyme ICE, which is a cysteine protease) of *ced-3* and *ced-4*. Therefore, to cause rapid cell death, steroids, azide, colchicine and other agents may activate the *ced-3* and *ced-4* homologues in addition to carrying out their immediate biochemical action (Sentman et al.,1991; Strasser et al.,1991; Alnemri et al., 1992). It is not known, however, wheter the activation of any *ced-3* and *ced-4* homologues is at transcriptional, translational or postranslational levels.

The physiological cell-death machinery that can be regulated by *bcl-2* may constitute a "default" pathway that can be activated in many different ways:

(i)Some cell types may not possess the cell death machinery and could only die by an alternative physiological cell death pathway or by necrosis.

(ii)Other cells may possess the machinery but it may remain inactive unless triggerd in some way. Inhibitors of RNA and protein synthesis inhibit apoptosis exclusively if the cell death machinery has to be synthesized ex novo (Schwartz et al., 1990).

(iii) Some other cells may have activated cell death machinery but keep it in check with *bcl-2*.

In conclusion, if a major default pathway for cells is to activate their cell-death machinery, then anything that threatens cell viability may lead to apoptosi.

Biochemical characteristics and role of Ca^{++}

The main morphological features of apoptosis are nuclear fragmentation and double-strand cleavage of DNA at linker regions between nucleosomes (Arends et al., 1991). Those180-200 base pair fragments obtained are readily shown by agarose gel electrophoresis of DNA and appear as a typical ladder pattern whereas in necrosis the DNA breakdown is random and it is seen as a smear after electrophoresis. The endonuclease(s) responsible for DNA fragmentation has (have) not been conclusively identified. To date, several studies have reported partial purification and characterization of apoptosis-linked nucleases. Most have a Ca^{2+} and Mg^{2+} requirement (Arends et al., 1990), others seem to have mainly Ca^{2+} (Gaido et al., 1991) or Mg^{2+} sensitivity (Kawabata et al., 1993). DNAse I (Peltsch et al., 1993) and acid nucleases such as DNAse II (Barry et al., 1993) have also been implicated. These studies supported the idea that Ca^{2+} mobilization could be a key event in the activation of the death program and eventually internucleosomal cleavage.

A sustained intracellular Ca^{2+} overload can trigger several lethal processes. Mytochondrial damage and ATP depletion, activation of proteases, activation of phospolipases, disruption of cytoskeleton structure and organization can all be caused by intracellualr Ca^{2+} overload and can rapidly lead to necrosis (Nicotera et al, 1992). Disturbances in Ca^{2+} signalling can also trigger apoptosis. Activation of Ca^{2+} dependent endonuclease activity, modification of chromatin conformation (Arends et al, 1990) and its susceptibility to cleavage by the former or other nucleases, alteration in gene expression can lead to apoptosis. The findings that removal of extracellular Ca^{2+} or buffering intracellular Ca^{2+} increases prevented DNA degradation and apoptotic body formation (Nicotera et al, 1992), that transfection of WEHI7.2 thymoma cells with calbindin (a Ca^{2+} binding protein) prevented apoptosis induced by glucocorticoids, cAMP and A23187 (Dowd et al., 1992) demonstrate the Ca^{2+} requirement in apoptosis . Although intracellular Ca^{2+} overload seems to trigger endonuclease activation in many systems, other observations suggest that the signalling involved in the induction of apoptosis is more complex and that additional mechanisms are involved. Thus, in some experimental models, apoptosis can occur also in the absence of detectable increases in intracellular Ca^{2+}. Brune et al (1991) have shown that agents known to modify chromatin structure (polyamines, namely spermine) can prevent DNA fragmentation and apoptosis in thymocytes exposed to glucocorticoids, Ca^{2+} ionophores, or organotin compounds. The mechanism responsible for the inibition of DNA fragmentation appears to be related to the ability of the polyamines to modify the chromatin structure. Conversely, agents which cause chromatin unfolding seem to stimulate DNA fragmentation and polyamine-depleted cells appear to be more susceptible to the onset of apoptosis. This suggests that endonuclease-mediated internucleosomal cleveage is secondary to chromatin decondensation or local reduction in histone-DNA interaction. Thus, changes in chromatin structure may play a determinant role in triggering apoptosis. Moreover, according to the working hypothesis outlined above, modifications of ion homeostasis may also favour those changes in chromatin conformation required for gene activation. A role for Ca^{2+} and Ca^{2+}-dependant proteins in the regulation of gene expression has been known from long time, in fact, Ca^{2+} can directly regulate the expression of certain genes (Morgan et al., 1986) and control the expression of other genes during cell activation (Crabtree, 1989).

Physiological and chemical stimuli can induce apoptosis.

Differents works on apoptosis features have shown its wide distribution in species and tissues, as well as the existence of multiple effectors.

Stimuli	Drugs
ATP	Amsacrine
Actinomycin D	Aphidicolin
A23187 $Ca^{2+}Mg^{2+}$ionophore	1-β-D-Arabinofuranosylcystonine
Cytochalasin B	BCNU
Calcium	Camptothecin
Cycloheximide	Cisplatin
Anti-CD3/T-cell receptor antibo	Etoposide
Epipodophyllo-toxins	5-Fluorodeoxyuridine
Gliotoxin	5-Fluorouracil
Glucocorticoids	Methotrexate
Hyperthermia	Melphalan
Irradiation (soft beta or gamm	Teniposide
Lymphotoxin	Vincristine
RU486	
TCDD	
Tumor necrosis factor	
TGF-β1	

Table II: Example of stimuli and drugs reported to induce apoptosis

Originally it was thought that in cells there may be a gene or set of genes that may be programmed to be expressed spontaneously at a given time resulting in the synthesis of lethal products. Neverthless, existing information shows that many exogenous stimuli can induce apoptosis. This concept implies the presence of target cells and appropriate levels of specific stimuli. In fact a wide spectrum of physiological or chemiotherapic agents have been shown to induce apoptosis in cells, particularly in thymocytes (Arrends et al., 1991; Lennon et al., 1991; Smith et al., 1989; Wyllie et al., 1980). There must be cell-specific and stimulus specific pathways, as well as a final common pathway of apoptosis. Some of these agents are physiologycal such as glucocorticoids, lymphotoxin, or TNF-α while others are strictly non-physiological such as γ-irradiation and certain xenobiotics.

Several independent observations have supported a critical role for calcium ion in thymocyte apoptosis. In glucocorticoid-treated thymocytes an early and sustained increase in cytosolic Ca^{2+} concentration occurs before DNA fragmentation or loss of cell viability (McConkey et al, 1989). Intracellular Ca^{2+} buffering or chelation of extracellular Ca^{2+} blocks both endonuclease activation and cell death (Nicotera et al, 1992). In addition, the finding that DNA fragmentation was sensitive to inhibition by antagonist of calmodulin provides further evidence for the involvement of Ca^{2+}-dependent processes in apopoptosis (Mc Conkey et al., 1989). A dose response relantionship between the cytosolic Ca^{2+} level, and DNA fragmentation was observed in thymocytes treated with Ca^{2+} ionophore (McConkey et al., 1989b) suggesting that the cytosolic Ca^{2+} increase may be rate-limiting for endonuclease activation. Ca^{2+} ionophore induced identical Ca^{2+} increases in immature thymocites and mature T cells (McConkey et al., 1989c) . Triggering of comparable biochemical processes has also been demonstrated following incubation of thymocytes with ant-CD3 antibodies.

Agents that elevate cAMP stimulate DNA fragmentation in thymocytes (McConkey et al., 1990). Increases in the cAMP level are often inhibitory in mature T lymphocytes and may be involved in the development of tolerance to self Ag (Jenkins et al., 1987). McConkey (1990b) showed that agents inducing an increase in the cAMP level by independent mechanisms were able to stimulate DNA fragmentation typical of apoptosis. Data obtained with cAMP analogs known to act synergistically to stimulate PKA suggested that the latter directly mediated endonuclease activation. Agents previously shown to stimulate PKC and to inhibit Ca^{2+}-dependent, T Cell Receptor-mediated thymocyte apoptosis, including IL-1, blocked both DNA fragmentation and cell death in response to cAMP, suggesting interactions (cross-talk) between the two protein kinase systems (McConkey et al., 1990).

Prostaglandins E (PGEs) are macrophages and dendritic cell products that are potent inhibitors of T-cell activation and their inhibitory action has been linked to accumalation of cAMP (Kammer, 1988). Treatment of rat thymocytes with PGE2 or pharmacological agents that directly raise cAMP levels resulted in rapid DNA fragmentation (McConkey et al, 1990).

Several anticancer drugs induce apoptosis in different cancer cells and in thymocytes (Dyson et al., 1986; Kaufmann, 1989; Barry et al., 1990; Walker et al., 1991). This phenomenon has never been reported in cells taken from animal tumors or human biopsies, however the lack of in vivo data doesn't reduce the potential relevance of apoptosis for the anti-tumoral activity of these drugs, since the cells undergoing apoptosis are probably rapidly destroyed by phagocytic cells and their very short "half-life" make difficult their detection. Among the long list of drugs reported to induce apoptosis are:

(i) DNA-damaging agents (BCNU, melphalan and cis-platinum) wich cause several types of DNA damage, such as DNA interstrand breaks, intrastand crosslinks and DNA-protein crosslinks.

(ii) Drugs like camptothecin that cause protein-associated DNA strand breaks mediated by the enzime DNA topoisomerase I

(iii) The epipodophyllotoxins and intercalating agents that are poison of DNA topoisomerase II

(iv) Inhibitors of mitotic spindle apparatus, such as the vinca alkaloids

(v) Inhibitors of DNA synthesis , such as aphidicolin

(vi) Several antimetabolites

The fact that apoptosis can be induced by perturbing cell biochemistry with antineoplastic drugs suggest that all these drugs cause a block of macromolecule synthesis and of the progression of the cell cycle which is reported to occur before the citotoxicity is evident. It seems, therefore that the activation of apoptosis is mediated by a specific block in the cell cycle, probably related to a general metabolic damage of cell structures. This phenomenon could be relevant for the susceptibility of the tumors to chemiotherapeutic agents even if apoptosis is phenotypically determined and can be induced only in some cell types. In other terms, it is possible that the drugs can induce or enhance the biochemical pathways leading to apoptotic death only in the predisposed or primed cells. The drugs would probably act by enhancing stimuli wich normally activate the suicide programme.

Role of apoptosis

The general role of apoptosis, though has not been studied extensively, appears to be crucial in development and growth regulation. It is clear that derangement in its regulation in development could result in structural and functional abnormalities, whereas lack of its tight regulation in growth could result in tumors. Emerging evidences show that cell proliferation and apoptosis may be subject to coordinated but inverse regulation. In fact the two main functions of apoptosis are in development and growth regulation. In development apoptosis leads to a programmed deletion of cell populations while in adult tissue the growth modulation results from the balance cell proliferation minus apoptosis. An interesting investigation (Rotello et al., 1991) using rabbit uterine epithelium has shown that cell proliferation and apoptosis are regulated in an inverse and coordinated fashion by progesterone in animals and by trasforming growth factor β-1 as well as serum in primary cultures. It has also been shown recently that Epstein-Barr virus latent protein activates B cell proliferation and decreases apoptosis (Gregory et al, 1991). The effect appears to involve *bcl-2*.

The products of the tumor suppressor gene *p53* have been described to be involved in arresting growth by increasing apoptotic activity. These data suggest a potential application of *p53* gene therapy for the treatment of human malignancies characterized by alterations in this important cell regulatory element. Thus, the expression of anti-apoptotic genes such as *bcl-2* can lead to chemioresistance in tumor cells. One therapeutic strategy to overcome this resistance is the use of antisense oligodeoxynucleotides to suppress translation of these gene.

The rate of apoptosis in a population of cells is determined by a balance between the activities of cell death effectors and endogenous cell death inhibitors. Recent advances in understanding the molecular biology and biochemistry of these effectors and inhibitors offer the perspective of discovering compounds that could modulate the apoptotic pathway. Therapeutic compounds that alter the biochemical balance in favour of the cell death effectors offer a potential strategy for the use in several types of cancer, including lymphoma and prostate cancer.

Cell death can be a physiological process in cell number regulation in tissues or it can be the result of exogenous or endogenous injuries. However, intriguing hypotheses were recently raised about the role of injured/dying cells in neoplasia (Wenner et al.,, 1985) and human immunodeficiency virus infection (Capron et al., 1991). In fact, it was suggested that malignancy can be related to a defective apoptosis (or PCD) in which some repair mechanisms to restore the cell cycle can trigger neoplastic proliferation (Wenner et al., 1985). Furthermore, recent insights into genetic involvement in some cell death processes have been provided by some investigators. In humans, high levels of *bcl-2* expression have been associated with impaired apoptosis, leading to inhibition of cell loss (Williams, 1991), thus active apoptosis could provide an additional means of regulating cell number, and unlike simple degeneration (necrosis) it could also depend on some intracellular components that can be exogenously activated. Hence, aberrant cell survival due to inhibition of cell death can effectively contribute to oncogenesis and, conversely, diverse anticancer drugs that induce apoptosis can be used to specifically antagonize the process (Dive et al., 1991). Thus, drug-induced tumor regression could result from a simple triggering in the commitment to apoptosis. Finally a role for apoptotic cell death in HIV infection has also been hypothesized, since viral particles appear able to induce apoptotic cell death in CD4 antigen-bearing lymphocytes (Capron et al., 1991).

Accumulating evidences have been obtained to support the involvement of apoptosis in several physiological and pathological processes. For instance, apoptosis is known to be the mechanism involved in the selective deletion of thymocytes during physiological regression of the thymic gland (McConkey et al, 1989b). Moreover, apoptosis seems to play a role in morphogenesis (Tata,1966), in the ageing process of blood cells in the bone marrow (Radley et al, 1982), in the cyclic involution of some tissues such as endometrium (Linch et al, 1986), and in regressive phenomena occuring in hyperplastic organs (Columbano et al., 1985). In vitro evidence has also been obtained to show that T-lymphocytes activate a DNA deavage process in target cells which result in an endogenous cell suicide, supporting the involvement of apoptosis in cell-mediated cytotoxicity (Ucker, 1987). Other data support the ability of some xenobiotics to kill target cells by apoptosis. In fact, as demostrated by McConkey et al (1988), dioxins cause thymocyte death by Ca^{2+}-mediated endonuclease activation.

References

Alnemri E. S., Fernandes T. F., Haldar S., Croce C., and Litwack G. (1992) Cancer Res. 52 : 491-495
Arends M.J., Morris R.G. and Wyllie A. H.. (1990) Amer. J. Pathol 136: 593-608
Arends M.J. and Wyllie A.H. (1991) Int. Rev. Exper.Pathol. 32 : 223-254
Barry M.A., Behnke C.A., and Eastman A. (1990) Biochem. Pharmacol. 40 : 2353-2362
Barry M.A. and Eastman A. (1993) Arch. Biochem. Biophys. 300 : 440-445
Bissonette R.P., Echeverri F., Mahboubi A. and Green D.R. (1992) Nature 359: 552-554
Borzillo G.V., Endo K. and Tsujimoto Y. (1992) Oncogene 7 : 869-876
Blackwood E.M., and Eisenman R.N. (1991) Science 251: 1211-1217
Brune B., Hartzell P., Nicotera P., and Orrenius S. (1991) Exp. Cell. Res. 195: 323-329
Capron A. and Ameisen J.C. (1991) Immunol. Today 4 : 102-105

Clarke A. R., Purdie C. A., Harrison D. J., Morris R. G., Bird C.C., Hooper M.L., and Wyllie A. H. (1993) Nature 362 :849-852
Cohen J.J. (1991) Adv. Immunol. 50 : 55-85
Columbano A., Ledda-Columbano G.M., Coni P.P., Faa G., Liguor C., Santa Cruz G. and Pani P. (1985) Laboratory Investigations 52 : 670-675
Cotter T.G., Lennon S.V., Glynn J.G. and Martin S.J. (1990). Anticancer Research 10 : 1153-1159.
Crabtree G.R. (1989) Science 243 : 335-362
Dyson J.E.D., Simmons D.M., Daniel J., McLaughlin J.M.,Quirke P., and Bird C.C. (1986) Cell Tissue Kinetics 19: 311-324
Dive C., Hickman J.A (1991) Br. J. Cancer 64: 192-196
Dive C., Gregory C.D., Phipps D.J., Evans D.L., Milner A.E. and Wyllie A.H. (1992) Biochim. Biophys. Acta 1133 : 275-285
Dowd D.R., Mac Donald P.N., Komm B.S., Haussler M.R and Miesfeld R.L. (1992) Mol. Endocrin. 6 : 1843-18487
Eilers M., Scirm S. and Bishop J.M. (1991) EMBO J. 10 : 133-141
Ellis H.M. and Horvitz H.R. (1986) Cell 44 : 817-829
Ellis R.E.Yuan J. and Horvitz,H.R. (1991) Annu.Rev.Cell.Biol. 7 : 663-698
Evan G.I., Wyllie A. H., Gilbert C.S, Littlewood T D, Land H., Brooks M., Waters C.M., Penn L.Z. and Hancock D.C. (1992) Cell 69 : 119-128.
Fanidi A., Harrington E.A. and Evan G.I. (1992) Nature 359 : 554-556
Fields J.K. and Jang S.K. (1990) Science 249 : 1046-1049
Fields J. K., and Spandidos D.A (1990b) Anticancer Reser. 10 : 1-22
Finlayson L.H. (1956) Q. J. Microsc. Sci. 97 : 215-234
Gaido M.L. and Gdlowski J.A. (1991) J. Biol. Chem. 266: 18580-18585
Gregory C.D., Dive C., Henderson S., Smith C.A. , Williams G.T., Gordon J., Rickinson A. B. (1991) Nature 349 : 612-614.
Hedgecock E.M., Sulston J.E. and Tomson J.N. (1983) Science 20 : 1277-1279
Heikkila R., Schwab G., Wickstrom E., Loke S.L., Pluznik D.H., Watt R. and Neckers L.M. (1987) Nature 328 : 445-449
Hengartner M.O., Ellis R.E., Horvitz H.R. (1992) Nature 356 : 494-499
Hollstein M., Sidransky D., Vogelstein B. and Harris C. (1991) Science 253 : 49-53
Jenkins M.J., Pardoll D.M., Mizuguchi J., Chused T.M. and Schwartz R.H. (1987) PNAS 84 : 5409-5413
Jenkinson E.J, Kingston R., Smith C.A., Williams G.T. and Owen J.J. (1989) Europ. J. Immunol. 19 : 2175-2177
Kammer G.M. (1988) Immunol. Today 9 : 222-226
Kastan M. (1992) Cell 71 : 587-597
Kaufmann S.H. (1989) Cancer Res. 49 : 5870-5878
Kawabata H., Anzai N., Masutani H., Hirami T., Yoshida Y. and Okuma M. (1993) Biochem. Biophys. Res. Commun. 191: 247-254
Kerr J.F.R. (1971) J. Pathol. 105 : 13-20.
Kerr J.F.R., Wyllie A.H. and Currie A.R. (1972) Br. J. Cancer 26 : 239-257
Kyprianou N. and Isaacs J.T. (1989) Biochem. Biophys. Res. Commun. 165 : 73-81
Lane D.P. (1992) Nature 358 : 15-16
Lennon S.V., Martin S.J. and Cotter T.G. (1990) Biochem.Soc.Trans. 18 : 343-345.
Levine A. J., Momand J. and Finlay C. (1991) Nature 352 : 453-456
Lennon S.V., Martin S.J. and Cotter T.G. (1991) Cell Prolif. 24 : 203-214
Linch M.P., Nawaz S. and Gerschenson L.E. (1986) PNAS 83 : 4784-4788
Livingstone L.R. (1992) Cell 70 : 923-935
Lockshin R.A. and William C.M. (1965) J. Insect. Physiol. 11 :123-133

Loke S.L., Stein C., Zhang X., Avigan M., Cohen J. and Neckers L.M. (1988) Curr. Top. Microbiol. Immunol. 141 : 282-289

Mapara M.Y., Bargou R., Zugck C., Dohner H., Ustaoglu F., Jonker R.R, Kramer P.H. and Dorken B. (1993) Eur.J. Immunol. 23 : 702-708

McConkey D.J., Hartzell P, Duddy S.K., Hakansson H., and Orrenius S. (1988) Science 242 : 256-259.

McConkey D.J., Nicotera P., Hartzell P., Bellomo G., Wyllie A.H. and Orrenius S. (1989) Arch. Biochem. Biophysic. 269 : 363-367

McConkey D J, Hartzell P., Nicotera P. and Orrenius S. (1989b) FASEB J. 3 : 1843

McConkey D.J., Hartzell P., Amador-Perez J.F., Orrenius S and Jondal M. (1989c) J. Immunol. 143: 1801-1803

McConkey D.J., Hartzell P., Chow S.C., Orrenius S. and Jondal M. (1990) J. Biol. Chem. 265 : 3009-3011

McConkey D.J., Orrenius S., and Jondal M. (1990b) The J. of Immunology 145 : 1227-1230

Mercep M., Weissman A.M., Frank S.J., Klausner R.D. and Ashwell J.D. (1989) Science 246 :1162-1165

Mercer W.E. (1990) PNAS 87 : 6166-6170

Michalovitz D., Halevy O. and Oren M. (1990) Cell 62 : 671-680

Morgan J.I. and Curran T. (1986) Nature 322 : 552-555

Nicotera P., Bellomo G. and Orrenius S. (1992) Annu. Rev. Pharmacol. Toxicol. 32 : 449-470

Nunez, G., London L., Hockenbery D., Alexander M., McKearn J.P., and Korsmeyer S.J. (1990) J.Immunol. 144 : 3602-3610

Ohyama H., Yamada T. and Watanabe I. (1981) Radiation Res. 85 : 333-339

Peltsch M.C., Polzar B., Stephan H., Crompton T., Mac Donald H.R., Mannherz H.G. and Tschopp J. (1993) EMBO J 12: 371-377

Prochownik E.V., Kukowska J and Rodgers C. (1988) Mol. Cell. Biol. 8 : 3683-3695

Radley J.M and Haller J. (1982) British Journal of Haematology 53 : 277-287

Raff M.C. (1992) Nature 356 : 397-400

Rotello R.J., Lieberman R.C., Purchio A.F. and Gerschenson L.E. (1991) PNAS 88 : 3412-3415

Schwartz L.M., Kosz L., and Kay B.K. (1990) PNAS 87 : 6594-6598

Schwartz L.M., Smith S.W., Jones M.E.E. and Osborne B. (1993) PNAS 90 : 980-984

Searle J., Kerr. J.F. and Bishop C.J. (1982) Phatol. Annu. 2 : 229-259

Sentman C.L., Shutter J.R, Hockenbery D., Kanagawa O. and Korsmeyer S.J. (1991) Cell 67 : 879-888

Shaw P. (1992) PNAS 89 : 4495-4499

Smith C.A., Williams G.T. Kingston R., Jenkinson E.J. and Owen J.J.T. (1989) Nature 337 : 181-184

Spencer C.A. and Groundine M. (1991) Adv. Cancer Res. 56 : 1-48

Strasser A., Harris A.W. and Cory S. (1991) Cell 67 : 889-899.

Tata J.R. (1966) Developmental Biology 13 : 77-85

Ucker D.S. (1987) Nature 327 : 62-64

Vaux D.L., Aguila H.L. and Weissman I.L. (1992) Int. Immunol 4 : 821-824.

Yin T., Tainsky M.A., Bischoff F.Z., Strong L.C. and Whal G.M. (1992) Cell 70 : 937-948

Yonish-Rouach E., Resnitzky D. Lotem J. Sachs L. Kimchi A. and Oren M. (1991) Nature 352 : 345-347

Yuan J.Y. and Horvitz H.R. (1990) Dev. Biol. 138 : 33-41

Walker P.R., Smith C., Youdale T., Leblanc J., Whitfield J.F. and Sikorsa M. (1991) Cancer Res. 51 : 1078-1085

Wenner C.E., Lesiter K.J. and Tomei L.D. (1985). Molecular Basis of Cancer. P.B. Farmer and J.M. Walker eds, Vol 175A: Macromolecular Structure, Carcinogens, and Oncogenes. A.R. Liss New York

Williams G.T., Smith C.A., Spooncer E., Dexter T.M. and Taylor D.R. (1990) Nature 343 : 76-79

Williams G.T. (1991) Cell 65: 1097-1098

Williams G.T, Smith C.A., McCarthy N.J., and Grimes E.A. (1992) Trends Cell Biol. 2 : 263-267

Wyllie A.H., Kerr J.F.R. and Currie A.R. (1980) Int. Rev. Cytol. 68 : 251-306

Wyllie A.H. (1980b) Nature 284 : 555-556

Wyllie A.H. and Morris R.G. (1982) Am. J. Pathol. 109 : 78-87

Wyllie A.H. and Morris R.G. (1984) Am. J. Pathol 142 : 67-77

Wyllie A.H. (1985) Anticancer Research 5 : 131-136

FREE RADICAL MEDIATED TOXICITY OF DRUGS OF ABUSE

P.F. Mannaioni, M.G. Di Bello, S. Raspanti, L. Mugnai, V. Romano and E. Masini.
Department of Preclinical and Clinical Pharmacology, Florence University, Viale G.B. Morgagni 65, 50134 Florence, Italy.

Author for correspondence:
Pier Francesco Mannaioni
Department of Pharmacology
University of Florence
Viale G.B. Morgagni 65
50134 Florence
Italy

ABSTRACT

1. Isolated purified rat serosal mast cells were incubated with given concentrations of drugs of abuse (morphine, cocaine, methadone), with oxidative enzymes (prostaglandin-H-synthetase, PHS; rat liver homogenate fraction - S-10 mix), and with the drugs of abuse in the presence of oxidative enzymes.
2. The release of mast cell histamine and the generation of malonyl-dialdehyde (MDA) is present only when mast cells were incubated with the drugs of abuse in the presence of oxidative enzymes.
3. The release of histamine and the generation of MDA were abated by the free radical scavengers, reduced glutathione (GSH) and α-tocopherol.
4. It is suggested that morphine, cocaine and methadone are activated into free radicals producing membrane lipid perturbation and histamine release.
5. The metabolic activation of morphine, cocaine and methadone into free radicals could entail pathophysiological relevance in the organic injuries of drug addiction.

NATO ASI Series, Vol. H 93
Modulation of Cellular Responses in Toxicity
Edited by C. L. Galli, A. M. Goldberg, M. Marinovich

INTRODUCTION

Heroin and cocaine addiction have been associated with cardiac, renal, and hepatic pathologies in humans. However, interpretation of many clinical studies examining the toxicology of cocaine and narcotic drugs is difficult, due to the polydrug use, nutritional status, and disease state of the individuals. Some investigations concluded that narcotics and cocaine were directly hepatotoxic (MARKS & CHAPPLE, 1967) eventhough changes which would reflect organic injury could not be produced experimentally with morphine or cocaine administration to humans or monkeys (BROOKS et al., 1963). Attempts to explain this disparity led others to conclude that the concomitant diseases seen in drug addicts were caused by infected needles and syringes, and by contamination of paraphernalia.

It appears, however, that the observed toxicity might be due to biologically transformed products of the drugs of abuse rather than the initial, non metabolized, compound. In the case of morphine, in addition to the major metabolite, morphine-3-glicuronide, morphine has been shown to be metabolized to morphinone 7,8-epoxide and to 2-hydroxymorphine which can be metabolized further to 2,3-quinone, all of which have the potential to generate the highly reactive free oxygen radicals, and to bind to glutathione (GSH), a naturally occurring free radical scavenger (Mc CARNTEY, 1989). In the case of cocaine, serum pseudocholinesterase and liver esterases rapidly hydrolyze cocaine to give ecgonine methylester and benzoic acid. However, the minor oxydative route appears to be responsible for the hepatotoxicity of this drug. It is now established that cocaine can be N-oxidised to give norcocaine, which can be further metabolised, producing N-hydroxy-norcocaine and norcocaine nitroxide. This free radicals may be regarded as ultimately responsible for the hepatotoxicity elicited by cocaine (KLOSS et al., 1984). Therefore, the reactive electrophils formed indirectly from the metabolism of morphine and cocaine could subsequently cause cellular toxicity through enzyme inactivation, DNA damage and lipid peroxidation.

We have recently proposed a free radical bioassay, in which the drug under study is incubated with isolated purified rat serosal mast cells, in the presence and in the absence of oxidative enzymes. The end point of the reaction is the free radical-driven release of histamine and of lactate-dehydrogenase (LDH), the generation of markers of membrane lipoperoxidation and the inhibition of these effects by free radical scavengers. Using this bioassay, we have recently shown that paracetamol, anthranilic antibiotics, arachidonic acid and linoleic acid evoke the release of mast cell histamine only in the presence of oxidative enzymes. The release of histamine was coupled with the generation of malonyldialdehyde (MDA) and abated by GSH and by α–tocopherol, thus fulfilling the criteria of a free-radical driven event (MANNAIONI et al., 1988; MASINI et al., 1990). Here we report on the generation of free radicals from commonly abused drugs, using the same bioassay system.

METHODS

Mast cells of serosal phenotype were obtained by pleural and peritoneal lavage in Wistar rats and purified by elutriation. The elutriation was carried out at 5° C at the rotor speed of 2400 ± 10 rpm for 1 hr and at a flow rate of 15 ml/min, using a Beckman elutriation system (Rotor JE-6, chamber size 4.5 ml). A final yeld of 90-95% pure mast cells was achieved. Mast cell incubations were conducted in 10 ml test tubes at 37° C in a metabolic shaker for 30 min. The final composition of the incubation was 5 x 10^4 mast cells and 2 ml of the incubation medium (Tyrode solution, pH 7.4, gas phase air) containing a) given concentrations of morphine, cocaine and methadone; b) various amounts of oxidative enzymes; c) morphine, cocaine and methadone in the presence of oxidative enzymes. The reaction was stopped by chilling the tubes in an ice-water buffer. Cells were then separated from the medium by centrifugation (400 g x 5 min) and histamine and LDH were measured in the supernatants and in the pellets. Histamine was measured fluorimetrically using the method of SHORE et al. (1959) as modified by KREMZNER and WILSON (1961). Histamine release (supernatant histamine) was expressed as a percentage of the total present in the cells plus supernatants. Spontaneous histamine release ranged between 1 and 8% and was subtracted. Release of cytoplasmic LDH was assayed by measuring spectrophotometrically the LDH catalysed reduction of pyruvate to lactate in the presence of NADH, as described by BERGMEYER and BERNT (1974). The determination of lipid peroxidation in mast cell homogenate was based on the reaction of MDA, the end product of lipid peroxidation, with 2-thiobarbituric acid (TBA) to form a pink coloured substance, measured at 532 nm, according to the method of DAHLE et al. (1962). The oxidative enzymes were prostaglandin-H-synthetase (PHS) purified from calf seminal vesicles purchased fro MILES, and the S-10 mix liver homogenate fraction (S-10 mix). S-10 mix was prepared from liver of phenobarbital treated rats (75 mg/kg i.p. for 6 days) acording to GARNER et al. (1972). The protein content of the enzymatic preparation was adjusted to 7.5 mg/ml and the activity was 0.15 units/g protein. One unit of the enzymatic activity was defined as the enzyme that catalyses the formation of 1 μmol PGE_2 per min. All the chemicals were from Merck, suprapur quality. The paired t-test was used as the statistical analysis.

RESULTS AND DISCUSSION

As shown in Tables I-III, the separate incubation of mast cells with morphine, cocaine and methadone did not evoke any significant cell activation at the concentration studied. Separate incubation of mast cells with the oxidative enzymes were similarly uneffective. However, a significant release of histamine was observed when the drugs of abuse were incubated with mast cells in the presence of both the oxidative enzymes.

The release of histamine afforded by the drugs of abuse in the presence of oxidative enzymes is non cytotoxic in nature. In fact, the release of histamine was not accompained by a substantial leackage of LDH as a marker of unspecific cell disruption. Moreover, electron microscopic analysis of mast cells exposed to the drugs of abuse in the presence of oxidative enzymes (data not shown) were consistent with a process of sequential exocytosis and not with a feature of cytotoxic cell disruption.

Tables I-III also show that a relationship exists between the release of histamine and the generation of MDA from mast cells incubated with the drugs of abuse in the presence of oxidative enzymes.

	MORPHINE (10^{-5} M)	PHS (25 mU)	S10Mix (400 μg)	MORPHINE (10^{-5}) PLUS PHS (25 mU)	MORPHINE (10^{-5}) PLUS S10Mix (400μg)
Histamine release	0*	3.0*	6.1*	47.6	39.7
LDH release	2.6*	3.1*	2.9*	9.5	15.1
MDA	8.9*	3.1*	7.1*	25.0	31.1

TABLE I. Release of histamine (%), leackage of LDH (%) and generation of MDA (nmol/mg protein) in rat mast cells induced by morphine in the presence of oxidative enzymes.
*Values not significantly different from untreated controls.
The values are the means ± s.e.m of 6 experiments performed in duplicate.

	COCAINE (10^{-5} M)	PHS (25 mU)	S10 Mix (400 μg)	COCAINE (10^{-5}) PLUS PHS (25 mU)	COCAINE (10^{-5}) PLUS S10Mix (400μg)
Histamine release	5.0*	3.0*	6.1*	38.7	51.3
LDH release	2.1*	3.1*	2.9*	6.5	8.1
MDA	10.1*	3.1*	7.1*	56.4	61.1

TABLE II: Release of histamine (%), leackage of LDH (%) and generation of MDA (nmol/mg protein) in rat mast cells induced by cocaine in the presence of oxydative enzymes.
*Values not significantly different from untreated controls.
The values are the means ± s.e.m of 6 experiments performed in duplicate.

	METHADONE (10^{-5} M)	PHS (25 mU)	S10 Mix (400 μg)	METHADONE (10^{-5}) PLUS PHS (25 mU)	METHADONE (10^{-5}) PLUS S10Mix (400μg)
Histamine release	0*	3.0*	6.1*	22.2	34.2
LDH release	3.0*	3.1*	2.9*	7.9	13.9
MDA	9.6*	3.1*	7.1*	19.9	25.4

TABLE III: Release of histamine (%), leackage of LDH (%) and generation of MDA (nmol/mg protein) in rat mast cells induced by methadone in the presence of oxydative enzymes.
*Values not significantly different from untreated controls.
The values are the means ± s.e.m of 6 experiments performed in duplicate.

The effect of endogenous free radical scavengers on the release of histamine in the same circumstances is reported in Table IV, showing an almost complete protection afforded by both GSH and α–tocopherol.

	INHIBITION OF HISTAMINE RELEASE (%) BY GSH (10^{-4} M)	INHIBITION OF HISTAMINE RELEASE (%) BY α-TOCOPHEROL (10^{-4} M)
MORPHINE (10^{-5} M)	65.2	92.7
COCAINE (10^{-5} M)	76.4	72.2
METHADONE (10^{-5} M)	69.8	90.9

TABLE IV: Effect of reduced glutathione (GSH) and α-tocopherol on mast cell histamine release induced by drugs of abuse in the presence of PHS (25 mU).
The values are the means ± s.e.m of 6 experiments performed in duplicate.

Taken together, these observations suggest that morphine, cocaine and methadone undergo a metabolic activation leading to a perturbation of membrane lipids and to a parallel release of histamine. Conceivably, the metabolic activation of the drugs of abuse could be identified with the generation of reactive radicals species, which are abated by endogenous free radical scavengers. The same conclusions have been drawn from experiments in which mast cells have been exposed to paracetamol, anthranilic antibiotics, polyunsaturated fatty acids, in the absence and in the presence of oxidative enzymes (MANNAIONI and MASINI, 1988; MASINI et al., 1990).

Cocaine and morphine have been reported previously as capable of producing reactive electrophils (KLOSS, 1984; Mc CARNTEY, 1989). We now add that the reactive electrophils may induce an oxidative stress in isolated rat mast cells. The knowledge that the commonly abused drugs may be activated into free radicals may entail significant pathophysiological consequences.

The induction of liver mixed function oxidases and other oxidative enzymes may be a consistent feature in drug addicts, due to the continous self-administration of drugs, often associated with high ethanol intake. Conceivably, the induction of microsomal enzymes could switch the unharmful metabolic pathways (N-demethylation of morphine and methadone; hydrolysis of cocaine) to the more harmful metabolic pathway leading to the oxidative activation into reactive electrophils, and to the final oxidative stress. In this way, morphine could be activated into a free radical by the P-450 system, producing the suicide inhibition of P-450 enzymes. This could explain the progressive decline of morphine N-demethylation observed in rats chronically treated with morphine (AXELROD, 1968). The cocaine free radical could explain not only the disruption of liver cells (Mc CARTNEY, 1989), but also the cardiac toxicity of cocaine. In this context, the sudden death of drug addicts, not necessarily related to drug overdose, could be also linked to the release of histamine. It is possible the hypothesis that a sudden activation of residential mast cells could produce a burst of free radical driven histamine release, leading to hypotension, cardiac arrhythmias, bronchoconstriction and endothelial permeabilization.

Finally, the possible use free radical scavengers in the chemotherapy of drug addicts could be worth of a clinical trial.

REFERENCES

Axelrod J (1968) Cellular adaptation in the development of tolerance of drugs. In: The Addictive States, Wikler A (ed), The Williams & Wilkins Company, Baltimore, pp. 247-265.

Bergmeyer HV, Bernt E (1974) Lactate dehydrogenase UV-assay with pyruvate and NADH. In: Methods for Enzymatic Analysis, Bergmeyer HV (ed),. Verlag Chemie, Academic Press, New York, Vol. 2, pp. 574-579.

Brooks FP, Deneau GA, Potter HP, Reinhold JG, Norris RF (1963) Liver function tests in addicted and non-addicted rhesus monkeys. Gastroenterology 44:287-290.

Dahle LK, Hill EG, Holman RT (1962) The thiobarbituric acid reaction and the autoxidation of polyunsaturated fatty acid methyl esters. Arch Biochem 98:253-261.

Garner RC, Miller EC, Miller JA (1972) Liver microsomal metabolism of aflatoxin B to a reactive derived toxic to salmonella typhimurium Ta1530. Cancer Res 32:2058-2066.

Kloss MW, Rosen GM, Rauckman EJ (1984) Cocaine-mediated hepatotoxicity. Biochem Pharmacol 33:169-173.

Kremzner LT, Wilson IB (1961) A procedure for the determination of histamine. Biochem Biophys Acta 50:364-367.

Mannaioni PF, Masini E (1988) The release of histamine by free radicals. Free Radical Biol Med 5:177-197.

Marks V, Chapple PA (1967) Hepatic disfunction in heroin and cocaine addicts. Br J Addict 62:189-196.

Masini E, Palmerani B, Gambassi F, Pistelli A, Giannella E, Occupati B, Ciuffi M, Bani Sacchi T, Mannaioni PF (1990) Histamine release from rat mast cells induced by metabolic activation of polyunsaturated fatty acid into free radicals. Biochem Pharmacol 39:879-889.

McCartney M (1989) Effect of glutathione depletion on morphine toxicity in mice. Biochem Pharmacol 38:207-209.

Shore PA, Burkhalter A, Cohn VR (1959) A method for the fluorimetric assay of histamine in tissues. J Pharmacol 127:182-186.

ALTERNATIVES IN TESTING FOR HEPATOTOXICITY.

Fabienne Goethals and Marcel Roberfroid
Unité de Biochimie Toxicologique et Cancérologique
Departement des Sciences Pharmaceutiques
Université Catholique de Louvain
BCTC 7369
B-1200 Brussels

INTRODUCTION

Toxicology studies the properties of chemicals and their effects on living organisms with the aim to assess their potential adverse effects so as to help preserve and protect human health.

The science of toxicology characterizes the adverse effects (the biological reactivity) of a chemical by integrating all available scientific informations But, presently, the scientific toxicity data base consists mainly of the results of studies carried out in two or three species of animals. These studies are termed "safety evaluation test" with the questionable implication that a chemical can be judged safe for human if it produces negative results in a serie of "scientifically" conducted protocols, most exclusively in experimental animals.

Toxicologists need to ask the question whether these protocols, really, characterise the toxicity of a chemical in terms of identifying all the potential adverse effects for humans and domestic animals and establishing the relevant data base. Furthermore, they have to ask whether, and most importantly, how the science can be improved (Kroes and Hicks, 1990). In that respect it is disquieting to realize that toxicologists have still only a limited understanding of the mechanisms through which most chemicals exert their toxic effects, and as yet,

NATO ASI Series, Vol. H 93
Modulation of Cellular Responses in Toxicity
Edited by C. L. Galli, A. M. Goldberg, M. Marinovich

have only a limited capacity to predict the adverse effects of a given chemical in an intact animal particularly in human.

The question thus becomes whether toxicologists are using the correct models (Rodricks, 1986) and what they can do to improve these models (Roberfroid and Goethals, 1990; Balls et al, 1990). With regard to the so-called alternative models, the question needs to be asked : can they contribute to the progress in toxicology or are they simply useless ?

THE LIVER, A MAJOR TARGET FOR XENOBIOTIC-INDUCED TOXICITY

The liver being the major site of xenobiotic metabolism, is often a target for xenobiotic-induced toxicity. Screening for hepatotoxicity of drugs is thus one of the main investigations carried out during the preclinical development of a new drug. Biotransformation and effects of xenobiotics on specific liver functions are difficult to study in the whole organisms because other organs as well as exogenous and endogenous factors may interfere. Therefore, in vitro liver systems have been developed as a major tool in pharmacotoxicology. In vitro methods offer a series of advantages : (1) they allow to evaluate the mechanism of action of chemical toxicants on the target tissue and to identify short-lived toxic metabolites; (2) if human material is used, they provide a very specific and direct information about potential effects on human; (3) from an ethical point of view, they help reducing the use of laboratory animals.

In vitro liver preparations include isolated perfused organs, liver slices, isolated hepatocytes and subcellular fractions. It is not the scope of this paper to describe all these models. Only hepatocytes and liver slices will be discussed. Suspensions or cultures of hepatocytes are certainly the most popular in vitro systems to study the metabolism of chemicals and their hepatotoxicity. Since the last decade, liver slices have also received much attention and their use is becoming very popular.

The quality of the information provided by such in vitro models depends on various factors : (1) the in vitro system has to retain to a large extent the typical liver functions; (2) the parameters used for evaluating hepatotoxic effects in vitro have to include not only cell lysis but also biochemical indicators of hepatic functions.

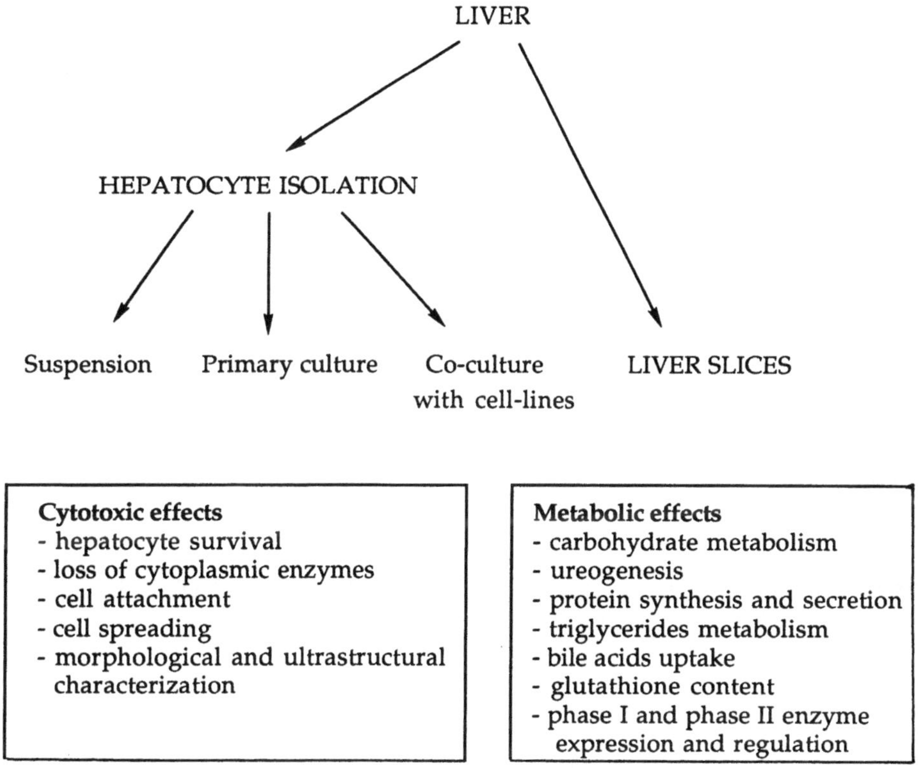

Fig. 1.
In vitro hepatotoxicity screening protocol.

As summarized in Fig. 1, the strategy of an in vitro screening protocol for hepatotoxicity involves exposing hepatocytes to the xenobiotic at different concentrations for different periods of time. Two types of effects can be evaluated : (1) cytotoxic effects for the purpose of determining the highest concentration of xenobiotic compatible with hepatocyte survival, and (2) effects of the xenobiotic on specific liver functions at concentrations which do not cause cell death. Leakage of cytoplasmic enzymes in the culture medium, due to an alteration of cell membrane permeability, is the most often used parameter for a rapid and sensitive evaluation of cytotoxicity. The most common metabolic indicators which are also used include protein synthesis and secretion, carbohydrate metabolism (glycogen content, gluconeogenesis, production of lactate), lipid metabolism (triglycerides synthesis and secretion, fatty acids oxidation), and xenobiotic metabolism including phase I (cytochrome P_{450} dependent monooxygenases) and phase II enzymes.

To be fully confident of the possible effect a compound might have on hepatic function in vivo, it may be necessary to examine most of the metabolic indicators. If alteration of one metabolic pathway is observed, more accurate investigation on the toxic mechanism can be done. Such early effects of chemicals on liver metabolic functions are indicative of its biological reactivity leading to cellular dysfunctions which may lead to function specific toxicity and related disease e.g. steatosis if the chemical causes early impairment in lipid metabolism.

LIVER SLICES AND ISOLATED HEPATOCYTES, TWO MAJOR IN VITRO MODELS TO STUDY LIVER TOXICITY

Liver slices

Tissue slices have a number of potential advantages over other in vitro systems such as primary cell cultures. Slices maintain tissue architecture so that all cell types are present and intercellular communications between the various cell types are preserved. This should permit the detection of both selective toxicity (where one or more cell types are affected) and interactive toxicity (where one cell type contributes to the toxicity observed in another cell type). In addition, tissue slicing avoids the damage to cells which may occur during the cell isolation procedure which uses proteolytic enzymes like collagenase. Liver slices are quite appropriate for studies on human liver since they can easily be prepared from organ pieces. The recent development of a new instrument providing precision cut liver slices (Krumdieck et al, 1980; Smith et al, 1985; 1986; Brendel et al, 1987) has led to new interest and new promising developments of this system.

With this technique, slices of uniform size (1 cm diameter, 250 µm thickness) may readily be prepared, under conditions that result in minimum tissue trauma not only from the liver but also from other organs : kidney (Ruegg et al, 1987; 1989; Phelps et al, 1987), lung (Stefaniak et al, 1988) and intestine (Vickers et al, 1992). Xenobiotic metabolism and toxicity have been studied on liver slices prepared from the liver of laboratory animals such as rat (Sipes et al, 1987; Wolfgang et al, 1990; Goethals et al, 1990; Dogterom, 1993; Lake et al, 1993), mouse (Connors et al, 1990), hamster (Miller et al, 1993), rabbit (Connors et al, 1990), pig (Connors et al, 1990; Fischer et al, 1991a), dog (Dogterom and Rothuizen, 1993), monkey (Stearns et al, 1992) or human (Connors et al, 1990; Fischer et al, 1991 b and c; Gunawardhara et al, 1991; Vickers et al, 1992; Stearns et al, 1992).

A critical step in the technique is the conditions of incubation of slices. Some incubations systems have been developed that guarantee an adequate nutrient and gas exchange necessary for cell survival in slices during incubation. One of these systems is the dynamic organ culture system (Smith et al, 1985; 1986): individual slices are supported on stainless-steel screen in rotating vials and cyclically go through the medium and atmosphere of the vial. An alternative system is described by Connors et al (1990) : slices are placed individually on baskets in 24-well plastic tissue culture plates on a gyratory shaker. We have developed a simple incubation system in which slices are deposited in cassettes which are submerged in the culture medium and gently agitated in a water-bath (37°C). The culture medium is the Waymouth medium supplemented with 0.3% serum albumin bovin, 100 IU penicillin/ml and 0.1 mg streptomycin/ml. The medium is oxygenated by bubbling O_2/CO_2 (95/5) during the whole incubation period. Under these conditions, liver slices can be kept viable for at least 8h.

We have demonstrated that the thickness of the slices is a critical factor for maintenance of functional parameters. Slices, 220 µm thick, maintain their glycogen content whereas in slices thicker than 250 µm, glycogenolysis occurs (Fig. 2). Using PAS staining on microtome sections of slices fixed in Carnoy solution, we have demonstrated that glycogen degradation takes place in the central part of the slice which is PAS negative whereas the outer cell layers have a well preserved glycogen content. A possible explanation is that glucose and oxygen cannot diffuse to the center of the slice if slices are more than 250 µm thick.

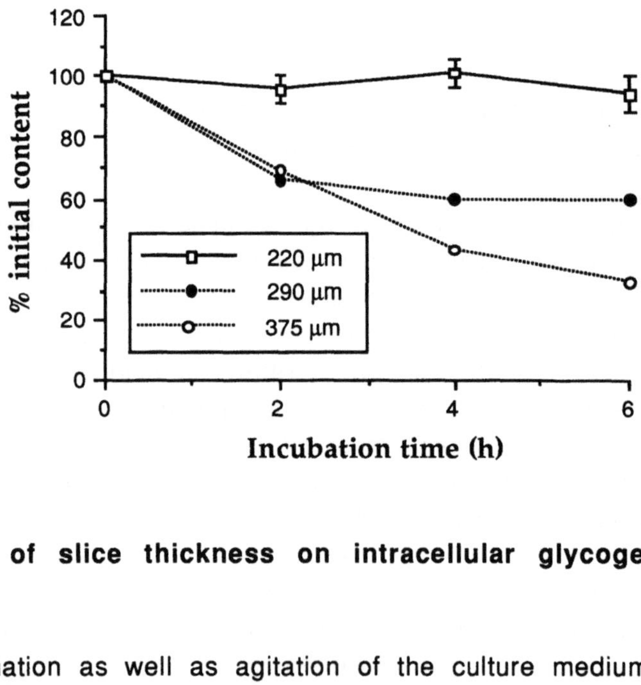

Fig. 2
Effect of slice thickness on intracellular glycogen content

Oxygenation as well as agitation of the culture medium are additional key factors in keeping viability of liver slices. As illustrated in Fig. 3, incubation of slices under static conditions (no stirring of the culture medium and thus no slice movement) results in an increase in LDH release in the incubation medium as well as a decrease in intracellular glycogen content. Agitation of the culture medium reduces the LDH leakage from the slices and favors the maintenance of the intracellular glycogen content. Dogterom (1993) has reported the influence of slice thickness and mixing of the culture medium on tolbutamide and diazepam metabolism by rat liver slices, confirming our own observation.

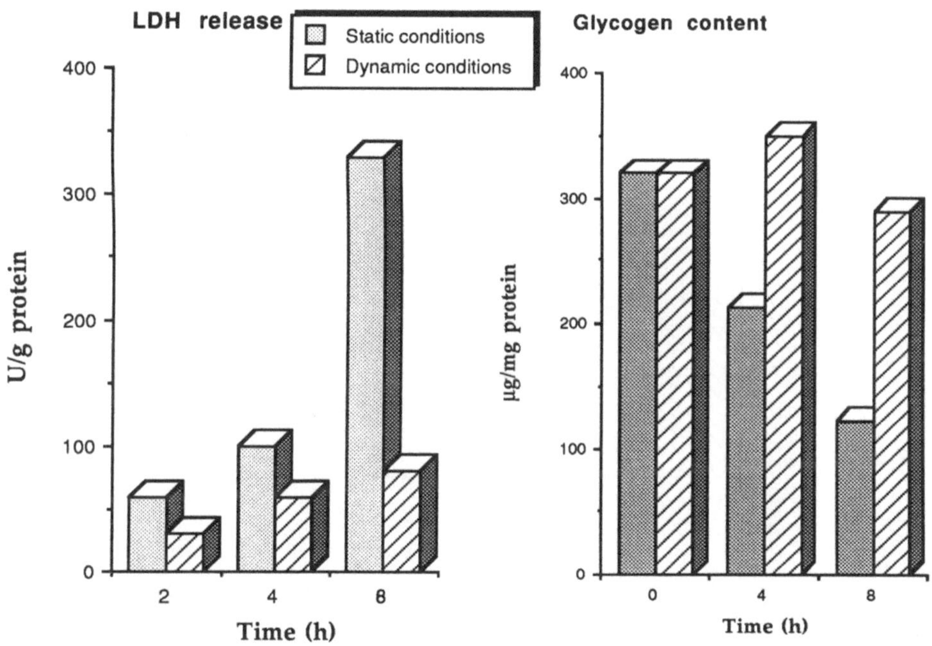

Fig. 3
Effect of culture medium agitation on LDH release and glycogen content.

The measurement of other biochemical parameters further demonstrates the maintenance of functional integrity of liver slices over 8h incubation period. ATP level which reflects the energy status of the hepatic cells is maintained (Fig. 4). Protein synthesis and secretion, as measured by the incorporation of ^{14}C-leucine into both intra- and extracellular $HClO_4$ precipitable material, remain constant over the whole incubation period (Goethals et al, 1990). The maintenance of these two processes which involves continued transport of amino acids, incorporation of these amino acids into proteins and the export of newly synthetized proteins from the cells further demonstrates the biochemical integrity of the cells in the slices.

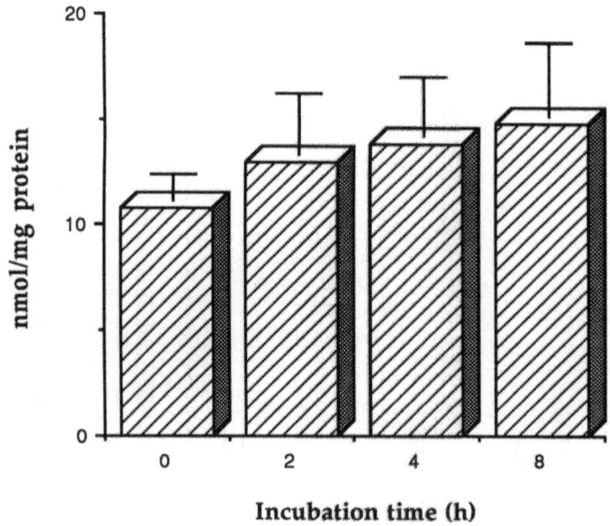

Fig. 4
Evolution with incubation time of the ATP content from rat liver slices

After 8h, cultured rat liver slices have a reduced level of cytochrome P_{450} i.e. 54% of initial value (Table 1). The activity of aminopyrin demethylase and 7-ethoxyresorufin O-deethylase after 8h of incubation is reduced to 65% and 79% of their initial levels respectively (Table 1). GSH content also progressively decreases with duration of incubation (Table 1). Even though many studies have investigated the biotransformation and cytotoxic effects of chemicals on liver slices, only few have controlled the level of cytochrome P_{450} and the activity of monooxygenases. Wright and Paine (1992) reported that cytochrome P_{450} level in rat liver slices declined to 26% of the initial level after 24h of culture. Lake et al (1993) have also observed a decrease in mixed function oxidase activities after 48h or 72h of culture, but they have also demonstrated that known enzyme inducers like phenobarbitone, ß-naphtoflavone or Aroclor induce, in vitro, the activity of 7-ethoxycoumarin O-deethylase, 7-benzoxyresorufin O-debenzylase and 7-ethoxyresorufin O-deethylase.

	Freshly cut slices	**8h incubated slices**
Cytochrome P_{450} (nmol/mg protein)	0,62 ± 0,046	0.33 ± 0.07
Aminopyrin demethylase (nmol/min . mg protein)	1.80 ± 0.21	1.17 ± 0.08
Ethoxyresorufin O-deethylase (nmol/min . mg protein)	12.1 ± 1.5	9.6 ± 1.4
GSH (nmol/mg protein)	27 ± 0.9	24 ± 1.7

TABLE 1
Levels of cytochrome P_{450}, mixed-function oxidase enzyme activities and GSH content in freshly cut and cultured rat liver slices.

Isolated hepatocytes

Isolated hepatocytes from various animal species, either in suspension or in primary cultures, are extensively used for foreign compound metabolism and drug-induced hepatotoxicity. In drug discovery, they are an important tool in the screening of metabolite profiles of potentially new drugs. In that way, they should contribute to a rational selection of the proper animal species for drug safety studies.

We will illustrate in the present paper the use of isolated hepatocytes to predict and analyse mechanisms of a common form of drug-induced hepatotoxicity, i.e. steatosis. The high incidence of fatty liver is attribuable to the central role of the liver in the metabolism of lipids and lipoproteins. The accumulation of lipids in the liver may result from many causes : a defect in triglycerides secretion as VLDL (alteration in apoprotein synthesis, in the assembly of proteins and lipids, or in the transport of the plasma membrane through the Golgi apparatus), an increase in fatty acid synthesis (lipogenesis) or esterification in triglycerides (triglycerides synthesis), a decrease in fatty acids oxidation (ß-oxidation).

Although isolated hepatocytes have been extensively used to investigate lipid metabolism, they have less frequently been used to investigate the mechanisms of drug-induced fatty liver. We report here data obtained in vitro with five compounds of known steatogenic potential. For some of them, the use of an in vitro model has contributed to the elucidation of the mechanisms involved in causing fatty liver.

All the chemicals tested induced, in vitro, an accumulation of triglycerides in hepatocytes (Fig. 5) confirming their potential steatogenic in vivo effect.

Cycloheximide 7.10^{-6} M	13.1 ± 3.5
Colchicine 10^{-} M	99.3 ± 6.4
Tetracycline 10^{-4} M	97.7 ± 8.1
Tetracycline 10^{-3} M	71.3 ± 4.5

TABLE 2
Effect of cycloheximide, colchicine and tetracycline on protein synthesis after 2h of treatment
Data are presented as percentages of the control value and expressed as means ± SEM (n = 3)

Cycloheximide strongly inhibits protein synthesis (Table 2), an effect which has been assumed to be the major cause of fatty liver induced by that chemical (Gravela et al, 1977). The accumulation of triglycerides induced by colchicine can be explained by an inhibition of triglyceride secretion (Fig. 5) which is related to the well-known interference of colchicine with the microtubular system (Stein and Stein, 1973; Le Marchand et al, 1973). We have indeed demonstrated that colchicine does not modify protein synthesis (Table 2) but drastically impairs both protein and triglyceride secretion (Deboyser et al, 1989).

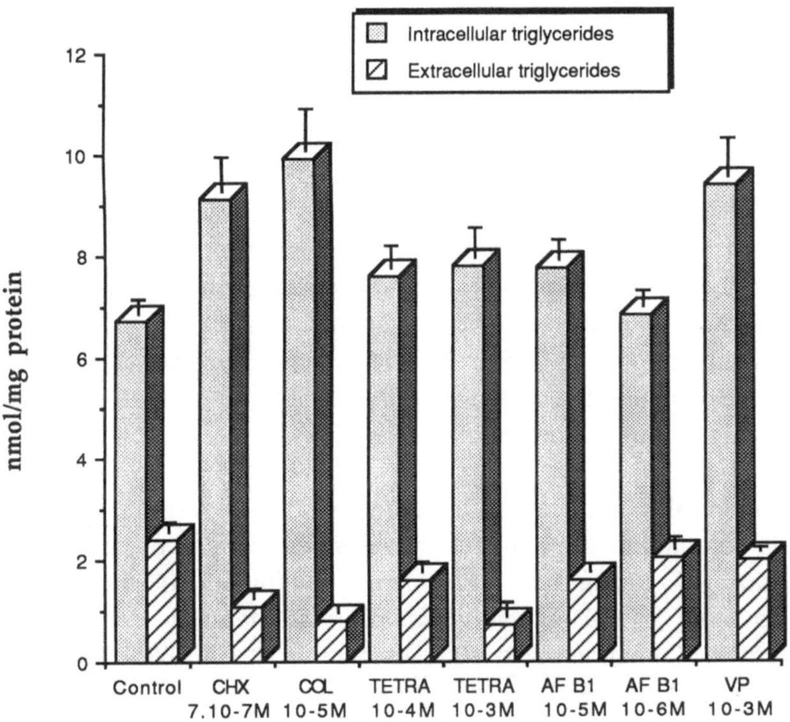

Fig. 5
Effect of cycloheximide (CHX), colchicine (COL), tetracycline (TETRA), aflatoxin B_1 (AF B_1) and sodium valproate (VP) on intracellular and extracellular content of ^{14}C-triglycerides. Isolated hepatocytes were incubated with ^{14}C-palmitate (710 μCi/mmol, 0.04mM) for 2h in the presence of drug, except for sodium valproate where it was for 3h.

High-doses of tetracycline have been reported to induce severe microvesicular steatosis in man (Zimmerman, 1979; Stricker and Spoelstra, 1985). Inhibition of protein synthesis induced by this antibiotic has been reported to be an important mechanism responsible for fatty liver. But, in our experimental conditions, the inhibition of triglyceride secretion occurs early and before protein synthesis inhibition (Fig. 5 and Table 2), suggesting that the inhibition of protein synthesis is not the key event in the pathological effect of the drug (Deboyser et al, 1989).

The addition of aflatoxin B_1 to hepatocytes induces an accumulation of intracellular triglycerides (Fig. 5) which is in concordance with fat accumulation observed in vivo in rats (Clifford and Rees, 1967). Aflatoxin B_1 causes, in isolated liver cells, a reduction of the secretion of triglycerides which is time- and dose- dependent. Moreover, we have demonstrated that the inhibition of triglyceride secretion is only partly due to inhibition of protein synthesis (Blaude et al, 1990; 1992). Since total esterification of fatty acids is not affected, we conclude that the intracellular accumulation of triglycerides is not due to an increase in lipogenesis or to an inhibition of ß-oxidation.

As shown in Fig. 5, valproate induces an important intracellular triglyceride accumulation in hepatocytes which cannot be explained by an inhibition of triglyceride secretion. Inhibition of ß-oxidation is indeed the main cause for the microvesicular steatosis described after administration of valproic acid (Becker and Harris, 1983; Coudé et al, 1983; Turnbull et al, 1983).

CONCLUSION

Risk assessment of drugs, cosmetics, food additives, is undoubtely the most difficult task for toxicologists. It implies characterization of the effects of these compounds on all tissues and functions. Moreover, metabolism and pharmacokinetics should be taken into account. Over the last two decades, toxicology, as a science, has made extensive alternative progress, and extremely powerful in vitro tests are available. These tests, as an adjuncts to in vivo tests, or as complementary or simply new tests provide fruitful information which is likely to improve risk assessment. Since the final goal of toxicology is to improve safety evaluation/risk assessment, it has become evident that such information cannot be ignored simply because in vivo tests are the only accepted or required tests.

In that respect, the liver, as a target for xenobiotic-induced toxicity, is a leading example. Models have been developed and scientifically validated. They are easily available and can be adapted so as to help answering important questions such as interspecies variation in biotransformation, specificity of metabolic target, reversibility of effects, dose effect relationship, and chemical-chemical interactions.

One important step forward in the scientific validation of in vitro hepatic test system has been the development of an approach based on the analysis of the effect of a chemical on specific liver metabolic function. Our group has significantly contributed to such a development. It must however be made clear that the identification of an acute, and early effect of a chemical on a metabolic function of liver cells, in vitro, is not the demonstration of a toxic effect. To demonstrate that a xenobiotic impairs lipid metabolism in isolated hepatocytes or in liver slices, does not prove that chemical is steatogenic. But rather it demonstrates that the xenobiotic has, in liver cells, a biological reactivity, the target of which is (are) in this particular metabolic pathway.

When evaluating the risk that might be associated with the use of that particular chemical, toxicologists should integrate the information when designing further tests. Such a scientific information together with interspecies comparison, knowledge on biotransformation pathways and results of pharmacological and other toxicological tests must become the basis for a more scientific toxicology we strongly advocate (Roberfroid and Goethals, 1990; 1992; Roberfroid, 1991). But for such a new toxicology to become implemented, present regulations have to become more flexible so as to permit, and even stimulate, more scientific approaches to risk assessment. The present rigidity which imposes a fixed set of tests for every chemical, must definitiely be abandoned (Balls et al, 1990; Roberfroid and Goethals, 1992).

References

Balls M., Blaauboer B., Brusick D., Frazier J., Lamb D., Pemberton M., Reinhardt C., Roberfroid M., Rosenkranz H., Schmid B., Spielman H., Stammati A.L. and Walum E. (1990) Report and recommandations of an international workshop on promotion of the regulatory acceptance of validated non-animal toxicity test procedures. ATLA, 18: 317-337.

Becker C.M. and Harris R.A. (1983) Influence of valproic acid on hepatic carbohydrate and lipid metabolism. Arch. Biochem. Biophys., 223: 381-392

Blaude M.N., Goethals F.M., Ansay M.A. and Roberfroid M.B. (1990) Interaction between aflatoxin B_1 and oxytetracycline in isolated rat hepatocytes. Cell. Biol. Toxicol., 6: 339-351

Blaude M.N., Goethals F.M., Ansay M.A., and Roberfroid M.B. (1992) Synergism between aflatoxin B_1 and oxytetracycline on fatty acid esterification in isolated rat hepatocytes. Toxicol. Lett., 61: 159-166

Brendel K., Gandolfi A.J., Krumdieck C.L. and Smith P.F. (1987) Tissue slicing and culturing revisited. Trends Pharm. Sci., 8: 11-15

Clifford J.I. and Rees K.R. (1967) The action of aflatoxin B_1 on the rat liver. Biochem. J., 102: 65-75

Connors S., Rankin D.R., Gandolfi A.J., Krumdieck C.L., Koep L.J. and Brendel K. (1990) Cocaine hepatotoxicity in cultured liver slices : a species comparison. Toxicology, 61: 171-183

Coudé F.X., Grimber G., Pelet A., Benoit Y. (1983) Action of the antiepileptic drug, valproic acid on fatty acid oxidation in isolated rat hepatocytes. Biochem. Biophys. Res. Commun., 115: 730-736

Deboyser D., Goethals F., Krack G. and Roberfroid M. (1989) Investigation into the mechanism of tetracycline-induced steatosis : study in isolated hepatocytes. Toxicol. Appl. Pharmacol., 97: 473-479

Dogterom P. (1993) Development of a simple incubation system for metabolism studies with precision-cut liver slices. Drug Metab. Disp., 21: 699-704

Dogterom P. and Rothuizen J. (1993) A species comparison of tolbutamide metabolism in precision-cut liver slices from rats and dogs. Qualitative and quantitative sex differences. Drug. Metab. Disp., 21: 705-709

Fischer R., Barr J., Zukoski C.F., Putnam C.W., Sipes I.G., Gandolfi A.J. and Brendel K. (1991b) In vitro hepatotoxicity of three dichlorobenzene isomers in human liver slices. Human Exp. Toxicol., 10: 357-363

Fischer R., Nau H., Gandolfi A.J. and Brendel K. (1991a) Toxicity of valproic acid in liver slices from Sprague-Dawley rats and domestic pigs. Toxicol. in Vitro, 5: 201-205

Fischer R., Nau, H., Gandolfi A.J., Putnam C.W. and Brendel K. (1991c) Valproic acid hepatotoxicity in human liver slices. Drug. Chem. Toxicol., 14: 375-394

Goethals F., Deboyser D., Lefebvre V., Decoster I. and Roberfroid M. (1990) Adult rat liver slices as a model for studying the hepatotoxicity of vincaalkaloids. Toxicol. in Vitro., 4: 435-438

Gravela E., Poli G., Albano E. and Dianzani M. (1977) Study on fatty liver with isolated hepatocytes. I. The action of colchicine, phalloidin, cytochalasin B and cycloheximide on protein and triglyceride synthesis and secretion. Exp. Mol. Pathol., 27: 339-352

Gunawardhara L., Barr J., Weir A.J., Brendel K. and Sipes I.J. (1991) The N-acetylation of sulfamethazine and p-aminobenzoic acid by human liver slices in dynamic organ culture. Drug. Metab. Disp., 19: 648-654

Kroes R. and Hicks M. (1990) Introduction. Reevaluation of current methodology of toxicology testing including gross nutrients. Proceedings of a symposium held on 5-7 April 1989, Chateau de Limelette, Belgium. Food and Chem. Toxicol., 28: 733-734

Krumdieck C.L., Dos Santos J.E. and Ho K.J. (1980) A new instrument for the rapid preparation of tissue slices. Anal. Biochem., 104 : 118-123

Lake B.G., Beamand J.A., Japenga A.C., Renwick A., Davies S. and Price R.J. (1993) Induction of cytochrome P_{450}-dependent enzyme activities in cultured rat liver slices. Fd. Chem. Toxicol., 31: 377-386

Le Marchand Y., Singh A., Assimacopoulos-Jeannet F., Orci L., Rouiller C. and Jeanrenaud B. (1973) A role for the microtubular system in the release of very low density lipoproteins by perfused mouse livers. J. Biol. Chem., 248: 6862-6870

Phelps J.S., Gandolfi A.J., Brendel K. and Dorr R.T. (1987) Cisplatin nephrotoxicity : in vitro studies with precision-cut rabbit renal slices. Toxciol. Appl. Pharmacol., 90: 501-512

Ruegg C.E., Gandolfi A.J., Nagle R.B. and Brendel K. (1987) Differential patterns of injury to the proximal tubule of renal cortical slices following in vitro exposure to mercuric chloride, potassium dichromate or hypoxic conditions. Toxicol. Appl. Pharmacol., 90: 261-273

Ruegg C.E., Grushenka H.I., Wolfgang G.H., Gandolfi A.J., Brendel K. and Krumdieck C.L. (1989) Preparation and utilization of positional renal slices for in vitro nephrotoxicity studies. In : In vitro Toxicology : Model Systems and Methods (Ch.A. Mc Queen, ed), 197-230, The Telford Press Inc., New Jersey

Roberfroid M. (1991) Long term policy in toxicology. In : animals in Biochemical Research (H. Koeter and Ph. Anderson, eds),pp 35-48, Elsevier Sci. Publ., New York, London

Roberfroid M. and Goethals F. (1990) In vitro toxicology : a challenge for the 21st century. ATLA, 18: 19-22

Roberfroid M. and Goethals F. (1992) In vitro toxico-pharmacology : past, present and future. In : In vitro Alternatives to Animal Pharmaco-Toxicology (J.V. Casterll and M.J. Gomez-Lechon, eds), pp 17-41, Farmaindustria, Madrid

Rodricks J.V. (1986) Improving the use of risk assessment in regulation. Comments on Toxicol., 1: 75-96

Smith P.F., Gandolfi A.J., Krumdieck C.L., Putnam C.W., Zukoski C.F., Davis W.M. and Brendel K. (1985) Dynamic organ culture of precision cut liver slices for in vitro toxicology. Life Sci., 36: 1367-1375

Smith P.F., Krack G., McKee R.L., Johnson D.G., Gandolfi A.J., Hruby V.J., Krumdieck C.L. and Brendel K. (1986) Maintenance of adult rat liver slices in dynamic organ culture. In vitro Cell Develop. Biol., 22: 706-712

Stearns R.A., Miller R.R., Doss G.A., Chakravarty P.K., Rosegay A., Gatto G.J. and Chiu S.H.L. (1992) The metabolism of DuP753, a nonpeptide angiotensin II receptor antagonist, by rat, monkey, and human liver slices. Drug. Metab. Disp. 20: 281-287

Stefaniak M.S., Gandolfi A.J. and Brendel K. (1988) Adult rat lung slices in dynamic organ culture : a new tool in pharmacology. Proc. West. Pharmacol. Soc., 31: 149-151

Stein O. and Stein Y. (1973) Colchicine-induced inhibition of very low density lipoprotein release by rat liver in vivo. Biochim. Biophys. Acta, 306: 142-147

Stricker B.H. and Spoelstra P. (1985) Tetracyclines. In : Drug-Induced Hepatic Injury (M.N.G. Dukers, ed), p 157, Elsevier, Amsterdam.

Turnbull D.M., Bone A.S., Bartlett K., Koundakjian P.P. and Sheratt H.S.A. (1983) The effects of valproate on intermediary metabolism in isolated rat hepatocytes and intact rats. Biochem. Pharmacol., 32: 1887-1892

Vickers A.E.M., Fisher V., Connors S., Fischer R.L., Baldeck J.P., Maurer G. and Brendel K. (1992) Cyclosporin A metabolism in human liver, kidney, and intestine slices. Comparison to rat and dog slices and human cell lines. Drug. Metab. Disp. 20: 802-809

Wolfgang G.H.I., Donarski W.J. and Petry T.W. (1990) Effects of novel antioxidants on carbon tetrachloride induced lipid peroxidation and toxicity in precision cut rat liver slices. Toxicol. Appl. Pharmacol., 106: 63-70

Wright M.C. and Paine A.J. (1992) Evidence that the loss of rat liver cytochrome P_{450} in vitro is not solely associated with the use of collagenase, the loss of cell-cell contacts and/or the absence of an extracellular matrix. Biochem. Pharmacol., 43: 237-243

Zimmerman H.S. (1979) Drug-induced chronic hepatic disease. Med. Clin. North Amer, 63: 567-582

In Vitro Skin Irritation Assays: Relevance to Human Skin

Sunita Patil, Ph.D.;Jeff Harvell, M.D.; Howard I. Maibach, M.D.
Department of Dermatolgy,
Surge 110
University of California, San Francisco,
San Francisco, CA - 94143. USA.

Introduction

The events occurring in primary skin irritation in vivo represent a complex series of chemical and physiological changes. Animals have been used to assess dermal irritation by observation of visible changes ranging from erythema and edema to corrosion and ulceration in the in vivo Draize rabbit skin test accepted by many regulatory agencies [Draize et al., 1944]. These responses, easily observed, are produced by diverse mechanisms.

Investigators [Kastner, 1977; Patrick and Maibach HI, 1989] have found that different species exhibit widely varying reactivity under identical test conditions especially in substances with only minor irritant potential. Thus, the accuracy of the Draize test and other animal testing as it relates to humans has been called into question [Helman et al., 1986]. Also, the results from the animal methods currently used differ due to the subjective visual test scoring [Oliver and Pemberton,1988]. These differences occur most frequently in the assessment of the toxicity of mild irritants and colored material [Oliver and Pemberton, 1988]. Thus, it has been postulated that animals should not be the exclusive means of evaluation.

Humane considerations have galvanized efforts to find alternative testing methods. The development of an in vitro irritation assay may be a means of improving upon the accuracy of the animal test while reducing the number of animals needed to test a given compound. An in vitro irritation assay could be designed to provide insight into the actual mechanism of damage produced by a given toxicant which could be addressed in future formulations.

In vitro skin irritation tests are being developed in the hope that methods of analysis can be determined that are more humane, more predictive of actual human response, and will provide an objective, quantifiable means of determining the irritancy potential of a substance - a major advantage compared to subjectively assessed animal tests [Oliver and Pemberton, 1988].

This review attempts to summarize previous studies, and proposes a strategy for validating such assays.

NATO ASI Series, Vol. H 93
Modulation of Cellular Responses in Toxicity
Edited by C. L. Galli, A. M. Goldberg, M. Marinovich

Assays

Proposed in vitro tests for irritation are based on cell cytotoxicity, inflammatory or immune system response, alterations of cellular or tissue physiology, cell morphology, biochemical endpoints, and structure activity analysis [SOT Position paper]. A reliable skin irritation test should provide a method for differentiating the varying degrees of irritation or corrosion of the skin [Gfeller et al., 1985]. In vitro test systems should allow for the measurement of the damage and the signals which initiate the physiological response to an applied irritant [Bell et al., 1989].

Many in vitro systems have been explored; these methods vary greatly in the parameters utilized to correlate with irritation. The most recently published methods currently under investigation examine toxic effects on keratinocyte and other cell cultures, multilayered "skin equivalents", the chorioallantoic membrane of fertilized eggs, irritant-sensitive microorganisms, and a keratin/collagen-reagent system membrane that mimics skin response to irritants (Table 1) [Bell et al., 1989; Bagley et al., 1988; Bloom et al., 1989; Borenfreund and Puerner, 1985; 1984; Bulich et al, 1981; Gordon et al., 1990; Leupke and Kemper, 1986; Parce et al., 1989; Silverman and Pennist, 1987; Bason et al; 1992].

One method using human skin elements is "Testskin", a multilayer epidermal system composed of cultured human keratinocytes that cover a layer of human dermal fibroblasts in a collagen matrix. It is used to detect chemically induced cytotoxicity by assessing the release of inflammatory mediators. The irritant exposed tissue can also be evaluated histologically for signs of damage [Bason et al., 1992].

TABLE 1. In Vitro Irritation Systems

Method	System	Principle	Toxicity
"Testskin"	synthetic human epithelium	Histologically, physiologically similar to skin	morphologic inflammatory mediators
Microphysio-meter	cell-culture	pH monitoredas an indicator of metabolism	decreased metabolism
Neutral red dye assay	cell-culture	dye uptake by viable cells measured spectro-photometrically	cell death
"Microtox" Photobacterium phosphorium	luminescent bacteria	only viable bacteria luminesce	organism death

Tetrahymena thermophila	protozoa	motility	decreased motility
Chorioallantoic membrane	fertilized egg membrane	vasculature changes	damage to vasculature
"Skintex" [MB/ PM] system	keratin/collagen matrix linked to dye	dye released and tubidity are measured spectrophotometrically	increased dye/ turbidity reflec ts membrane disruption

An evaluation utilizing cell culture includes a device called the microphysiometer. This method employs a silicon based biosensor known as a light activated potentiometric sensor (L.A.P.S) that can detect small changes in the pH of culture media surrounding a single layer of confluent cells by determing the rate at which cells excrete the acidic products of cellular metabolism, such as lactic acid and CO2. The endpoint of the assay is the MRD50 , or that concentration of test material resulting in a 50% reduction of cellular acidification rate [Parce et al., 1989].

These metabolic changes can be observed on a time scale of seconds to minutes, and therefore, this assay system allows one to observe the processes of irritation/toxicity in a dynamic fashion (Bruner, et al., 1991) Shortly after exposure to a known irritant, one typically observes a decrease in metabolic rate; and, provided that the insult is not lethal, one can then observe the recovery of the monolayer cell system to baseline. In a preliminary experiment, the device was utilized to determine its ability to reproduce the in vivo rank order irritancy/toxicity of fifteen cutaneously applied basic compounds. The fifteen basic compounds, which spanned a range of pKa's from 1.4 to 11.2, were tested both in vivo, in human volunteers, and in vitro, in the microphysiometer. The Spearman in vitro/in vivo rank order correlation coefficient was $r=0.829$ ($p<0.001$), suggesting the ability of this in vitro system to reproduce the irritation response seen in the human. (Bason, Miller, Nangia, Harvell, Owicki, and Maibach, UCSF Department of Dermatology, unpublished data). This system can thus be used to assess subtle perturbations of cellular metabolism and thereby assess cellular damage and recovery after irritant exposure [Parce et al., 1989]. This is the only in vitro system that can assess recovery of the cell monolayer after toxicological insults and thereby adds another dimension to in vitro toxicity prediction.

Such a comparative approach using *in vivo* human data will no doubt enhance the ability to assess the true relevance of a particular *in vitro* system; however, this approach is not without its limitations. Probably the most significant weakness of this approach is the fact that one is limited to using fairly innocuous compounds in the *in vivo* arm of the comparison. We may therefore be skewing the relevance assessment to less irritating/toxic compounds. On the other hand, it is generally thought that if a test is sensitive enough to detect differences in the toxic potential of compounds which are in a narrow range of irritancy, it is most likely capable of making more broad-ranged assessments. The veracity of the latter statement will be borne out by future validation projects.

The neutral red dye assay utilizes human keratinocytes in a serum free medium [Borenfreund and Puerner, 1985; 1984]. Studies have shown that serum in cell culture can alter the toxicity potential of some materials. This assay may be useful for those materials in which serum-binding is a confounding factor when analyzing toxicity. After cells are exposed to an irritant, they are washed and incubated with neutral red dye. The dye is taken up by viable cells only and the amount incorporated by these cells is measured spectrophotometrically. This information can be used to quantify the toxicity of a test substance [Borenfreund and Puerner, 1985; 1984].

A similar system, "Microtox", has been used for years in environmental studies. This luminescent bacteria (Photobacterium phosphorium) toxicity assay examines the reduction of fluorescence normally emitted by a suspension of luminescent bacteria [Photobacterium phosphoreum] after exposure to toxins [Bulich et al., 1981]. This process can be used to assess irritation because the amount of viable remaining cells corresponds to the degree of fluorescence [Bulich et al., 1981].

Another system utilizing microorganisms [Tetrahymena thermophila] evaluates the ability of a test substance to decrease the normal motility of this ciliated protozoan [Silverman and Pennist, 1987; Silverman, 1983]. The movement of these organisms is examined after irritant exposure, and compared to the motility of untreated T. thermophila to detect any decrease that results from toxicity [Silverman and Pennist, 1987; Silverman, 1983].

Fertilized egg chorioallantoic membrane (CAM) is a system incorporating a vascular network [Bagley et al., 1988; Leupke and Kemper, 1986]. A small window is cut into the shell of fertilized chicken eggs to expose the CAM. Test substances are placed directly onto the surface of the CAM and the response is evaluated exactly 30 min after dosing. Serial dilutions of a toxicant are studied and the CAM is visually graded with a positive result being the appearance of hemorrhaging blood vessels [Leupke and Kemper, 1986]. Thus far, this method has been used mainly to examine eye irritancy, however, its basic principles could also be employed to assess skin irritancy.

"Skintex", one of the systems utilizing non-human substrates, can be described as a membrane barrier/protein matrix (MB/PM) system [Gordon et al., 1990]. This method detects changes in the intact barrier matrix with an indicator dye attached to the matrix that is released with exposure to an irritant. The amount of dye released is quantified and correlates with expected protein disruption and denaturation. A second compartment is a reagent system which responds to irritants by producing turbidity. This response provides an internal detection for materials which perturb conformations after permeating the membrane. This provides a quantitative response to materials which may produce irritation by membrane damage, protein-binding, enzyme-inactivation and a variety of other pathways where macromolecule conformation is altered in the initiation of dermal irritation [Gordon et al., 1990].

One example of such an approach compared the irritant capabilities of the irritants benzalkonium chloride, trichloroacetic acid, phenol, and hydrochloric acid in the human and in the Skintex™ (In Vitro International, Irvine, CA) dermal assay system [Bason et al., 1992]. The Skintex™ system was fairly sensitive in its ability to predict the irritant potential of these compounds in man. (sensitivity 82%, specificity 71%, positive predictive value 82%) [Bason et al., 1992]. Additionally, the *in-vivo* dose response curves for each of the 4 substances was compared to the *in-vitro* dose response curves, and correlation coefficients calculated. The *in-vitro* dose response for benzalkonium chloride (R^2=0.987) and phenol (R^2=0.994) were strikingly similar to those generated in-vivo, possibly indicating that the mechanisms of action *in-vivo* and *in-vitro* are similar for these two compounds.

Since susceptibility to irritants changes with increasing numbers of passages using primary keratinocyte cultures; one more human keratinocyte cell line, HaCaT, was recently reported to quantitate and assess cytotoxicity by measuring the absorption of the vital dye neutral red (Wilhelm et al 1994) using a scanning spectrophotometer in an 'ELISA' reader. Concentrations resulting in 50% inhibition of the dye uptake (IC 50) have been claimed to highly resemble the in vivo observations. Evaluation of cell toxicity by changes in mitochondrial metabolic activity (MTT assay) and plasma membrane integrity (LDH leakage) utilizing either skin keratinocytes or oral mucosa has been proposed (Eun et al., 1994) . Oral mucosa is also subjected to contact with various primary irritants and since both cell types express similar sensitivity to irritants either one may be applied for in vitro irritancy assays.

Other epitheloid cell lines like adult rat lung (ARL), human lung carcinoma (A549), human cervix epitheloid carcinoma (HeLa) cell lines, rabbit corneal epithelial cells, and keratinocytes have been employed to assess the potential of irritancy of different compounds. In these cells various cellular parameters were used as markers to identify possible mechanisms of cytotoxicity in a dermal irritancy model. Four different fluorometric assays were applied to quantitate any changes in levels of reactive oxygen intermediates, intracellular glutathione, intracellular calcium and mitochondrial membrane (MM) potential of these cells after exposure to falcarinol (Avalos et al., 1994). Use of these probes provide a rapid and sensitive tool to study in vitro cell irritancy with negligible toxic side effects on the cells. Furthermore, the response to standard irritants like sodium lauryl sulfate and benzalkonium chloride were observed in mouse keratinocytes using these fluorescent probes (Acosta et al., 1994; Levin et al., 1994; Yang and Acosta, 1994). Both the compounds were shown to be cytotoxic and caused destruction of MM potential and dissipation of intracellular calcium in a dose dependent manner as indicated by a decresae in rhodamine 123 and Fluor-3 fluorescence respectively.

Intracellular calcium concentrations, MM potential, generation of reactive oxygen species, and changes in intracellular pH were also measured using fluorescent probes in rabbit corneal epithelial cells after an exposure to surfactants (Yang and Acosta, 1994; Jiang et al., 1994). The probes used were fura-2 and fluo-3, rhodamine 123 , 2',7'-bis(2-carboxyethyl)- 5(6)-carboxyfluorescein for calcium, MM potential, ROS, and intracellular pH, respectively. SDS caused a rapid dose-dependent rise in calcium and dissemination of MM potential in this primary culture of corneal epithelial cells (Yang and Acosta, 1994). In addition, the levels of ROS did not increase with treatemnt of SDS while SDS (40μg/ml) and benzalkonium (20μg/ml) acidified the cells (Yang and Acosta, 1994; Jiang et al., 1994).

These methods have evolved from years of laboratory and clinical research in determining the basic features of irritation of the skin. Factors predisposing individuals to irritation have also been evaluated [Maibach et al., 1989]. No single parameter has yet to emerge as definitive of irritation. The targets of the toxin vary so much that the effect of toxic substances on the structure of the skin is poorly understood.

Studies have elucidated considerable information about the mechanisms of damage and repair that occur in skin. Typical events identified in dermal irritation include protein denaturation, epidermal cell lysis, cytotoxicity, enzyme leakage, and production of epidermal antigens and cytokines [Gordon et al., 1990; Imokawa, 1980]. The means of evaluating the evidence of damage include examining morphology, signs of the inflammatory reaction initiation, cellular toxicity, changes in electrical properties, and decreased metabolism [Serban et al.,1981]. Also, synthetic models of epithelium have been designed to mimic irritant damage characteristics [Blake-Haskins et al., 1986]. Some investigators have combined two or more of these modalities and compared them to assess the differences.

Helman [1988], comparing the morphologic responses of in vitro and in vivo skin exposed to chemicals with light microscopy, found that the absence of an intact vascular system in in vitro skin specimens did not interfere significantly with the ability to detect graded microscopic epidermal lesions and concluded that the morphologic response of skin maintained in organ culture is an accurate indicator of skin toxicity. In addition to the altered histology seen with light microscopy, electron microscopic analysis of irritant-damaged skin reveals characteristic changes including spongiosis of epidermis, disappearance of tonofilament-desmosome complexes, and dissolution of the horny cells [Nagao et al., 1972; Kanerva and Lauharanta, 1986].

Enzyme leakage may provide a means for detecting sublethal cell injury which might not be observed histologically. Skin in organ culture has been analyzed to determine quantifiable parameters to assess injury such as cellular enzyme leakage, glucose metabolism, DNA synthesis, water loss, and changes in electrolyte concentration [Helman et al., 1988]. Rat skin in vivo exposed to toxicants causes release of acid phosphatase, lactate dehydrogenase and N-acetylglucosaminidase which is associated histologically with epidermal edema and an increase in dermal leukocytes [Gibson and Teal, 1983].

Irritation has been evaluated by analyzing epidermal edema with other techniques. Sodium lauryl sulfate produced swelling in in vitro skin discs prepared from excised human skin and dermal calf collagen [Choman, 1963]. In an in vitro system without skin, the tritiated water uptake [i.e. swelling] of a collagen film substrate correlated with the irritation potential of anionic surfactants [Blake-Haskins et al., 1986]. The swelling response was concentration-dependent and increased substrate water uptake indicated greater irritation potential - results showed that the C12 homolog produced the most swelling.

The presence of the cell products made in response to inflammation is a potential means of quantitating irritation. An in vitro system could be designed to quantitate the substances liberated during the irritant process such as prostaglandins, glucocorticoids, complement factors [Malten, 1981]. Bell [1989] has reported production of interleukin 1 α, PGE 2, and prostacyclin in the dermal equivalent testing system after the application of a number of irritants.

Several investigators evaluated the changes in electrical properties of damaged skin. In human volunteers, skin exposed to toxicants displayed gradual loss of barrier properties of the stratum corneum reflected by an increase in electrical conductance and visible signs of damage [Serban et al., 1981]. Changes in electrical conductances occurred before visible irritation signs and thus may be a sensitive monitor to predict skin irritation. Oliver [SOT Position paper, 1989] utilized in vitro skin discs from rats to measure stratum corneum integrity by analyzing the change in the electrical resistance.

Our experience with the validation process of dermal irritancy assays has allowed the identification of the above limitations when using solely human *in vivo* data; however, the experience has also allowed for the identification of other limitations of the validation process which are more general, and are irrespective of the species used. Common questions are (1) What parameter do we use to define irritation? (2) How can we make an assessment of a particular *in vitro* assay's ability to predict both acute and chronic irritation? (3) Similarly, what *in vivo* exposure period should be used to compare results.? 24 hours? 72 hours? (4) Should the *in vivo* exposure period be an open or closed test? and (5) How do we make the *in vivo* data more objective. By answering these questions, we begin to form a "gold standard" *in vivo* data base, and thus improve our ability to make sound assessments of an *in vitro* assay's true relevance.

Thus far, no single means of evaluation of epithelial damage has been acknowledged as the most important indicator. The fact that different components of the epithelium vary in their vulnerability to chemicals and the ways in which they are damaged may make this impossible. An ideal in vitro test would give information about the specific action of a toxicant on the epithelium and the mechanism of activation, inflammation, and repair of skin in response to the chemical as well as correlate with the relative ranking of animal and human irritation tests. A system could be developed to exhibit a series of biosynthetic properties like those of epithelium and provide a broad range of pharmacologic responses including the induced release of cytokines [Bell et al., 1989].

We believe that any validation study comparing in vitro and in vivo responses should take into account general principles of toxicology including dose response, intended application method to the skin, the anatomic site, and the delivery system. It is likely that predictions from homologous series will be easier than from unrelated moieties. Using pure chemicals rather than complex mixtures may simplify the learning process. As our knowledge of skin irritation biology increases, we should become more facile in designing appropriate validation studies. A recent text provides a more indepth overview of this rapidly emerging area (Rougier et al., in press).

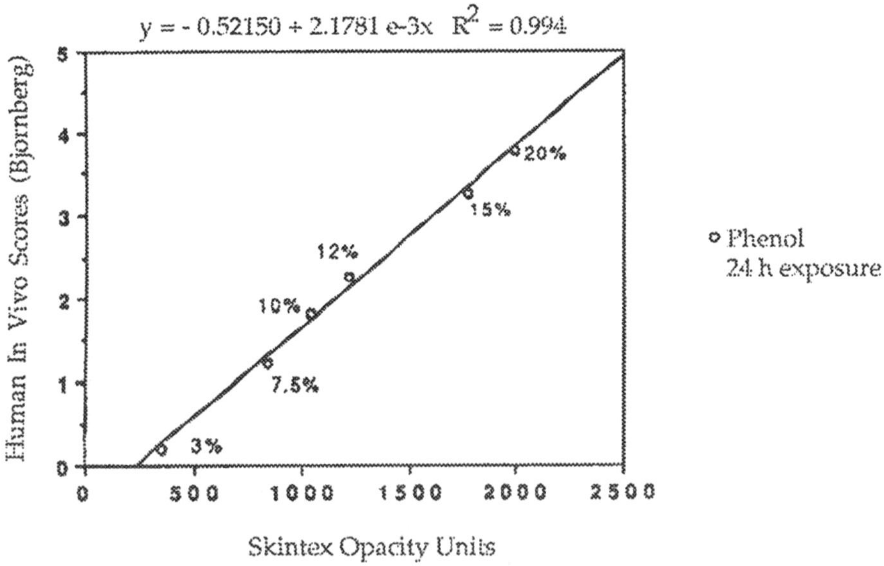

Figure 1. Human In Vivo Data vs Skintex HIE
Phenol diluted in distilled water to concentration listed, applied to the thigh in 100 volunteers and allowed to dry. Reading taken 24 h after exposure, and arithmetic mean determined.
0 = no reaction; 1 = slight erythema, test area; 2 = pronounced erythema, test area; 3 = brown-red superficial epithelial defects, test area; 4 = brown-red

Figure 1. Human In Vivo Data vs Skintex HIE

Phenol diluted in distilled water to concentration listed, applied to the thigh in 100 volunteers and allowed to dry. Reading taken 24 h after exposure, and arithmetic mean determined.

0 = no reaction; 1 = slight erythema, test area; 2 = pronounced erythema, test area; 3 = brown-red superficial epithelial defects, test area; 4 = brown-red superficial defects outside test area; 5 = defects outside test area and deep necrosis inside test area. See reference [Bason et al., 1992] for details.

Recent Studies

Recently, the MB/PM method was evaluated for irritation prediction using human in vivo data for validation [Bason et al., 1992]. Four compounds were analyzed, some at different concentrations for a total of 18 evaluations. The results were reproduced with multiple test runs.

Actual data are presented for one compound (phenol) in Figure 1; note that details of human testing conditions are in the legend. The nature of the data generated by in vitro and in vivo testing can make correlations difficult. In the in vivo experiments, investigators tested substances in a variety of ways (exposure time, application, etc.), making a single system of correlation of all test materials impractical. Due to these differences in the in vivo testing methods, the results of compounds that were examined in a similar manner were compared. Thus, substances were grouped by study conditions to provide a means of correlation and likewise graphed with their in vitro results.

Detailed human data were available for different concentrations of phenol and these results were correlated with the opacity unit readings of the irritation produced in the in vitro system at the 100 mL dose (see Figure 1).

The extraordinary correlation of in vivo data with in vitro results suggests that the damage mechanism in vivo may be similar to that in producing disruption in the synthetic membrane barrier/protein matrix system.

Discussion

In order to protect humans from substances, skin irritation tests must be performed. At a minimum, the new tests must be sensitive enough to characterize the potential degree of irritation. In the future, the developing skin systems may also provide information about the mechanism by which a substance causes irritation.

There are many possible uses for an in vitro system of irritation: it could be a worthwhile tool in the assessment of ranking groups of compounds for raw ingredients. It is unlikely that an in vitro system could ever be developed to mimic the complex cascade of reactions that occur in the human skin. However, an in vitro system utilized as an initial screening device would permit the use of a minimum number of animals for skin tests and simplify the process by which new compounds are developed.

Appropriate use of human irritancy data with standard compounds, utilizing occlusive and non-occlusive dosing, should permit a facile and realistic correlation between the in vitro assays and likely human experience. At present, in vitro irritation systems are entering a validation phase. Criteria must be established permitting identification of standards, i.e., how irritating are model compounds in man (or animal)? Our personal impression suggests that the data must be clearly related to method of application (occlusive or open), single dose (primary irritant) or multiple dose (cumulative irritation); and anatomic site (face responding differently than back).

References

Bagley DM, Rizva PY, Kong BM, et al.(1988) An improved CAM assay for predicting ocular irritation potential. In: Alterna~ive Methods in Toxicology, Vol 6, Goldberg AM, ed., New York: Mary Ann Liebert, Inc, 131-138.

Bason MM, Harvell J, Realica B, Gordon V, Maibach HI (1992) Comparison of In vitro and In vivo derrmal irritancy data forr four primary irritants. Toxic. in vitro. 521-526.

Bell E, Gay R, Swiderek M, et al. Use of fabricated living tissue and organ equivalents as defined higher order systems for the study of pharmacologic responses to test substances. Manuscript presented at the NATO advanced research workshop, Pharmaceutical Application of Cell and Tissue Culture to Drug Transport, 9/4/89-9/9/89.

Blake-Haskins JC, Scala D, Rhein LD, et al.(1986) Predicting surfactant irritation from the swelling response of a collagen film. J Soc Cosmet Chem 37:199-210.

Bloom E, Maibach HI, Tammi R (1989) In vitro models for cutaneous effects of glucocorticoids using human skin organ and cell culture. In: Models Dermatology, Vol 4, Maibach HI, Lowe NJ, eds., New York: Karger 12-19.

Borenfreund E, Puerner JA (1985) Toxicity determined in vitro by morphological alterations and neutral red absorption. Toxicol Lett 124: 119-124.

Borenfreund E, Puerner JA (1984) A simple quantitative procedure using monolayer cultures for cytotoxicity assays. J Tissue Culture Methods 9:7-9.

Bulich AA, Greene MW, Isenberg DL (1981) Reliability of bacterial luminescence assay for determination of the toxicity of pure compounds and complex effluents. In: Aquatic Toxicology and Hazard Assessment. Branson DR, Dickson KL, eds., 4th Conference of the American Society for Testing and Materials (ASTM 737) Philadelphia, 338-347.

Choman BR(1963) Determination of the response of skin to chemical agents by an in vitro procedure. J Inv Dermatol 44: 177-182.

Draize JH, Woodard G, Calvery HO (1944) Methods for the study of irritation and toxicity of substances applied topically to the skin and mucous membranes. Division of Pharmacology, Food and Drug Administration.

Gfeller W, Kobel W, Seifert G (1985) Overview of animal test methods for skin irritation. Food Chem Toxicol 23: 165-168.

Gibson WT, Teall MR (1983) Interactions of C12 surfactants with the skin: Changes in enzymes and visible and histological features of rat skin treated with sodium laurel sulfate. Food Chem Toxicol 21:587-593.

Gordon VC, Kelly CP, Bergman HC. Evaluation of "Skintex", an in vitro method for determining dermal irritation. Toxicologist. 1990;10:78.

Helman RG, Hall JW, Kao JY (1986) Acute dermal toxicity: In vivo and in vitro comparisons in mice. Fundam Appl Toxicol 7:94-100.

Imokawa G (1980) Comparative study on the mechanism of irritation by sulfate and phosphate type of anionic surfactants. J Soc Cosmet Chem 31:45-66.

Kanerva L, Lauharanta J (1986) Variable effects of irritants (Methylmethacrylate, Terphenyls, Dithranol, and Methylglyoxalbis-Guanylhydrazone) on the fine structure of the epidermis. Arch Toxicol 9 Suppl:455.

Kastner D (1977) Irritancy potential of cosmetic ingredients. J Soc Cosmet Chem 28:741-754.

Luepke NP, Kemper FH (1984) The HET-CAM test: An alternative to the Draize eye test. Food Chem Toxicol 24:495-496.

Maibach HI, Lammintausta K, Berardesca E, Freeman S (1989) Tendency to irritation. Sensitive skin. J Am Acad Dermatol 21:833-835.

Malten KE (1981) Thoughts on irritant contact dermatitis. Contact Dermatitis 7:238-247.

Nagao S, Stroud JD, Hamada T, et al (1972) The effect of sodium hydroxide and hydrochloric acid on human epidermis: an EM study. Acta Derm Venereol (Stockh) 52: 11-23.

Oliver GJ, Pemberton MA (1988) An in vitro model for identifying skin-corrosive chemicats. I. Initiat vatidation. Toxicol in Vitro. 2:7-17.

Parce JW, Owicki JC, Kercso KM, et al. (1989) Detection of cell-affecting agents with silicon biosensor. Science 246:243-247.

Patrick E, Maibach HI (1989) Comparison of the time course, dose response, and mediators of chemically induced skin irritation in three species. In: Current Topics in Contact Derrnatitis. Frosch PJ, et al, eds., New York: Springer-Verlag, 399-403.

Serban GP, Henry SM, Cotty VF, et al. (1981) In vivo evaluation of skin lotions by electrical capacitance: I. The effect of several lotions on the progression of damage and healing after repeated insult with sodium laurel sulfate. J Soc Cosmet Chem 32:407-419.

Silverman J (1983) Preliminary findings on the use of protozoa (Tetrahymena thermophila) as models for ocular irritation testing rabbits. Lab Anim Sci 33:56-58.

Silverman J, Pennist S (1987) Evaluation of Tetrahymena thermophilia as an in vitro alternative to ocular irritation studies in rabbits. J Toxicol Cut Ocular Toxicol. 6:33-42

SOT Position paper, Comments on the LD50 and acute eye and skin irritation tests. Fundam Appl Toxicol 1989;13:621-623.

Stephens TJ, Silber PM, Recce B, et al. (1990) "Testskin": An in vitro model for detecting cytoxicity and inflarnmation. Toxicologist 10:78.

Wilhelm KP, Maibach HI (1990) Factors predisposing to cutaneous irritation. Contact Dermatitis 8:17-22.

(Adapted from Harvell J, Bason MM , Maibach HI. In vitro skin irritation assays: Relevance to Human skin. Clinical Toxicology, 30 (3): 359-369.1992.

In vitro responses to immunotoxicants.

H. Van Loveren, R.J. Vandebriel
National Institute of Public Health and Environmental Protection
PO Box 1, 3720 BA Bilthoven, the Netherlands

KEYWORDS/ABSTRACT: immunotoxicology/ immunotoxicity testing/ immune suppression/ allergy/ autoimmunity/ in vitro testing/ extrapolation/ risk assessment

This paper describes in general terms the function of the immune system, and to what consequences adverse effects of chemical exposure on the immune system can lead. It introduces approaches to test for potential immunotoxicity, and describes in more detail what in vitro test methods are available. The benefits and limitations of in vitro immunotoxicity testing are discussed.

1. Basic and cellular mechanisms of the immune system

The major function of the immune system is defense against bacteria, viruses, parasites, fungi, and certain tumors. The system is located in primary lymphoid organs (bone marrow, thymus) and secundary tissues (lymph nodes, spleen, lymphoid tissues along mucosal surfaces, blood, lymph). Many cell types and their mediators in concert execute the function of the system. They originate in the bone marrow, and home to their target tissues. Lymphocytes migrate to primary and secundary lymphoid tissues, and also recirculate via extravascular tissue to these secundary lymphoid tissues to mature and execute their function.

A first line of defense to microorganisms is executed by a physical and chemical barrier, such as on the skin, the respiratory tract and the alimentary tract. This barrier is helped by non-specific cells that are able to ingest and/or lyse pathogens. Such cells are phagocytes (macrophages, polymorphonuclear leukocytes), that ingest and lyse foreign material, and also natural killer cells belong to this first line of defense. These latter cells lyse tumor cells and virus-infected cells in a non-specific fashion.

After initial contact of the host with the pathogen, specific immune responses are induced. The hallmark of this second line of defense is specific recognition of determinants, so called antigens, of the pathogens. These specific immune responses help the non-specific defense presented to the pathogens by stimulating the efficacy of the non-specific responses; for instance, more non-specific cells can be recruited, or the phagocytic or lytic activity is enhanced. In addition, the specific immune system comprises highly efficient effector cells (such as cytotoxic T cells) that are not present within the non-specific system. A second

NATO ASI Series, Vol. H 93
Modulation of Cellular Responses in Toxicity
Edited by C. L. Galli, A. M. Goldberg, M. Marinovich

effect of induction of specific immunity is that memory develops. If a host is exposed to a pathogen to which it has been exposed at an earlier time point, the specific immune system reacts and in this instance the pathogen will be terminated much more quickly and efficiently than a pathogen that the host had never encountered before.

Specific cells are lymphocytes. They recognize foreign material specifically, by antigen receptors. Antigen receptors on B lymphocytes are immunoglobulins. Mature B cells (plasma cells) produce immunoglobulins that act as antibodies in serum or along mucosal surfaces. This constitutes the humoral arm of the specific immune system. There are 5 major classes of immunoglobulins. IgM circulates in blood. When it recognises antigens on a pathogen, it fixes complement. A cascade of processes is induced, in which ultimately the cell membrane of the pathogen is damaged, leading to lysis. IgG also circulates in blood, but this immunoglobulin is smaller than IgM, and can also be found in tissues. It neutralizes toxins, and is also able to opsonize pathogens, that can then be more easily ingested by phagocytes. IgG is able to cross the placenta, thus conveying immune responsiveness to the fetus. IgA, although also found in the circulation, is predominantly present on mucosal surfaces, where it protects them. IgE is present in serum as well as in tissues (on tissue mast cells). It is thought to play a role in defense to parasites, but is also involved in allergic and inflammatory processes. IgD is a membrane determinant of some B cells; the function is not well known.

Immunoglobulins consist of 2 heavy and 2 light chains, that are linked to each other. One end of the immunoglobulin molecule is the Fc part, that can interact with Fc receptors, that are found on cells such as B cells, macrophages, and mast cells. This part of the molecule is rather constant. The other end comprises is the variable region, that recognizes the antigens, and hence determines the specificity of the molecule. The different classes of immunoglobulins differ in their molecular weight. This stems from the fact that IgG for instance is a monomer, whereas IgM is a pentamer, i.e. 5 basic structures linked together form one IgM molecule. IgA is most often present as a dimer, linked together by a secretory component (produced by epithelial cells) that protects the immunoglobulin from protection by proteolytic enzymes, such as they occur in the gut.

The cellular arm of the specific immune system is mediated by T lymphocytes. These cells also have antigen receptors on their membranes. They recognize antigen if presented by antigen presenting cells in the context of histocompatibility antigens. Hence, these cells have a restriction in addition to the antigen specificity. T cells function as helper cells for various (including humoral) immune responses, mediate recruitment of inflammatory cells, and can be active as cytotoxic effector cells in a specific fashion. A dichotomy exists in T cell activities. Based on profiles of production of certain proteins (cytokines), these T cells are called T-helper 1 cells or T helper 2 cells. The former produce interleukin 1 and interferon gamma, and mediate inflammatory responses, whereas the latter produce interleukin 4 and interleukin 5, and help IgE production and recruitment of eosinophils. These immunoglobulins but also eosinophilic granulocytes are implicated to play a role in defense to parasites, but are also known in allergic reactions, o.a. extrinsic asthma.

The various components communicate with one another, and with other relevant cell types such as endothelial cells and fibroblasts, through soluble mediators such as immunoglobulins and cytokines, and molecules such as prostaglandins, leukotrienes, etc. They also communicate through receptors, such as antigen receptors, cytokine receptors, adhesion molecules, etc.

The number of antigens that can be encountered exceeds the capacity of the genome to encode antigen receptors to recognize this array. The repertoire of specificity develops by a process of gene rearrangements. This is a random process, during which various specificities are brought about. This includes specificities for self components, that are undesirable. A selection process, that takes place in the thymus (T cells), or bone marrow (B cells) operates to delete these undesirable specificities.

2. Immunotoxicity

Chemicals can affect the immune system. Direct effects of chemicals result in dysfunction of the system. This can lead to decreased resistance to infections or tumors, as is most evident after treatment with immunosuppressive drugs. Epidemiologic data that exposure to environmental immunotoxicants leads to increased incidence of infections are scant. In contrast, it is well known that treatment with immunosuppressive therapeutics, such as is the case with patients receiving organ transplants, show increased vulnerability to infectious diseases. Moreover, it is known that such patients suffer from higher incidences of certain types of cancer.

Autoimmunity, a situation in which immune reactions to self components occur, can also be brought about by chemicals. They may alter self determinants in such a way that these components are recognized as foreign. In addition, selection processes may be altered, so that undesired specificities occur. Data exist indicating that wildlife populations suffer from higher incidences of infectious diseases, associated with exposure to environmental immunotoxicants. This concerns fish in coastal waters of the Netherlands, and to seals in the Waddensea.

Certain chemicals are potent sensitizers. They induce immune responses that are as such physiologic, but that cause more harm to the tissue in which they take place than the chemical itself would have done. This situation is called allergy or hypersensitivity. There are two main subtypes of hypersensitivity. One is mediated by a specific class of immunoglobulins, (i.e) IgE, that triggers release of mediators from mast cells upon antigen recognition, leading to adverse reactions. The onset of these reactions is soon after antigen recognition, and is therefore named immediate type hypersensitivity. The other type of hypersensitivity is also named after its time course upon antigen recognition: delayed type hypersensitivity. Delayed type hypersensitivity is mediated by T lymphocytes, that produce chemoattractive mediators, recruiting mononuclear inflammatory cells to the site of antigen contact. This leads to a swelling reaction, 24 hr or longer after antigen contact. T cells that mediate delayed type hypersensitivity reactions produce interleukin 2 and interferon gamma which are both chemoattractants. These T cells are called Th1 cells. Also in immediate type hypersensitivity T cells play a role. These latter T cells produce interleukin 4, and 5, and function as helper cells for IgE synthesis. They are called Th2 cells.

In addition to altering the capacity to resist infections, direct immunotoxicity may alter the course of autoimmunity and of allergy.

3. Examples of immunotoxicants.

Examples of environmental immunotoxicants that induce immunosuppression are among others: dioxins, PCB's, benzidine, oxidant gases, benzene, toluene, xylol, asbestos (for which chemicals besides animal data there are also human data), and dimethylnitrosamine, hexachlorobenzene, organotin compounds, lead, cadmium, mercury, benzo(a)pyrene, for which there are only animal data. Many therapeutics exert immunotoxic activities, including alkylating agents, anti-inflammatory agents, anti-metabolites, natural products, estrogens, and opiates. Agents that induce allergic reactions are among others formaldehyde, phtalatic anhydrides, pesticides, food additives, anti-microbials, resins, toluene-diisocyanate, platinum salts, nickel, chromium, gold, mercury, beryllium, and many drugs. Of most cases for which exogenous causes of autoimmunity are known, these are predominantly drugs.

4. Assessment of immunotoxicity.

Assessment of potential direct immunotoxicity of chemicals can be well carried out in experimental animals. Histopathology of lymphoid organs, in addition to differential cell counting, determination of subpopulations of lymphocytes in the spleen, and measurement of total immunoglobulin levels in the serum have proven valuable screening parameters for potential immunotoxicity. An array of *in vivo* assays of the immune system, *in addition to ex vivo/in vitro* assays are available to study the functional consequences of such immunotoxic effects. Such parameters should address the different aspects of the immune system, i.e. non-specific responses (macrophage and natural killer activity), humoral responses (antibody responses to antigens, mitogen responsiveness), and cellular responses (mitogen responsiveness, mixed lymphocyte reactions, cytotoxic T cell responses, delayed-type hypersensitivity). In particular host resistance models are valuable in this respect, since these models offer endpoints that are clinically most relevant, and therefore relevant for estimation of risk to man. Since the defense mechanism that is put into action differs for different pathogens, different host resistance models need to be employed to study the consequences of immunotoxicity. Models that are used include bacterial, viral, parasitical, and tumor challenges.

An example of a compound that has been used as a model immunotoxicant to validate assessment methodologies is the immunosuppressive therapeutic azathioprine. This compound, that is used in the clinic, and has shown immunosuppressive activity in man, came out as a chemical with the immune system as its prime target also in rodent studies. This pertained to non-functional screening parameters of the immune system, i.e. weights of lymphoid organs, histopathology of the thymus, and numbers of circulating white blood cells. This immunotoxicity in rats was further corroborated with effects on functional parameters, i.e. effects on natural killer activity, mitogen responsiveness, and antibody responses to sheep erythrocytes and parasitic antigens.

Another compound that has been studied extensively is the organotin compound bis(tri-n-butyl)tinoxide. In the rat this compound induces involution of the thymus as its most sensitive parameter, along with autorosettes in mesenteric lymph nodes. In the spleen the cellularity of the periarteriolar lymphocyte sheath is decreased. Immunoglobulin levels in serum are altered, especially in animals exposed for longer periods of time increased concentrations of IgA are observed. Many functional parameters are affected by exposure to TBTO, such as mitogen responsiveness, natural killer activity, and delayed type hypersensitivity. In host resistance models, there appeared a significantly decreased resistance to *Trichinella spiralis* and Listeria monocytogenes. Especially if elder rats were exposed for longer periods of time to lower concentrations of TBTO, the host infectivity models were the only assays that showed altered immune capacity. This suggests that in these models the reserve capacity of the immune system was rather modest. In mice it is known that TBTO does not induce thymic involution. Because of this species difference, it is difficult to predict whether this important feature will occur in man after exposure to this chemical.

Hexachlorobenzene induces in rats increased cellularity of the marginal zone in the spleen. In addition, it induces focal accumulation of macrophages in the lungs and stimulated high endothelial venules in lymph nodes. Serum IgM levels were increased. Mitogen responses as well as antibody responses to antigens were stimulated. In two host infectivity models HCB failed to induce altered resistance. In the autoimmune model experimental allergic encephalomyelitis, in which autoimmunity is induced by immunization to myelin, HCB enhanced the course of the disease. In contrast, in another model, in which immunization with Freunds complete adjuvant leads to autoimmune arthritis, HCB diminished the responses. Although not fully understood, it can be concluded that HCB, by stimulating immune responses, has an immunotoxic activity. However, in mice HCB appears to suppress immune responses. Again, these contrasting results make it difficult to predict consequences of human exposure to HCB.

Ammonia caramel, a constituent of many foods, was shown in rats to be immunotoxic. It had a severe effect on thymus weight, decreased IgA serum levels, decreased numbers of circulating lymphocytes, decreased CD4/CD8 ratios in the periphery, had effects on the architecture of the thymus, and altered expression of macrophage differentiation markers. Functionally, this compound had a severe effect on mitogen responsiveness, decreased natural killer cell activity, diminished resistance to Trichinella spiralis, but enhanced resistance to Listeria monocytogenes. So far, no immunotoxic effects in man have been described.

Models for the assessment of potential induction of allergy or autoimmunity are also available. For skin sensitizers a number of skin tests are available. One promising test seems to be the auricular lymph node assay, that seems not only to be valuable for assessing (detecting) skin hypersensitivity, but also for respiratory allergy. The cytokine profile and antibody isotype induction profile, induced in the auricular lymph node after application of the test compound on the ear seems to predict skin hypersensitivity (if a Th1 response is induced), or respiratory allergy (if a Th2 response is induced). The swelling response of popliteal lymph nodes after injection of a test compound in the footpad of mice seems to correlate well with induction of autoimmunity.

Assessment of immunotoxic processes in man is difficult. Although there are tests available to study immune responses in man, it is often not easy to link the outcome of such assays to exposure to immunotoxicants, especially in the case of moderate changes as they may occur after exposure to environmental immunotoxicants. Only few biomarkers are available for the immune system, that provide specific information on the extent of exposure, or (susceptibility for) disease, associated with chemical exposure. Allergy induced by chemicals is relatively easy assessed in man, once it has developed. It is far more difficult to predict that in a given person a certain chemical will result in allergy.

5. *In vitro* immunotoxicology

Assessment of immunotoxicity in man is difficult, as often detailed information on exposure is lacking, and the immune system can be studied only to a limited extent. In experimental animals assessment of immunotoxicity is far more easy, but it is not easy to extrapolate immunotoxic data from animals to man. One tool for better extrapolation of animal data may be *in vitro* testing. Sources for human material that are immunologically relevant, and that are accessible to study (in various degrees) are: blood; umbilical vein blood/endothelial cells; bone marrow; bronchoalveolar lavage; nasal lavage; skin biopsies; donor organs; cell lines. Most of such cell populations can also be derived from animals. Cell lines that are available include: macrophages; T cells (Th1 and Th2), antigen specific); B cells (antigen specific); granulocytes (mast cells, basophilic cells); keratinocytes; and thymus epithelial cells. For the cells from these sources many tests are available. These produce among others parameters, such as viability, phagocytosis, respiratory burst, calcium influx, cytotoxicity, antigen presentation, antibody production, cytokine production, proliferation, apoptosis, and expression of cell surface markers.

Interleukins include IL-2, IFN-gamma, TNF-alpha (Th1 cells); IL-4, IL-5, IL-10 (Th2 cells), IL-1, TNF-alpha, IL-12 (macrophages), IL-1, IL-6, TNF-alpha (epithelial cells), IL-1, IL-2, IL-4, IL-6, IL-7 (thymus). Interleukins can be detected at various levels, i.e. functionally (using *in vitro* bioassays), at the level of protein (by ELISA), and molecular biologically (using nothern blotting, dot blotting RT/PCR, or *in situ* hybridisation). Surface markers for T cells are CD2, CD3, CD4, CD8, MHC Class I (constitutive), and CD25, CD26, MHC Class II (activation markers). Surface markers for B cells include CD19, FcR, Ig, MHC (constitutive), CD23, CD25, CD26 (activation markers). Surface markers for NK cells are CD16, and for macrophages/monocytes FcR, Complement receptors, MHC, CD14, CD16 (constitutive), and CD23, various enzymes (activation markers).

Adhesion molecules include LFA-1, VLA-4, ICAM-3, L-selectin (lymphocytes), and MAC-1, LFA-1, VLA-4, ICAM-1, L-selectin (macrophages).

In vitro exposure to chemicals may lead to alterations in any of the above. Indeed, examples of such studies have been presented. A beautiful example is the study of dioxin on thymus epithelium. Coculture of bone marrow cells with thymus epithelium induces maturation of the bone marrow precursor cells, that can be measured by mitogen responsiveness. Pretreatment of thymus epithelium with dioxin, that has these cells as one of their targets, led to decreased maturation. This was found not only with cells from rodent origin, but also with cells from human origin.

It must be noted, that in *in vitro* systems, the complexity of the immune system as it exists *in vivo*, is not present. This pertains to interactions between different cell types, such as is the case with *in vitro* studies of effects of ammonia caramel. Whereas *in vivo* exposure led to a severely decreased proliferative activity of lymphocytes, *in vitro* exposure had no such effects. Beside different cell types, that may be involved in immune responses and (different) effects of chemicals on each of them, also metabolizing of the test compound is a complexity that does not occur *in vitro*. Coculture of different relevant cell types, or addition of metabolizing systems (microsomal fraction of hepatocytes) to the *in vitro* culture may, to a certain extent, overcome these problems. Yet, it is clear that such *in vitro* systems are far away from the *in vivo* situation. For these reasons it is doubtful that *in vitro* testing will be very powerful in screening immunotoxicity of novel compounds in general. However, for prescreening for immunotoxicity of products such as those derived by recombinant technology, or monoclonal antibodies, *in vitro* testing may prove an adequate option. *In vivo* toxicity testing of such products may, in some cases, result in effects that are intolerable inconvenience for the experimental animals. The strength of *in vitro* immunotoxicity testing lays also in studies aimed at unravelling mechanisms of immunotoxicity.

Another important benefit may be that it offers a mode to compare sensitivity and mechanisms of immunotoxicity in experimental animals to man, TBTO for instance, induces thymus involution in rats, resulting in suppressed T lymphocyte functions. In mice effects on the thymus are not observed. *In vitro* studies have shown that TBTO also affects human lymphocytes. HCB induces immunostimulation in rats, and immunosuppression in mice. *In vitro* studies with human cells may shed more light on what consequences HCB exposure in man would have. Unfortunately, such studies on HCB have not yet been performed.

Such comparison of *in vitro* data, in addition to other available data both in the experimental animal and in man, may form a solid basis for extrapolating the larger data base as it exists in experimental animals, in particular those stemming from *in vivo* host resistance models using experimental infections or tumors.

Suggested further reading

- Sharma RP (Ed) (1981). Immunologic Considerations in Toxicology. CRC Press, Boca Raton

- Gibson G, Hubbard R, Parke DV (Eds) (1983). Immunotoxicology. Academic Press, London

- Dean JH, Luster MI, Munson AE, Amos H (Eds) (1985). Immunotoxicology and Immunopharmacology. Target Organ Toxicology Series. Raven Press, New York

- Descotes J (1986). Immunotoxicology of Drugs and Chemicals. Elsevier, Amsterdam

- Berlin A, Dean JH, Draper MH, Smith EMB, Spreafico F (Eds) (1987). Immunotoxicology. Martinus Nijhof Publishers, Dordrecht

- Kammüller ME, Bloksma N, Seinen W (Eds) (1989). Autoimmunity and Toxicology. Immunoregulation and disregulation by drugs and chemicals. Elsevier, Amsterdam

- Van Loveren H, Vos JG. Immunotoxicological considerations: A practical approach of immunotoxicity testing in the rat. In: Dayan AD and Paine AJ (Eds) (1989) Adv. Appl. Toxicol., Taylor and Francis Ltd, London, pp 143-163

- Dayan AD, Hertel RF, Hesseltine E, Kazantis G, Smith EM, Van Der Venne MT (Eds) (1990). Immunotoxicity of Metals and Immunotoxicology. Proceedings of an International Workshop. Plenum Press, New York

- Descotes J. Drug-induced Immune Diseases. Dukes MNG (Ed) (1990) Drug-induced Disorders Vol. 4. Series. Elsevier, Amsterdam

- Schuurman HJ, Krajnc-Franken MAM, Kuper CF, Van Loveren H, Vos JG. Immune System. In: Haschek-Hock WM, Rousseaux CG (Eds) (1991). Handbook of Toxicologic Pathology. Academic Press, Washington

- Newcombe DS, Rose NR, Bloom JC (Eds) (1992). Clinical Immunotoxicology. Raven Press, New York.

- Biological Markers in Immunotoxicology. National Research Council. National Academic Press, Washington, 1992.

- Miller K, Nicklin S (Eds) (1992). Principles and Practice of Immunotoxicology. Blackwell Scientific Publication, London

- Rogers V, Sonck W, Shephard E, Vercruysse A (Eds) (1993). Human Cells in in vitro pharmacotoxicology. Present Status within Europe. Vubpress, Brussels

- Burleson G (Ed) (In press). Methods in immunotoxicology. Wiley & Sons, New York.

- WHO Environmental Health Criterium Document. Priciples and Methods for Assessing Direct Immunotoxicity associated with Exposure to Chemicals. In Preparation.

NEPHROTOXICITY

G.Gordon Gibson
Molecular Toxicology Group
School of Biological Sciences
University of Surrey
Guildford, Surrey GU2 5XH
England, U.K.

INTRODUCTION

The mammalian kidney is a very complex organ, both anatomically and functionally, and plays an important role in the control and regulation of homeostasis. The cortex forms the major part of the kidney and receives 80 % of the tissue blood supply, which itself receives 25 % of the cardiac output, and in conjunction with the organic molecule concentrating abi;ity of this organ, predisposes the tissue to the toxicity of blood borne xenobiotics, and the cortex in particular. In addition, xenobiotics are delivered to the medulla and because of the anatomy of the vasa rectae and the loop of Henle, can become trapped by the countercurrent system in this region of the nephron.

TYPES OF NEPHROTOXITY AND MECHANISMS

Many structurally diverse xenobiotics exhibit nephrotoxicity, and some examples of these are given in Table 1.

LESION	EXAMPLE
Proximal tubule necrosis	Cadmium, mercury, hexachlorobutadiene
Distal tubule necrosis	Cisplatin, gentamycin
Changes in renal blood flow	Cyclosporin A
Medullary	Paracetamol
Mutagenesis and carcinogenesis	Hexachlorobutadiene

Table 1. Types of Nephrotoxicity

NATO ASI Series, Vol. H 93
Modulation of Cellular Responses in Toxicity
Edited by C. L. Galli, A. M. Goldberg, M. Marinovich

The above examples may arise via a varity of mechanisms in different regions of the kidney, including an effect on renal blood flow, glomerular filtration rate, cellular cytotoxicity, immunologically-mediated injury, obstruction, interaction with DNA, influence on the cell cycle or apoptosis (Walker and Duggin, 1992).

TECHNIQUES TO STUDY NEPHROTOXICITY

Just as there many different mechanisms of nephrotoxicity, equally there are several different techniques to study these mechanisms (Table 2).

In vivo techniques
Renal perfusion
Kidney slices
Renal homogenates
Renal cells in culture
Molecular techniques

Table 2. Techniques to Study Nephrotoxicity.

The choice of model system to use in the assessment of nephrotoxicity is a compromise between the complexity of in vivo integrity and the simplicity and sometimes misleading simplicity of cellular and molecular biology approaches. With this in mind, there is no one ideal system to study nephrotoxicity, and the choice of the method depends on the toxicological question being asked. Obviously, whole animal or human studies are the most relevant, but what does one measure as an index of nephrotoxity ? Blood and urine analysis of constitutive components seems an obvious answer, and indeed, blood urea nitrogen values are extensively used as an index of gross toxicity, but suffers from the disadvantage that this type of analysis yields no additional information on either renal site or mechanisms of toxicity. A more promising approach to analysing nephrotoxicity in vivo is by high resolution NMR urine analysis of endogenous metabolite patterns (Bales et al, 1984). For example, this approach has demonstrated that proximal tubule toxins produce marked glycosuria, aminoaciduria and lacticaciduria, whereas papillary toxins cause early increases in trimethylamine N-oxide and dimethylamine (Gartland et al, 1889).

The perfused kidney preparation has the advantage that it retains a closely integrated structure and function and is a useful model system in that it is free from systemic hormonal or adrenergic nervous regulaion. Whereas this may be a disadvantage with respect to comparison with the in vivo situation, it does allow a more focussed experimental approach and much valuable information has been gleaned from such studies. The drawbacks of perfusion studies are that they require a substantial amount of experimental expertise, they are only viable for up to around four hours with a marked loss of renal concentrating cpacity and it is relatively difficult to identify the specific site of xenobiotic-induced toxicity.
Renal tissue slices are becoming increasingly used in that cortical or medullary slices are readily produced and can be used to monitor the uptake and efflux of both endogenous compounds and xenobiotics and the influence of the latter on the former (Wolfgang et al, 1989). Then disadvantages of using slices are that they are rarely viable after 4 - 6 hours and they do not yield much information on segment-specific toxicity. Freshly isolated tubule cells are also used in nephrotoxicity studies, but suffer from the disadvantage that they have a short life span of 1 - 2 hours. Accordingly, much effort has been expended in assessing the utility of established cell lines in cell culture and the pig kidney-derived LLC - PK1 cell line has proved very useful as it retains several features of a typical proximal tubular epithelium, including the retention of several transport and enzyme-dependent bioactivation systems (Chen et al, 1990). This cell line has the additional advantage that, if grown on a porous support, it also retains its epithelial polarity, in that it allows both apical and basolateral exposure to xenobiotics (thus reflecting the in vivo situation) and hence further insight into site-specific nephrotoxicity.

CASE STUDY : HEXACHLOROBUTADIENE

Several nephrotoxicants require metabolic activation and specific kidney uptake to express their toxicity and the above in vivo and in vitro approaches have yielded a coherent understanding of the mechanism of action of several xenobiotics, particularly halogenated alkenes such as hexachlorobutadiene (HCBD, Dekant et al, 1990 ; Lock et al, 1984 andReichert et al, 1985). HCBD is an industrial chemical produced as a by product of several manufacturing processes, including high volume trichloroethylene production, and therefore has a potential occupational and environmental exposure. Acute exposure of experimental animals to HCBD results in a marked and distinct necrosis of the kidney, and the chemical is additionally mtagenic and carcinogenic. Initially, HCBD is conjugated to glutathione in the liver, excreted in the bile and into the gastrointestinal tract where it is cleaved to the N-acetyl cysteine conjugate of pentachlorobutadiene (N-AcCYS-PCBD). This organic anion is selectively taken up by the proximal tubular cells, where it is N-deacetylated to form the CYS-PCBD metabolite. This latter metabolite is subsequently cleaved by an enzyme termed cysteine conjugate beta-lyase (or C-S lyase), forming a biologically reactive thiol that is thought to produce the kidney necrosis, as summarised in Figure 1. It is salutary to note that the HCBD-induced renal necrosis is very site specific to the P_3 segment of the proximal convoluted tubule, because this is where both the organic anion concentrating mechanism and the cysteine conjugate beta-lyase activating enzyme are located (MacFarlane et al, 1989).

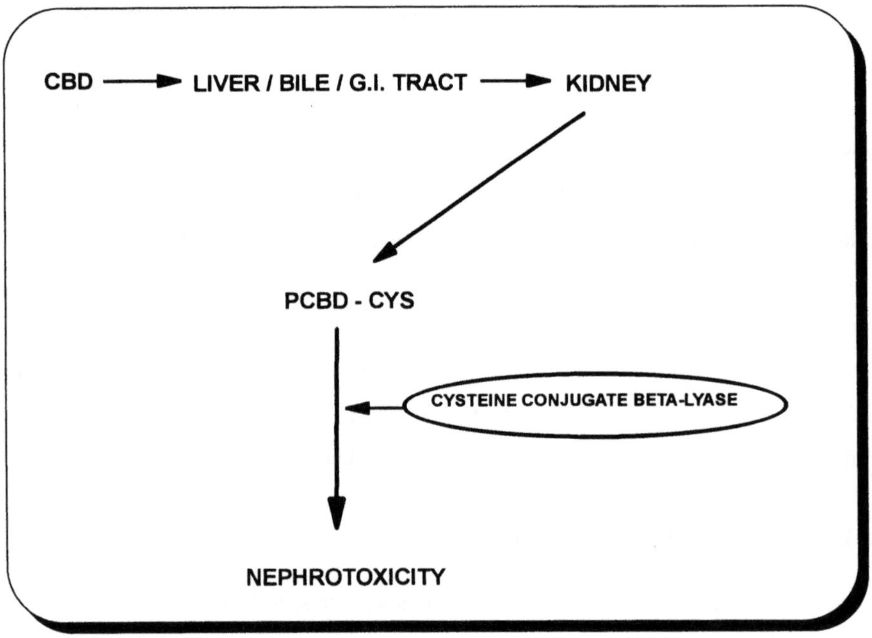

Figure 1. Role of Metabolism in the Nephrotoxicity of HCBD.
HCBD, hexachlorobutadiene ; PCBD, pentachlorobutadiene ; CYS, cysteine

Thus it is very clear that the cysteine conjugate beta-lyase enzyme plays a pivotal role in the expression of haloalkene-induced nephrotoxicity in animal models. But what relevance, if any, is this to man ? This is a difficult question to answer by traditional experimental approaches, but our research group has taken a molecular biology approach to understanding the human dimension, our ultimate goal being to clone and isolate the human cysteine conjugate beta-lyase gene, and to compare and contrast the substrate specificities of the expressed rat and human genes (Figure 2). In this way, we can directly assess the activity and substrate specificity of the cognate human enzyme, which can therefore make a substantial contribution to the human risk assessment of the nephrotoxicity of HCBD.

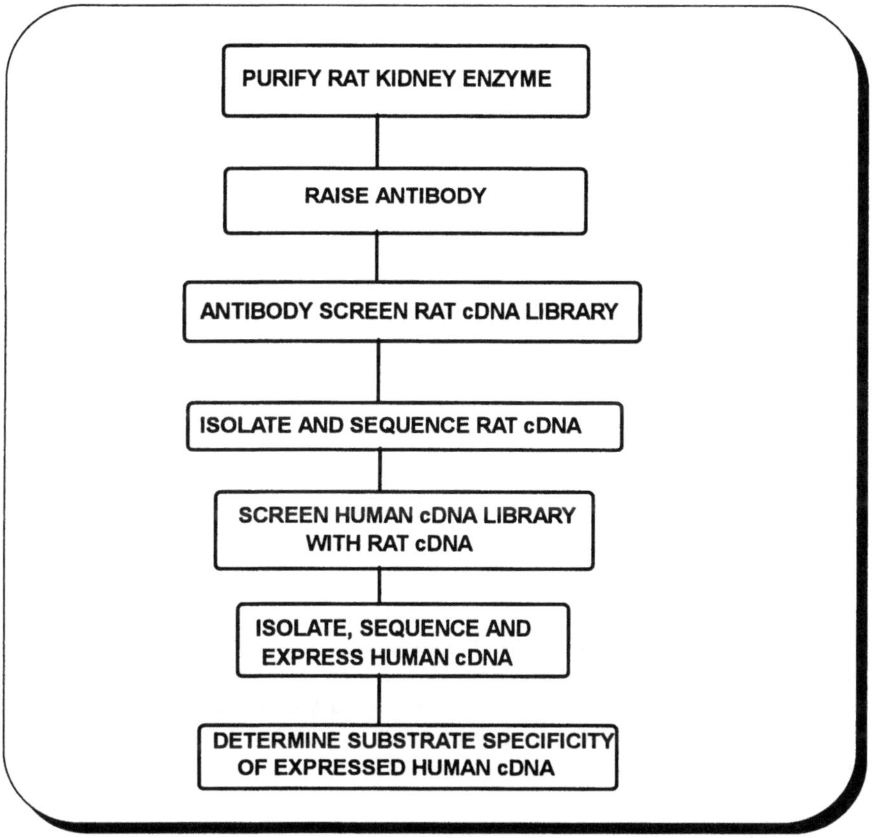

Figure 2. Strategy for the Isolation, Expression and Characterisation of the Human Cysteine Conjugate Beta-lyase Gene.

To summarise our studies to date, we have isolated the rat kidney cysteine conjugate beta-lyase enzyme to electrophoretic homogeneity and used this protein to raise a desorbed, monospecific antibody (MacFarlane et al, 1989), which in turn, was used to screen a rat kidney cDNA library. The isolated, full length rat cDNA has been completely sequenced, the identity of the correct gene established and a functional enzyme expressed by transient transfection of COS cells with the rat cDNA (Perry et al, 1993, Figure 3.)

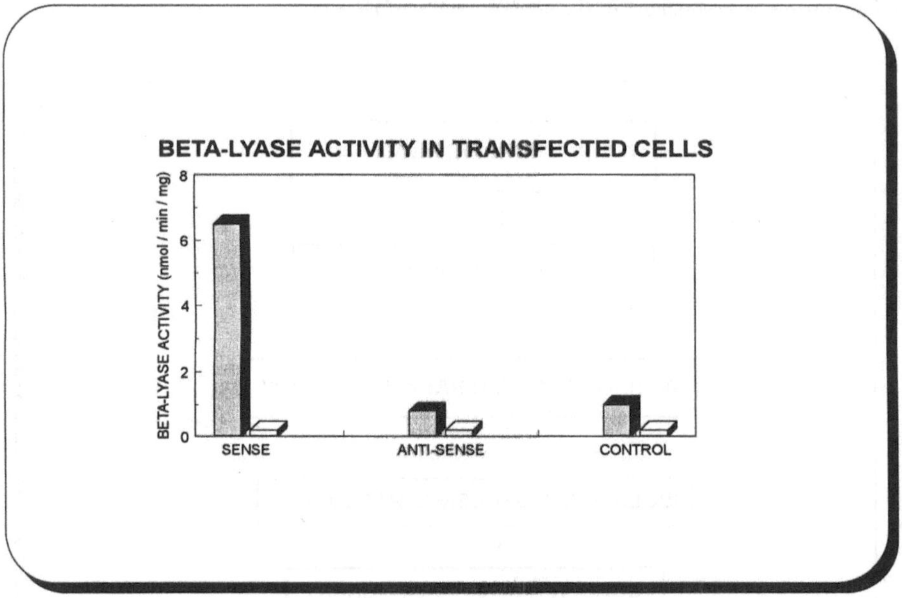

Figure 3. Functional Expression of the Rat Kidney Cysteine Conjugate Beta-lyase Gene. TFEC, the cysteine conjugate of 1,1,2,2-tetrafluoroethylene ; AOAA, aminooxy acetic acid. The open bars (second of the pair of bars) represents the resultant enzyme activity in the presence of AOAA. Note that no enzyme activity was expressed when the anti-sense cDNA was transfected, the expressed sense cDNA was essentially inhibited by AOAA (a beta-lyase inhibitor) and the control transfection (plasmid with no cDNA insert) resulted in almost negligible activity.

Current studies in our laboratory have used the rat cDNA to screen a human kidney cDNA library, and we have just recently completed the isolation and sequencing of the full length human cDNA, which shares approximately 90 % sequence similarity to the rat gene. Although our studies on the functional expression of the human gene and its substrate specificity are as yet incomplete, this molecular strategy provides a means to directly assess the role of a human kidney toxin activating enzyme, which would be very difficult to assess by any other means.

A further spin off from this molecular approach to understanding xenobiotic nephrotoxicity is that enables us to stably integrate the human cDNA into the genome of LLC-PK1 cells. This will allow the development of a genetically engineered cell line that stably expresses a human enzyme of toxicological significance and to concomitantly use the engineered cells themselves as a reporter end point of toxicity, after exposure to appropriate cysteine conjugate beta-lyase substrates.

REFERENCES

Bales, J.R., Higham,D.P., Howe,I., Nicholson, J.K. and Sandler, P.J. (1984). Use of high resolution proton nuclear manetic resonance spectroscopy for rapid multi-component analysis of urine. Clin.Chem. 30 : 426 - 432.

Chen, Q., Jones, T.W., Brown, P.C. and Stevens, J.L. (1990). The mechanism of cysteine conjugate cytotoxicity in renal epithelial cells. Covalent binding leads to thiol depletion and lipid peroxidation. J.Biol.Chem. 265 : 21603 - 21611.

Dekant, W., Vamvakas, S. and Anders, M.W. (1990). Bioactivation of hexachlorobutadiene by glutathione conjugation. Fd. Chem. Toxicol. 28 : 285 - 293.

Gartland, K.P.R., Bonner,F.W. and Nicholson, J.K. (1989). Investigations into the biochemical effects of region-specific nephrotoxins. Mol. Pharmacol. 35 : 242 - 250.

Lock, E.A., Ishmael, J. and Hook, J.B. (1984). Nephrotoxicity of hexachloro-1,3-butadiene in the mouse : the effect of age, sex, monooxygenase modifiers and the role of glutathione. Toxicol.App.Pharmacol. 72 : 484 - 494.

MacFarlane, M., Foster, J.R., Gibson, G.G., King, L.J. and Lock, E.A. (1989). Cysteine conjugate beta-lyase of rat kidney cytosol : characterisation, immunocytochemical localisation and correlation with hexachlorobutadiene nephrotoxicity. Toxicl. App. Pharmacol. 98 : 185 - 197.

Perry, S.J., Schofield, M.A., MacFarlane, M., Lock, E.A., King, L.J., Gibson, G.G. and Goldfarb, P.S. (1993). Isolation and expression of a cDNA coding for rat kidney cytosolic cysteine conjugate beta-lyase. Mol.Pharmacol. 43 : 660 - 665.

Reichert, D., Schutz, S. and Metzler, M. (1985). Excretion pattern and metabolism of hexachlorobutadiene in rats. Evidence for metabolic activation by conjugation reactions. Biochem.Pharmacol. 34 : 499 - 505.

Walker, R.J. and Duggin, G.G. (1992). Cellular mechanisms of drug nephrotoxicity. In, The Kidney : Physiology and Pathophysiology (D.W.Seldin and G.Giebisch, eds.), Raven Press, New York, pps.3571 - 3595.

Wolfgang, G.H.I., Gandolfi, A.J., Stevens, J.L. and Brendel, K. (1989). N-acetyl S-(1,2-dichlorovinyl)-L-cysteine produces a similar toxicity to S-(1,2-dichlorovinyl)-L-cysteine in rabbit renal slices : Differential transport and metabolism. Toxicol.App.Pharmacol. 101 : 205 - 219.

Neurotoxicity *in Vitro*

Cultured Neuronal Cell Lines and Electrophysiological Approach

H.P.M. Vijverberg, M. Oortgiesen, T. Leinders, R. Zwart
Research Institute of Toxicology
Utrecht University,
P.O. Box 80.176
NL-3508 TD Utrecht
The Netherlands

Introduction.

Traditionally, neurotoxic effects have been investigated in a number of model systems that were considered to express properties representative for specific functions of the nervous system. The recent discovery of a variety of genes coding for ion channel proteins and the investigations of the properties of these proteins has resulted in the concept of families of voltage- and receptor-operated ion channels. This development urges for reevaluation of existing and acquisition of novel model systems for the study of neurotoxic mechanisms of action. In parallel with these developments refined electrophysiological techniques have emerged to investigate the interactions of drugs and neurotoxins with ion channels in the neuronal membrane both at the cellular (whole-cell voltage clamp) and at the molecular (single channel patch clamp) level.

The characteristics of neuronal cell lines and their potential as *in vitro* models for neuropharmacological and neurotoxicological research have been topic of recent investigations. These studies have demonstrated that cultured neuroblastoma cells express a variety of voltage- and receptor-operated ion channels. In various research projects we have characterized these ion channels by investigating their basic physiological, pharmacological and toxicological properties. The vulnerability of some of these ion channels to modification by selective and non-selective neuroactive agents has been studied using voltage clamp techniques under strictly defined experimental conditions. The results thus far show that *in vitro* neuronal cell lines are useful models for the unequivocal

NATO ASI Series, Vol. H 93
Modulation of Cellular Responses in Toxicity
Edited by C. L. Galli, A. M. Goldberg, M. Marinovich

characterization of selective and non-selective mechanisms by which neuroactive agents modify specific neuronal targets.

Neuronal cell lines.

Since 1969, when various clonal cell lines were isolated from the mouse C-1300 tumor (for review see: Kimhi, 1981), neuroblastoma cells have developed into important model systems for basic research. Mouse neuroblastoma cells of the clone N1E-115 (Amano et al., 1972) have been studied extensively over the past decades. Developmental, morphological, electrophysiological, neurochemical and molecular biological investigations of these clonal cells have demonstrated the diversity and the complex interrelations of processes involved in excitability and signal transduction in these cells.

Cultured mouse N1E-115 neuroblastoma cells serve as a model to investigate cell growth and differentiation (Prasad, 1975, 1977; de Laat and van der Saag, 1982). They possess many neuronal properties which may be expressed upon or enhanced by various chemical stimuli that induce differentiation of these cells into morphologically mature neurons. Differentiated neuroblastoma cells are excitable and respond to electrical stimulation with one or more action potentials, that are mediated by voltage-dependent ion channels (Nelson et al., 1969, 1971; Chalazonitis and Greene, 1974; Tuttle and Richelson, 1975; Hamprecht, 1977). Neuroblastoma cells also express a variety of receptors and receptor-operated ion channels and some complete neurotransmitter systems are present. The relative ease by which large quantities of neuroblastoma cells are obtained has facilitated investigation of the properties of ion channels, receptors and signal transduction pathways by radioligand binding and other neurochemical detection techniques (Breakefield, 1976; Hamprecht, 1977; Pfenning and Richelson, 1990).

Electrophysiological properties are generally studied at the level of single cells. Fast and slow components are observed in the action potential of neuroblastoma cells, depending on membrane resting potential, ionic composition of the extra- and intracellular solutions and on stimulus parameters. The action potential may be followed by an afterhyperpolarization and is sometimes regenerative. Like in other neurons ion currents through voltage-dependent sodium, potassium and calcium channels as well as through calcium-dependent potassium channels underlie the electrical excitability of neuroblastoma cells. Various types of ligand-gated ion channels expressed by neuroblastoma cells (Table 1) may modulate electrical excitability. Ion currents passing through the excited membrane are highly interdependent, because the kinetic properties as well as the state of voltage-dependent ion channels depend on membrane potential. In addition, excitation may cause the release of intra- and intercellular messengers which modulate voltage- and receptor-operated ion channels. As a consequence of this complex interdependence, action potential parameters, and changes of these parameters, are generally poor indicators of the type(s) and size(s) of ion currents involved.

Table 1. Ion channels and receptors in mouse N1E-115 neuroblastoma cells

VOLTAGE-DEPENDENT ION CHANNELS		
name	*blocker*	*reference*
sodium channel	tetrodotoxin	Moolenaar and Spector, 1978
potassium channels		
slow	TEA[1]	Moolenaar and Spector, 1978
fast	TEA/4-aminopyridine	Quandt, 1988
calcium channels		Moolenaar and Spector, 1979
L-type	Cd^{2+}/streptomycin	Narahashi et al., 1987
T-type	Ni^{2+}	Yoshii and Tsunoo, 1988
chloride channel	9-anthracene carboxylic acid	Forshaw et al., 1993
RECEPTOR-OPERATED ION CHANNELS		
name	*antagonist*	*reference*
nicotinic ACh receptor	n-bungarotoxin	Oortgiesen and Vijverberg, 1989
muscarinic ACh receptor	atropine	Kato et al., 1983
serotonin 5-HT_3 receptor	ICS 205-930	Neijt et al., 1986, 1988
Ca^{2+}-activated channels		Moolenaar and Spector, 1979
BK	TEA	Romey et al., 1984
SK	apamin	Leinders and Vijverberg, 1992
cation	unknown	Yellen, 1982
metal ion-activated channel	unknown	Oortgiesen et al., 1990b

[1]tetraethylammonium ions

Voltage clamp techniques.

Under accurate and fast control of the membrane potential using appropriate voltage clamp techniques membrane currents can be measured directly and many of the drawbacks encountered with freely varying membrane potential are removed. At present any variants of the voltage clamp technique, initially developed for the squid giant axon (Hodgkin and Huxley, 1952), are available. All of the techniques rely on accurate measurement of membrane potential and a rapid feed-back amplifier to pass current through the membrane in order to maintain the membrane potential at the desired value. For the voltage clamping of neurons the two microelectrode technique (Hagiwara and Saito, 1959), the large suction pipette voltage clamp technique (Kostyuk et al., 1975; Lee et al., 1980) and the whole-cell patch clamp technique (Hamill et al., 1981) are applied. The latter two techniques are basically the same in that they use small glass pipettes that are sealed to the membrane of the neuron after which the electrical contact with the cell interior is established by rupture of the membrane inside the pipette. These suction pipette techniques are highly suited to voltage clamp isolated cells as they cause smaller leakage than microelectrodes and allow control of the internal ion concentration through dialysis. The smaller diameter pipettes used for patch clamping form extremely high-resistant seals with biological membranes and allow low-noise recordings of membrane currents in small cells and of single channel currents in membrane patches (Hamill et al., 1981). Under voltage clamp various types of ion channels can be investigated in separation after elimination of unwanted current components by ion substitution and/or addition of selective blockers to either external or internal solution, and by imposing the proper membrane potential or membrane depolarizing steps.

Selective neurotoxic effects.

Effects of selective neurotoxic compounds on ion channels in cultured neurons have been investigated using the whole-cell voltage clamp and single channel patch clamp techniques. The target sites of pyrethroids and nitromethylene heterocyclic (NMH) compounds, two classes of neuroactive insecticides, have been defined.

The pyrethroids constitute a major class of synthetic and highly active insecticides with a fairly selective neurotoxic mechanism of action. They produce pronounced excitatory effects in insects as well as in mammals, including man, whenever they reach the nervous system in sufficient concentration. The principal effect of pyrethroids is the induction of repetitive nerve activity in sense organs and in other parts of both the peripheral and the central nervous system. In addition, they may cause excessive neurotransmitter release due to depolarization of presynaptic nerve endings and, eventually, nerve conduction block (for review see: Vijverberg and van den Bercken, 1990). All insecticidally active pyrethroids share the same mechanism of action on voltage-dependent sodium channels in the nerve membrane. In frog myelinated nerve fibers pyrethroids induce a prolonged inward sodium tail current, that follows the sodium current evoked by a step depolarization of the membrane (Vijverberg et al., 1982). A similar prolonged sodium tail current is observed in squid giant axons and other invertebrate nerve fibers after treatment with pyrethroids (Narahashi, 1986). The prolongation of the sodium current varies greatly with pyrethroid structure. Cyano substituted pyrethroids prolong the sodium current to a much greater extent than non-cyano pyrethroids (Vijverberg et al., 1983). The time course of decay of the sodium tail current in frog myelinated nerve fibers correlates well with the intensity of repetitive nerve activity in the lateral-line sense organ of *Xenopus laevis*. Dependent on its size and time course the prolonged sodium current causes a depolarizing afterpotential, all or not associated with repetitive activity, or more permanent membrane depolarization. This may lead to excessive neurotransmitter release and to complete block of membrane excitability. In N1E-115 neuroblastoma cells exposed to pyrethroids the sodium current is prolonged both during depolarization and after repolarization, resulting in a sodium tail current. In addition, pyrethroids also induce an increase in the peak amplitude of the inward sodium current in N1E-115 cells (Ruigt et al., 1987). On the basis of the prolongation of macroscopic whole-cell sodium current by pyrethroids it was hypothesized that these insecticides selectively affect the kinetic properties of the sodium channels and that they keep these channels open for a prolonged period of time. Single channel patch clamp experiments on 'outside-out' patches of N1E-115 cells have in fact confirmed that the open time of individual sodium channels is greatly prolonged by pyrethroids (Fig. 1) (Chinn and Narahashi, 1986; de Weille, 1986; Yamamoto et al., 1983). Moreover, the summation of a large number of single channel records yields the reconstruction of 'macroscopic' sodium currents typically observed in whole-cell voltage clamp experiments and the slowing of its kinetics by pyrethroids is clearly demonstrated (Fig. 1). Similar results were obtained in dissociated frog spinal ganglion cells (de Weille and Leinders, 1989).

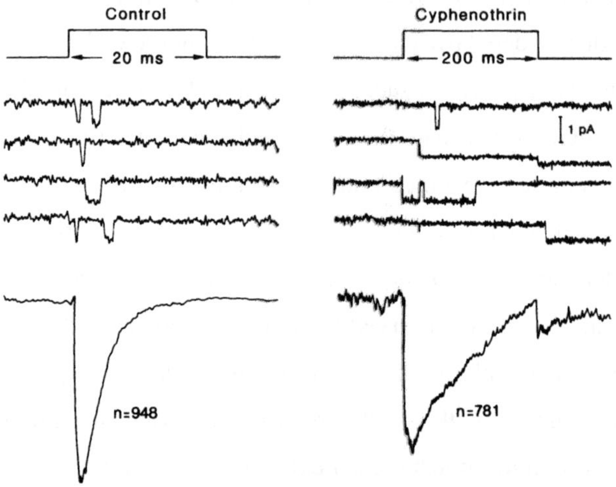

Figure 1. Opening and closing of individual sodium channels in excised (outside-out) membrane patches of cultured mouse neuroblastoma cells. Downward deflections of the current traces are discrete openings of single, voltage-dependent sodium channels, which are triggered by the membrane depolarization indicated on top of the traces.
In the presence of 10 μM of the pyrethroid (1*R,cis*)-cyphenothrin the open time of sodium channels is greatly prolonged and channel openings are also observed after termination of membrane depolarization. Summation of large numbers (n) of single channel records results in a rapid, transient sodium current in the control situation (left panel) and a prolonged sodium current followed by a sodium tail current in the presence of the pyrethroid (right panel). (adapted from de Weille, 1986)

The nitromethylene heterocyclic (NMH) insecticides directly interact with nicotinic acetylcholine receptors (nAChR) and produce agonistic and antagonistic effects (Benson, 1992; Sattelle et al., 1989; Leech et al., 1991). We have compared the effects of NMH insecticides on mammalian and non-mammalian nAChR (Zwart et al., 1992; 1994b). Effects of various NMH compounds on subtypes of nAChR have been investigated in locust thoracic ganglion neurons in primary culture, and in the mouse clonal N1E-115 neuroblastoma and BC3H1 muscle (Schubert et al., 1974) cell lines. These cells respectively express insect type neuronal nAChR of unknown subunit composition, mammalian type neuronal nAChR, most likely composed of α4 and ß2 subunits (Zwart et al., 1994a), and the fetal type of mammalian endplate nAChR composed of α1, ß1, γ and δ subunits (for review see: Sakmann et al., 1992). Whole-cell superfusion of voltage clamped neurons with ACh induced an inward current, which rapidly decays in the continued presence of the agonist, due to desensitization of the ligand-gated ion channels (Fig. 2). In locust neurons the NMH compound PMNI induced transient inward currents resembling ACh-induced inward current.

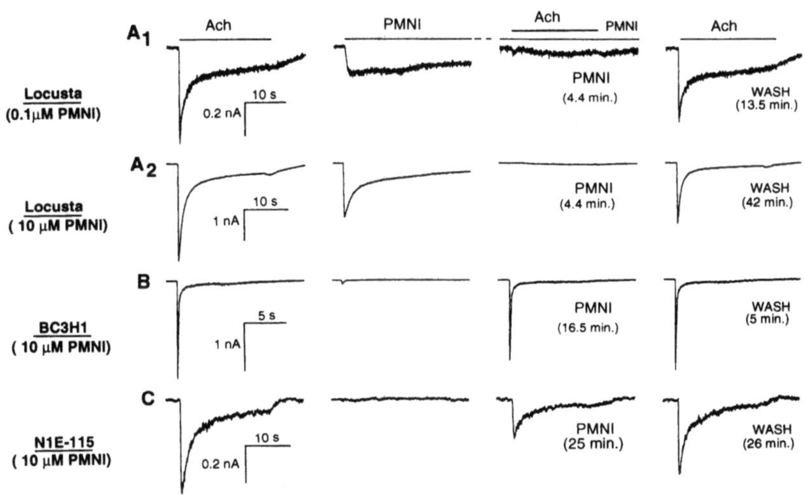

Figure 2. Effects of the nitromethylene heterocyclic insecticide PMNI on insect and mammalian subtypes of nAChR. Control inward currents (left column) were evoked by superfusion of the voltage clamped cells with saline solution containing 1 mM ACh at a holding potential of -80 mV. The decay of the ACh-induced inward currents is due to desensitization. Following recovery of desensitization by washing for 3-5 min, the same cells were superfused with PMNI at the concentrations indicated. In locust neurons (A1,2) PMNI also induced a transient inward current, which increased with PMNI concentration. In the continued presence of PMNI, the ACh-induced inward current was inhibited. This effect was reversed after removal of PMNI and ACh by prolonged washing (right column). Similar experiments on mouse BC3H1 muscle cells (B), and on mouse N1E-115 neuroblastoma cells (C) showed that the endplate and neuronal types of mammalian nAChR in these cells are much less sensitive to PMNI. (after Zwart et al., 1992)

The ACh-induced inward current was inhibited in the continued presence of PMNI (Fig. 2a1). Increasing the concentration of PMNI from 0.1 to 10 μM enhanced the amplitude of the PMNI-induced inward current as well as the PMNI-induced block of ACh-induced inward current (Fig. 2a2). Mammalian endplate type nAChR in BC3H1 cells and mammalian neuronal type nAChR in N1E-115 cells are much less sensitive to PMNI as compared to locust neuronal nAChRs. At 10 μM, PMNI partially inhibited the ACh-induced inward currents mediated by the two subtypes of mammalian nAChR, but failed to induce significant agonist effects (Fig. 2b,c).

Additional experiment showed that the inward currents induced by various NMH compounds were abolished in the presence of ACh, indicating that the effects of the endogenous agonist and NMH compounds cross-desensitize (Zwart et al., 1994b). From these results it has been concluded that the NMH compounds and ACh act at the same population of nAChRs in locust neurons. The NMH compounds appear to be agonists of the nAChR, which selectively interact with insect type nAChRs and cause activation and desensitization of nAChR-operated ion channels. The species comparison indicates that target site differences may contribute to the selective insecticidal activity of the NMH insecticides.

Non-selective neurotoxic effects.

As an example of non-selective neurotoxic action, we have investigated multiple concentration-dependent effects of inorganic lead (Pb^{2+}) on various types of ion channels in cultured N1E-115 mouse neuroblastoma cells. The effects of external Pb^{2+} on voltage-dependent calcium and sodium channels have been studied in whole-cell voltage clamped N1E-115 cells. In external solution containing 50 mM of barium ions membrane depolarization evokes a fast transient as well as a non-inactivating inward barium current in whole-cell voltage clamped neuroblastoma cells. These barium current components are carried by distinct transient and sustained types of voltage-dependent calcium channels (Narahashi et al., 1987). Superfusion with Pb^{2+} caused a reduction of the amplitude of both types of barium current within 5-10 min and the blocking effects were reversed after 5-10 min of washing with control external solution. The blocking effect of Pb^{2+} on the transient barium current component is concentration-dependent with an IC_{50} value of 4.8 ± 0.8 µM and a slope factor of -0.88 ± 0.14 (Fig. 3). The transient and sustained barium current components appear to have a similar sensitivity to Pb^{2+}. In contrast, voltage-dependent sodium current, evoked in cells bathed in normal external saline by membrane depolarizing steps, was not affected by 10-100 µM Pb^{2+} (Oortgiesen et al., 1990a). Effects of Pb^{2+} on calcium channels have been confirmed by a detailed investigation using N1E-115 cells (Audesirk and Audesirk, 1991) and similar results have been obtained with human neuroblastoma cells (Reuveny and Narahashi, 1991) and

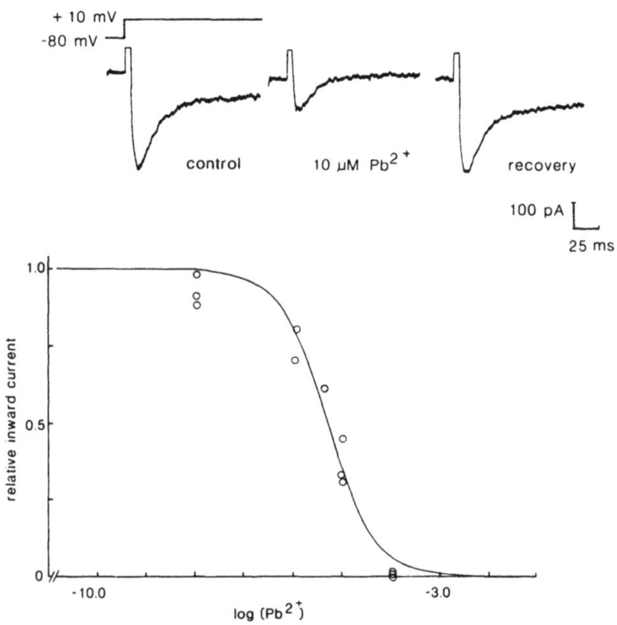

Figure 3. Pb^{2+} blocks voltage-dependent Ca^{2+} channels in N1E-115 cells. The upper panel shows Ba^{2+} currents evoked by step depolarizations to +10 mV in control external solution, after 5 min of superfusion with 10 µM Pb^{2+} and recovery after 5 min of washing with control external solution. The peak amplitude of the transient component of inward Ba^{2+} current was reduced to 34% of the control value by 10 µM Pb^{2+}. Holding potential was -80 mV. The lower panel shows the concentration dependence of the blocking effect of Pb^{2+}. Ordinate represents the peak amplitude of the transient inward current normalized to control value. The IC50 value and the slope factor estimated from the concentration-effect curve of block of the transient component of the Ba^{2+} current by Pb^{2+} are 4.8 ± 0.8 µM and -0.88 ± 0.14, respectively. (after Oortgiesen et al., 1990a)

with rat dorsal root ganglion cells (Evans et al., 1991). The IC_{50} obtained from N1E-115 cells is in the same order of magnitude as the value of 1 μM reported for the inhibition of calcium influx by Pb^{2+} in rat synaptosomes (Minnema et al., 1988; Suszkiw et al., 1984), and as the value of 1 μM of the apparent affinity of Pb^{2+} for presynaptic calcium binding in frog end-plate (Manalis et al., 1984). Thus the present experiments corroborate the hypothesis that presynaptic calcium channels constitute a target for the block of neurotransmission by micromolar concentrations of Pb^{2+} (Manalis et al., 1984; Pickett and Bornstein, 1984).

Receptor-activated ion currents in N1E-115 cells are also affected by external Pb^{2+}. The peak amplitude of the ACh-induced inward current is reduced by 32% during superfusion with 10 nM Pb^{2+}, while the time course of the ACh response remains unaffected at this very low concentration of Pb^{2+}. At 3 μM Pb^{2+} the amplitude of the ACh-induced inward current reaches only 10% of the control value. However, at higher concentrations of Pb^{2+} the blocking effect is reversed and after superfusion with 100 μM Pb^{2+} the peak amplitude of the ACh response amounts to almost 70% of the control value. In addition, the decay of the remaining ACh-induced inward current is markedly delayed by high concentrations of Pb^{2+}. Figure 4 shows the concentration dependence of the effect of Pb^{2+} on the inward current induced by 1 mM ACh in N1E-115 cells. The data can be fitted by the sum of a descending and an ascending concentration-effect curve with an IC_{50} value of 19 ± 6.3 nM and an EC_{50} of 21 ± 5.5 μM, respectively. The reversal of block of the ACh-induced current suggests a dual effect of Pb^{2+} on neuronal nicotinic ACh receptor-activated channels (Oortgiesen et al., 1990a). In frog end-plate 10 mM Ni^{2+} also causes a dual effect on ACh receptor-operated ion channels; a reduction of single channel conductance and a simultaneous prolongation of channel open time (Magleby and Weinstock, 1980). Recently, calcium ions have been shown to reduce single channel conductance and to increase opening frequency of ACh receptor-operated ion channels in rat central neurons (Mulle et al., 1992). Effects of Pb^{2+} on single cholinergic ion channels remain to be investigated.

Superfusion of neuroblastoma cells with 3 μM serotonin (5-HT) activates a transient inward current, mediated by an independent population of serotonin $5\text{-}HT_3$ receptor-operated ion channels (Neijt et al., 1986; 1989). In contrast to the nACh receptor-

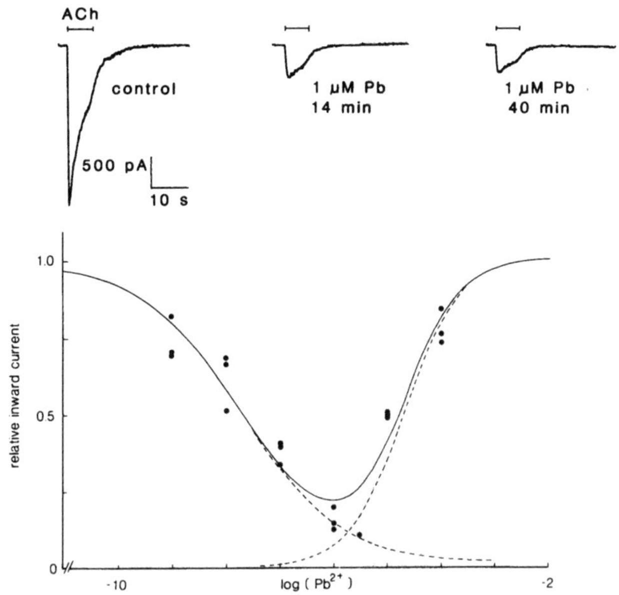

Figure 4. Effects of Pb^{2+} on nicotinic ACh-induced ion current. Upper panel left trace shows an inward current induced by 1 mM ACh in control external solution. During superfusion with 1 µM Pb^{2+} the peak amplitude of the ACh-induced inward current was reduced to a steady level which amounts 18% of the control value (middle and right traces). Note that 1 µM Pb^{2+} also delays the time course of decay of the ACh-induced inward current. Membrane potential was held at -80 mV. Superfusion periods are indicated by bars. Lower panel shows the concentration dependence of Pb^{2+} effects on the ACh-induced ion current. Ordinate represents the inward current peak amplitude normalized to control value. The data were fitted by the sum of two sigmoidal concentration-effect curves. The estimated parameters of the fitted curve (solid line) are: IC_{50} = 19 ± 6.3 nM; EC_{50} = 21 ± 5.5 µM and the slope factors are -0.45 ± 0.08 and 0.84 ± 0.15, respectively. The dashed lines represent the concentration-effect curve for the blocking effect of Pb^{2+} and for the reversal of block, according to the fitted parameters. (after Oortgiesen et al., 1990a)

operated ion current, serotonin 5-HT_3 receptor-operated ion current in N1E-115 cells was blocked only by high concentrations of Pb^{2+}. Application of 10 nM to 100 µM Pb^{2+} caused a concentration-dependent reduction of the 5-HT-induced inward current, without effects on the transient time course of the current. This blocking effect of Pb^{2+} was almost completely reversed by washing with external solution. The estimated values of the IC_{50} and the slope factor of the concentration-effect curve of block of the 5-HT-induced inward current by Pb^{2+} are 49 ± 18 μM and -0.32 ± 0.04, respectively (Oortgiesen et al., 1990a).

In addition to the effects described above, superfusion of N1E-115 cells with high concentrations of Pb^{2+} (> 10 μM) induced a slow, non-inactivating inward current, which was reversed by the removal of Pb^{2+} during washing. The amplitude of this inward current increased in the range of 1-200 μM Pb^{2+}. Exposure of excised outside-out membrane patches to Pb^{2+} revealed that the slow inward current was mediated by the opening of discrete ion channels (Fig. 5) with a single channel conductance of 24 pS. These single channel events were detected at Pb^{2+} concentrations ≥ 0.1 μM. From the reversal of the whole-cell membrane current and of the single channel currents at approximately 0 mV (not shown) it has been suggested that the Pb^{2+}-induced current is mediated by non-selective cation channels. The Pb^{2+}-induced ion channels could not be identified with any previously described type of ion channel, as they were neither

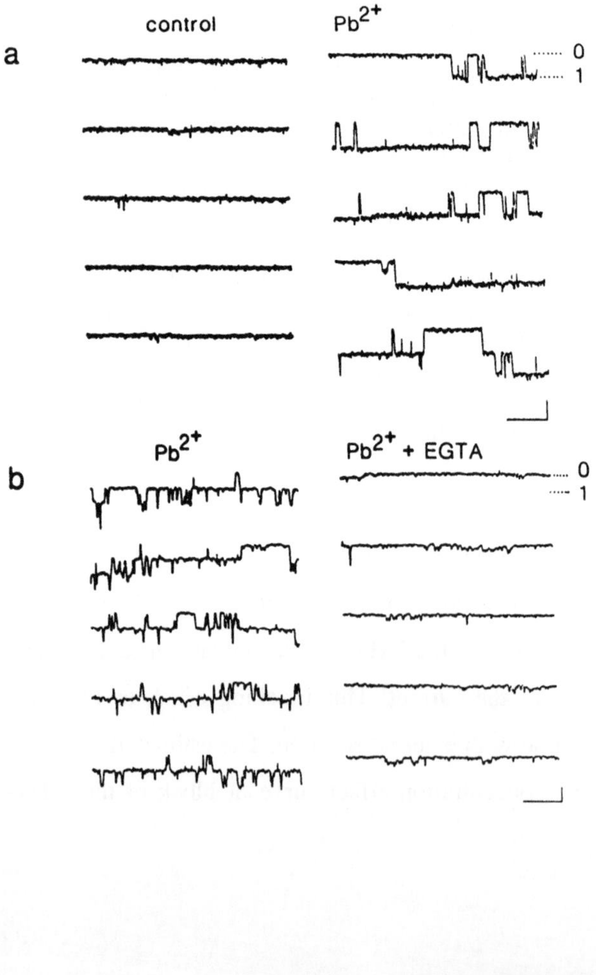

Figure 5. Single channel openings induced by Pb^{2+} in excised outside-out membrane patches of N1E-115 cells. (a) In control physiological solutions no channel openings were recorded. After addition of Pb^{2+} to the external solution at a final concentration of 10 μM discrete single channel openings occurred. (b) Multiple single channel openings induced by 10 μM Pb^{2+} disappear within 3-4 min after addition of 2 mM calcium-EGTA to the bathing solution. Calibration: horizontal 100 ms, vertical 2 pA. Membrane potential was held at -80 mV. (after Oortgiesen et al., 1990b)

blocked by external tetrodotoxin, TEA, d-tubocurarine, atropine, the potent 5-HT_3 antagonist ICS 205-930, nor by internal EGTA. Similar slow inward currents were induced by superfusion with Cd^{2+} and Al^{3+}. It has been concluded that the metal ion-induced ion currents are mediated by a novel type of metal ion-activated channel, which has no known physiological function at present (Oortgiesen et al., 1990b).

At the inside of excised membrane patches of neuroblastoma cells Pb^{2+} activates calcium-dependent potassium channels. Two types of calcium-dependent potassium channels have been identified in excised membrane patches of N1E-115 neuroblastoma cells. SK channels, which have a low single channel conductance of 5 pS, are potently blocked by the bee venom peptide apamin and show a relatively high sensitivity to calcium (Hugues et al., 1982; Leinders and Vijverberg, 1992). BK channels, which have a high single channel conductance of 98 pS, are sensitive to block by TEA and are less sensitive to calcium (Romey et al., 1984; Quandt, 1988; Leinders and Vijverberg, 1992). Figure 6 shows representative traces of single SK and BK channel recordings from two inside-out excised patches of N1E-115 membrane at a holding potential of 0 mV. In these experiments SK and BK channels were maximally activated by superfusion of the inside of the patches with solutions containing 14.4 and 115.2 μM buffered free calcium, respectively. During subsequent superfusion with calcium-free, EGTA-containing solution no single channel openings were observed. In the same membrane patches the superfusion of calcium-free external solution containing 1 μM free Pb^{2+} caused the activation of SK and of BK channels. At the given concentration of Pb^{2+} the open probability of the SK channel was the same as during superfusion with a maximally activating concentration of calcium. Conversely, BK channel open probability in the presence of 1 μM Pb^{2+} is only a fraction of the maximum attainable with calcium. Effects of several other metal ions on SK and BK channels in N1E-115 cells were also investigated. The open probability of the channels in the presence of the metal ions, as related to the maximum obtained with calcium, varied dependent on the metal ion species and concentration. Potency orders derived from effects measured at metal ion concentrations between 1 and 100 μM are for the SK channel: $Cd^{2+} \approx Pb^{2+} > Ca^{2+} > Co^{2+} >> Mg^{2+}$ and for the BK channel: $Pb^{2+} > Ca^{2+} > Co^{2+} >> Cd^{2+}$, Mg^{2+}. The sequences show that Pb^{2+} is more potent than Ca^{2+} in activating both SK and BK channels. Cd^{2+} is also a very potent activator of SK channels, but is unable to activate BK channels even at a concentration of 100 μM. Mg^{2+} is completely inactive at concentrations up to 100 μM (Leinders et al., 1992).

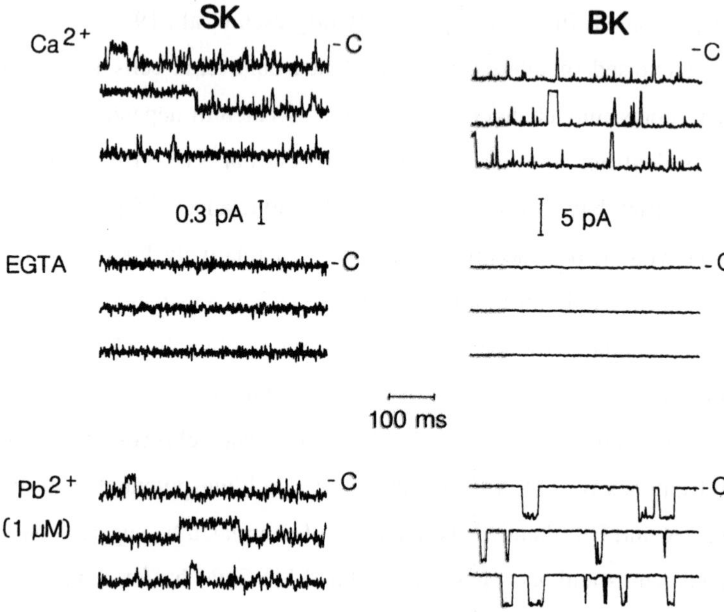

Figure 6. Effects of Pb^{2+} on calcium-dependent potassium channels. Maximum open probability of single small (SK, left) and big (BK, right) calcium-dependent potassium channels in two inside-out excised patches during superfusion with 14.4 and 115.2 μM buffered free calcium, respectively. Subsequent superfusion with calcium-free EGTA-containing solution abolished single channel activity. In the same membrane patches superfusion with 1 μM buffered free Pb^{2+} induced full activation of the SK and partial (10% of maximum open probability) activation of the BK channel. The cells were bathed in salines creating a physiological potassium gradient and the membrane potential was held at 0 mV. (after Leinders et al., 1992)

The results, which are summarized in Table 2, show that neuronal type nAChRs are the most sensitive to Pb^{2+} and are selectively blocked by nanomolar concentrations of Pb^{2+}. The same low concentrations of Pb^{2+} neither affect 5-HT_3 receptor-activated channels, nor voltage-dependent sodium and calcium channels, or big calcium-dependent potassium channels. Small calcium-dependent potassium channels may constitute another sensitive target of Pb^{2+} after entering the cytoplasmic compartment of neurons. Table

2 also demonstrates that, at low as well as at high concentrations, Pb^{2+} selectively modifies subtypes of channels of the various classes investigated. In the micromolar range external Pb^{2+} blocks voltage-dependent calcium channels selectively as compared to sodium channels and internal Pb^{2+} selectively activates calcium-dependent SK channels as compared to BK channels. The differential effects on neuronal type nicotinic ACh receptor-operated ion channels and 5-HT_3 receptor-operated channels also suggests selective and distinct interactions of Pb^{2+} with subtypes of ligand-gated ion channels. Thus, it appears that the effects of Pb^{2+} on the nervous system excitability depends on the concentration of the heavy metal as well as on the availability of and the differential affinity of Pb^{2+} to the various target sites.

Table 2. Effects of Pb^{2+} on various types of ion channels in cultured mouse N1E-115 neuroblastoma cells. Left hand column IC_{50} values are the external concentrations of Pb^{2+} required to inhibit 50% of the ion current. EC_{50} values denote the concentrations for 50% reversal the inhibitory effect or for 50% activation of the ion currents.

	inhibition (IC_{50})	*activation* (EC_{50})
voltage-dependent ion channels		
calcium channels	5 µM	
sodium channels	> 100 µM	
receptor-operated ion channels		
neuronal nicotinic ACh receptor	20 nM	21 µM
serotonin 5-HT_3 receptor	50 µM	
metal ion-activated channels		
Pb^{2+}-activated cation channels		> 30 µM
Ca^{2+}-dependent SK channels	> 100 µM	< 1 µM[1]
Ca^{2+}-dependent BK channels	> 100 µM	> 1 µM[1]

[1]buffered free internal Pb^{2+} concentrations. All others indicate total external Pb.

Concluding remarks.

The results illustrate the suitability of *in vitro* models to investigate specific mechanisms of action and target- as well as species-selectivity of neuroactive agents. N1E-115 neuroblastoma cells contain a variety of voltage-dependent, calcium-dependent and receptor-operated channels. The investigation of structure-related effects of selective agents at the target site provides information on their intrinsic potencies and such knowledge may assist in the prediction and the interpretation of the nature of neurotoxic effects in intact animals. The diversity of ion channels in these cells can be exploited to investigate effects of non-selective neurotoxicants over a wide concentration range and on a variety of potential targets. The example of Pb^{2+} demonstrates that this is not only confirmatory but also leads to surprising results as the finding that postsynaptic neuronal type nAChRs constitute a highly sensitive target for Pb^{2+} and the discovery of ion channels that are yet to be supplied with a counterpart *in vivo*. Since it is highly unlikely that a single cell type can ever be regarded as a universal model of the mammalian neuron, additional models with complementary properties are required to extend *in vitro* neurotoxicological research. The use of human cell lines, which are presently in the focus of interest, will allow to define the specific toxicological properties of human subtypes of receptors and ion channels. Investigation of basic physiological, biochemical and pharmacological properties of distinct cell types is time consuming, but seems to be essential to adequately define and select complementary models.

Acknowledgements.

The experimental results described in this chapter were obtained in research projects subsidized by the Foundation for Biological Research (BION/NWO) and by Shell Internationale Petroleum Mij. B.V.

References.

Amano T, Richelson E, Nirenberg PG (1972) Neurotransmitter synthesis by neuroblastoma clones. Proc Natl Acad Sci USA 69:258-263

Audesirk G, Audesirk T (1991) Effects of inorganic lead on voltage-sensitive calcium channels in N1E-115 neuroblastoma cells. Neurotoxicol 12:519-528

Benson JA (1992) Electrophysiological pharmacology of the nicotinic and muscarinic cholinergic responses of isolated neuronal somata from locust thoracic ganglia. J Exp Biol 170:203-233

Breakefield XO (1976) Neurotransmitter metabolism in murine neuroblastoma cells. Life Sci 18:267-278

Chalazonitis A, Greene LA (1974) Enhancement in excitability properties of mouse neuroblastoma cells cultured in the presence of dibutyryl cyclic AMP. Brain Res 72:340-345

Chinn K, Narahashi T (1986) Stabilization of sodium channel states by deltamethrin in mouse neuroblastoma cells. J Physiol (London) 380:191-207

De Laat SW, van der Saag PT (1982) The plasma membrane as regulatory site in growth and differentiation of neuroblastoma cells. Int Rev Cytol 74:1-54

De Weille JR (1986) The modification of nerve membrane sodium channels by pyrethroids. Dissertation, Utrecht University, The Netherlands

De Weille JR, Leinders T (1989) The action of pyrethroids on sodium channels in myelinated nerve fibres and spinal ganglion cells of the frog. Brain Res 482:324-332

Evans ML, Büsselberg D, Carpenter DO (1991) Pb^{2+} blocks calcium currents of cultured dorsal root ganglion cells. Neurosci Lett 129:103-106

Forshaw PJ, Lister T, Ray DE (1993) Inhibition of a neuronal voltage-dependent chloride channel by the type-II pyrethroid, deltamethrin. Neuropharmacol 32:105-111

Hagiwara S, Saito N (1959) Voltage-current relations in nerve cell membrane of *Onchidium verruculatum*. J Physiol (London) 148:161-179

Hamill OP, Marty A, Neher E, Sakmann B, Sigworth FJ (1981) Improved patch-clamp techniques for high-resolution current recording from cells and cell-free membrane patches. Pflügers Arch 391:85-100

Hamprecht B (1977) Structural, electrophysiological, biochemical, and pharmacological properties of neuroblastoma-glioma cell hybrids in cell culture. Int Rev Cytol 49:99-170

Hodgkin AL Huxley AF (1952) A quantitative description of membrane current and its application to conduction and excitation in nerve. J Physiol (London) 117:500-544

Hugues M, Romey G, Duval D, Vincent JP, Lazdunski M (1982) Apamin as a selective blocker of the calcium-dependent potassium channel in neuroblastoma cells: Voltage-clamp and biochemical characterization of the toxin receptor. Proc Natl Acad Sci USA 79:1308-1312

Kato E, Anwyl R, Quandt N, Narahashi T (1983) Acetylcholine-induced electrical responses in neuroblastoma cells. Neurosci 8:643-651

Kimhi Y (1981) Nerve cells in clonal systems. In: Nelson PG, Lieberman M (eds) Excitable cells in tissue culture. Plenum Press, New York, pp 173-245

Kostyuk PG, Krishtal OA, Pidoplichko VI (1975) Effect of internal fluoride and phosphate on membrane currents during intracellular dialysis of nerve cells. Nature 257:691-693

Lee KS, Akaike N, Brown AM (1980) Properties of internally perfused, voltage-clamped, isolated nerve cell bodies. J Gen Physiol 71:489-507

Leech CA, Jewess P, Marshall J, Sattelle DB (1991) Nitromethylene actions on in situ and expressed insect nicotinic acetylcholine receptors. FEBS Lett 290:90-94

Leinders T, van Kleef RGDM, Vijverberg HPM (1992) Divalent cations activate SK and BK channels in mouse neuroblastoma cells: Selective activation of SK channels by cadmium. Pflügers Arch 422:217-222

Leinders T, Vijverberg HPM (1992) Ca^{2+} dependence of small Ca^{2+}-activated K^+ channels in cultured N1E-115 mouse neuroblastoma cells. Pflügers Arch 422:223-232

Magleby KL, Weinstock MM (1980) Nickel and calcium ions modify the characteristics of the acetylcholine receptor-channel complex at the frog neuromuscular junction. J Physiol (London) 299:203-218

Manalis RS, Cooper GP, Pomeroy SL (1984) Effects of lead on neuromuscular transmission in the frog. Brain Res 294:95-109

Minnema DJ, Michaelson IA, Cooper GP (1988) Calcium efflux and neurotransmitter release from rat hippocampal synaptosomes exposed to lead. Toxicol Appl Pharmacol 92:351-357

Moolenaar WH, Spector I (1978) Ionic currents in cultured mouse neuroblastoma cells under voltage-clamp conditions. J Physiol (London) 278:265-286

Moolenaar WH, Spector I (1979) The calcium current and the activation of a slow potassium conductance in voltage clamped mouse neuroblastoma cells. J Physiol (London) 292:307-323

Mulle C, Léna C, Changeux JP (1992) Potentiation of nicotinic receptor response by external calcium in rat central neurones. Neuron 8:937-945

Narahashi T (1986) Mechanisms of actions of pyrethroids on sodium and calcium channel gating. In: Ford MG, Lunt GG, Reay RC, Usherwood PNR (Eds) Neuropharmacology and Pesticide Action, Ellis Horwood Ltd., Chichester, England, pp 36-60

Narahashi T, Tsunoo A, Yoshii M (1987) Characterization of two types of calcium channels in mouse neuroblastoma cells. J Physiol (London) 383:231-249

Neijt HC, Plomp JJ, Vijverberg HPM (1989) Kinetics of the membrane current mediated by serotonin 5-HT_3 receptors in cultured mouse neuroblastoma cells. J Physiol (London) 411:257-269

Neijt HC, te Duits IJ, Vijverberg HPM (1988) Pharmacological characterization of serotonin 5-HT_3 receptor-mediated electrical response in cultured mouse neuroblastoma cells. Neuropharmacol 27:301-307

Neijt HC, Vijverberg HPM, van den Bercken J (1986) The dopamine response in mouse neuroblastoma cells is mediated by serotonin 5-HT_3 receptors. Eur J Pharmacol 127:271-274

Nelson PG, Peacock JH, Amano T, Minna J (1971) Electrogenesis in mouse neuroblastoma cells *in vitro*. J Cell Physiol 77:337-352

Nelson PG, Ruffner W, Nirenberg M (1969) Neuronal tumor cells with excitable membranes grown *in vitro*. Proc Natl Acad Sci USA 64:1004-1010

Oortgiesen M, van Kleef RGDM, Bajnath RB, Vijverberg HPM (1990a) Nanomolar concentrations of lead selectively block neuronal nicotinic responses in mouse neuroblastoma cells. Toxicol Appl Pharmacol 103:165-174

Oortgiesen M, van Kleef RGDM, Vijverberg HPM (1990b) Novel type of ion channel activated by Pb^{2+}, Cd^{2+}, and Al^{3+} in cultured mouse neuroblastoma cells. J Membr Biol 113:261-268

Oortgiesen M, Vijverberg HPM (1989) Properties of neuronal type acetylcholine receptors in voltage-clamped mouse neuroblastoma cells. Neurosci 31:169-179

Pfenning MA, Richelson E (1990) Methods for studying receptors with cultured cells of nervous tissue origin. In: Yamamura HI, et al. (eds) Methods in neurotransmitter receptor analysis, Raven Press, New York, pp 147-175

Pickett JB, Bornstein JC (1984) Some effects of lead at mammalian neuromuscular junction. Am J Physiol 246:C271-C276

Prasad KN (1975) Differentiation of neuroblastoma cells in culture. Biol Rev 50:129-265

Prasad KN (1977) Role of cyclic nucleotide in the differentiation of nerve cells. In: Federof S, Hertz L (eds) Cell, organ and tissue culture in neurobiology, Academic Press, New York, pp 448-483

Quandt FN (1988) Three kinetically distinct potassium channels in mouse neuroblastoma cells. J Physiol (London) 395:401-418

Reuveny E, Narahashi T (1991) Potent blocking action of lead on voltage-activated calcium channels in human neuroblastoma cells SH-SY5Y. Brain Res 545:312-314

Romey G, Hugues M, Schmid-Antomarchi H, Lazdunski M (1984) Apamin: a specific toxin to study a class of Ca^{2+}-dependent K^+ channels. J Physiol (Paris) 79:259-264

Ruigt GSF, Neijt HC, van der Zalm JM, van den Bercken J (1987) Increase of sodium current after pyrethroid insecticides in mouse neuroblastoma cells. Brain Res 437:309-322

Sakmann B, Witzemann V, Brenner H (1992) Developmental changes in acetylcholine receptor channel structure and function as a model for synaptic plasticity. In: Fidia Research Foundation Neuroscience Award Lectures, Vol. 6, Raven Press, New York, pp 51-103

Sattelle DB, Buckingham SD, Wafford KA, Sherby SM, Bakry NM, Eldefrawi AT, Eldefrawi ME, May TE (1989) Actions of the insecticide 2(nitromethylene)-tetrahydro-1,3-thiazine on insect and vertebrate nicotinic acetylcholine receptors. Proc Roy Soc London B 237: 501-514

Schubert D, Harris AJ, Devine CE, Heinemann S (1974) Characterization of a unique muscle cell line. J Cell Biol 61:397-413

Suszkiw J, Toth G, Murawsky M, Cooper GP (1984) Effects of Pb^{2+} and Cd^{2+} on acetylcholine release and Ca^{2+} movements in synaptosomes and subcellular fractions from rat brain and *Torpedo* electric organ. Brain Res 323:31-46

Tuttle JB, Richelson E (1975) Ionic excitation of a clone of mouse neuroblastoma. Brain Res 84:129-135

Vijverberg HPM, van den Bercken J (1990) Neurotoxicological effects and the mode of action of pyrethroid insecticides. CRC Crit Rev Toxicol 21:105-126

Vijverberg HPM, van der Zalm JM, van den Bercken J (1982) Similar mode of action of pyrethroids and DDT on sodium channel gating in myelinated nerves. Nature 295:601-603

Vijverberg HPM, van der Zalm JM, van Kleef RGDM, van den Bercken J (1983)

Temperature and structure-dependent interaction of pyrethroids with the sodium channels in frog node of Ranvier. Biochim Biophys Acta 728:73-82

Yamamoto D, Quandt FN, Narahashi T (1983) Modification of single sodium channels by the insecticide tetramethrin. Brain Res 274:344-349

Yellen G (1982) Single Ca^{2+}-activated nonselective cation channels in neuroblastoma. Nature 296:357-359

Yoshii M, Tsunoo A (1988) Pharmacological characterization of two types of calcium channels in neuroblastoma cells. Biomed Res 9 Suppl. 2:29-30

Zwart R, Abraham D, Oortgiesen M, Vijverberg HPM (1994a) α4ß2-subunit combination specific pharmacology of neuronal nicotinic acetylcholine receptors in N1E-115 neuroblastoma cells. Brain Res in press

Zwart R, Oortgiesen M, Vijverberg HPM (1992) The nitromethylene heterocycle 1-(pyridin-3-yl-methyl)-2-nitromethylene-imidazolidine distinguishes mammalian from insect nicotinic receptor subtypes. Eur J Pharmacol 228:165-169

Zwart R, Oortgiesen M, Vijverberg HPM (1994b) Nitromethylene heterocycles: Selective agonists of nicotinic receptors in locust neurons as compared to mouse N1E-115 and BC3H1 cells. Pestic Biochem Physiol 48:202-213

In vitro Toxicology of the Respiratory System

Robert A. Roth
Department of Pharmacology and Toxicology
Institute for Environmental Toxicology
Michigan State University
E. Lansing, MI 48824 USA

INTRODUCTION

The lungs represent an important target organ for toxicants and present particular challenges regarding the development of *in vitro* methodologies and their use in screening of pneumotoxicants and in understanding mechanisms by which chemicals injure the respiratory tract. The development of *in vitro* lung methodologies for use in mechanistic studies has received considerable attention over the past several years. Some methods, for example, lung organ culture and culture of specific airway epithelial cells, presently are best described as still under development. Others, such as isolated, perfused lung preparations, have been used for many years and are well characterized. Attention among toxicologists to developing methods for *in vitro* screening of potential pneumotoxicants has been less than that devoted to certain other tissues, such as the eye and skin. Indeed, there seem to be relatively few investigators interested in development of *in vitro* lung screening methodologies as a major focus of study. Perhaps this is due to less external pressure of the types that have driven efforts in areas such as ocular and dermal testing.

To appreciate the particular challenges of the lungs in the development and use of *in vitro* methodologies, it is necessary to understand the organization of the respiratory system and the role of its various cells as well as how the lungs as an organ respond to chemically induced injury. A detailed description of lung anatomy and function is beyond the scope of this commentary, but a variety of useful texts is available (Netter, 1980; Gil, 1982; Spencer, 1985; 1989). After a brief introduction, various *in vitro* methodologies for evaluating

NATO ASI Series, Vol. H 93
Modulation of Cellular Responses in Toxicity
Edited by C. L. Galli, A. M. Goldberg, M. Marinovich

injury to the respiratory system will be discussed. Details of the various methods have been described elsewhere and will be treated only briefly. Advantages and limitations of specific methodologies will be listed, but emphasis will be placed on how selected methods have been used to advance knowledge about mechanisms of lung injury, using specific examples primarily from the author's laboratory. Finally, a comment will be made about the use of cultured cells in predicting toxicity to the lungs and other organs.

The lungs are a selective target for chemical toxicity, in part because of their anatomic position in the body and their functions as an organ. The most critical function of the lungs is that of gas exchange. In this role, cells of the lungs are exposed directly to the ambient gaseous environment and therefore to any toxic constituents therein. Indeed, they are the first point of contact with airborne toxicants. The lungs have a very large surface of epithelial cells exposed directly to the atmosphere (Murray, 1976). These cells not only receive higher doses of airborne toxicants relative to other cells in the body but normally experience oxygen concentrations higher than those to which cells in other organs are exposed.

In addition to epithelial lining cells exposed directly to the inspired air, the lungs possess a large vascular surface that is critical to maintaining gas exchange function. Thus, it is important to remember that lung tissue can be exposed to toxic chemicals not only via inhalation, but by the circulating blood too. Chemicals absorbed through the skin or the gastrointestinal tract are likely to contact the lungs via the circulation. For example, there are numerous dietary toxicants that target the lungs selectively. Indeed, the vascular bed of the lungs is the first to be encountered by toxicants that are absorbed into the circulation after intravenous, dermal or intramuscular administration, as well as by toxic metabolites that are produced by the liver then released into the circulation. The lung vasculature is distinguished from that of other organs in that it is a low pressure system (ie, pulmonary arterial pressure is about 1/5 that of the systemic circulation), the gradients for oxygen and carbon dioxide from the arterial to the venous beds are reversed, and it is the only organ in the body which has a capillary bed that receives all of the cardiac output. The vasculature of the lungs performs metabolic functions that also distinguish it from other organs and that may be important in protecting organs from untoward effects of numerous circulating substances (Roth and Vinegar, 1990; Ryan and Li, 1993). Thus, several qualities of the lungs render them not only critical to the functioning of other organs, but a selective target for toxic agents as well.

One can think of the lungs as comprising three anatomic and functional compartments; the conducting airways, the conducting blood vessels and a parenchymal or gas exchange (respiratory) region. Each of these anatomical subdivisions has its own role(s) in gas exchange, metabolism, control of pathogens, etc., yet the functioning of each is dependent on the others. To perform these roles requires cells that are highly specialized, and the nature of these cells differs considerably from one region to another. For example, the conducting airways consist of the nasopharynx, trachea, large and small bronchi and bronchioles. Each of these has cell types that differ from the others in structure and function. Indeed, the respiratory system comprises approximately forty different types of cells, no single one of which predominates either in number or mass (Gil, 1982). Some lung cell types demonstrate innate sensitivity to certain toxicants. This sensitivity usually derives from a particular functional characteristic of a differentiated cell of the lung. For example, the selective sensitivity of airway Clara cells to toxicants requiring metabolic bioactivation derives from the considerable capacity of these cells to metabolize xenobiotic agents (Boyd, 1977; Devereux et al., 1981; Devereux et al., 1989). Similarly, Type II alveolar cells are sensitive to toxicants such as paraquat that are accumulated by a polyamine transport system peculiar to these cells (Waddell and Marlowe, 1980).

For the reasons noted above, it is clear that the lungs posess certain anatomical and functional characteristics that are important determinants of responses to toxicants. Any experimental system or model is only as useful as the extent to which it mimics or predicts functional responses in vivo. In a clinical setting, radiography and pulmonary function tests are the usual methods for evaluating lung injury. Most in vitro methods are not amenable to these endpoints, and so one challenge of in vitro lung toxicology is the matching of outcomes of toxic exposure in vitro with effects measured in the clinic. On the other hand, most tests commonly used in the clinic are poor markers of early lung injury and yield limited information about mechanisms of toxicity. In vitro methodologies coupled with whole animal studies in which a variety of endpoints can be measured have provided valuable information about mechanisms of injury when evaluated in the context of clinical findings.

IN VITRO METHODS IN PULMONARY TOXICOLOGY

A large number of in vitro systems are used currently for the study of pulmonary biology. For use in toxicological studies, it is important that a system is chosen that is appropriate for the questions that are to be addressed. The complexities of lung function and structure in terms of the number of cells, important interactions among them and differences in function and susceptibility to toxic insult should be kept in mind when choosing, employing and interpreting results from a system in vitro. To choose or design the most appropriate in vitro methodologies for the study of pulmonary toxicity, one must have a conception of what is to be modeled, that is, the responses of the lungs to injury and, if possible, the particular cells involved. Early knowledge of these aspects often arises from pulmonary function measurements and morphological evaluation of lung tissue after exposure of animals or people to toxicants in vivo. This knowledge is clearly a necessary prerequisite for the development, refinement and validation of systems in vitro to be used for exploring mechanisms of toxicity and as tests to predict pulmonary toxicity in vivo.

A number of factors influence the usefulness of in vitro methodologies, either in the exploration of mechanisms of pulmonary toxicity or in toxicity testing. These include the nature of specific cellular receptors, signal transduction pathways and differentiated cellular functions, the number of different cell types and how they communicate with each other, the ways in which cells are organized into a functioning tissue, the breadth and complexity of lung functions, the nature and effectiveness of cellular and tissue protective mechanisms, and how the toxicant is delivered. Another determinant of utility is the degree of complexity in the mechanism by which a particular toxicant causes its effects: more complex mechanisms (eg, inflammatory responses) commonly require more complex systems than do simpler mechanisms (eg, cell membrane destruction).

In vitro systems currently in use and/or under development include isolated, perfused lung preparations, lung slices or explant cultures and isolated cells or mixed cell populations in short or long term culture. In the text that follows, some but not all of these will be discussed. Particular emphasis will be placed on isolated lung preparations and cultured cells. The discussion will be limited to a brief description of methodology, a listing of advantages and limitations and a specific example illustrating how a technique has been useful in extending knowledge or in constructing hypotheses about the responses of lungs to a toxic agent.

Isolated, perfused lungs. Isolated lung preparations have been used extensively over the years for the study of both inhaled and circulating toxicants. Preparations have been developed for several animal species and for numerous purposes, many of them relating to toxicologic evaluation. The use of isolated, perfused lung preparations in toxicological studies has been reviewed in detail, and the reader is referred to these for an in depth discussion of methodology and use (Roth and Bassett, 1989; Smith and Bend, 1981; Niemeier, 1984; Rhoades, 1984; Fisher, 1985). Usually, the pulmonary artery and trachea of an anesthetized animal are cannulated, and the lungs are placed in a warmed, humidified chamber (Fig 1).

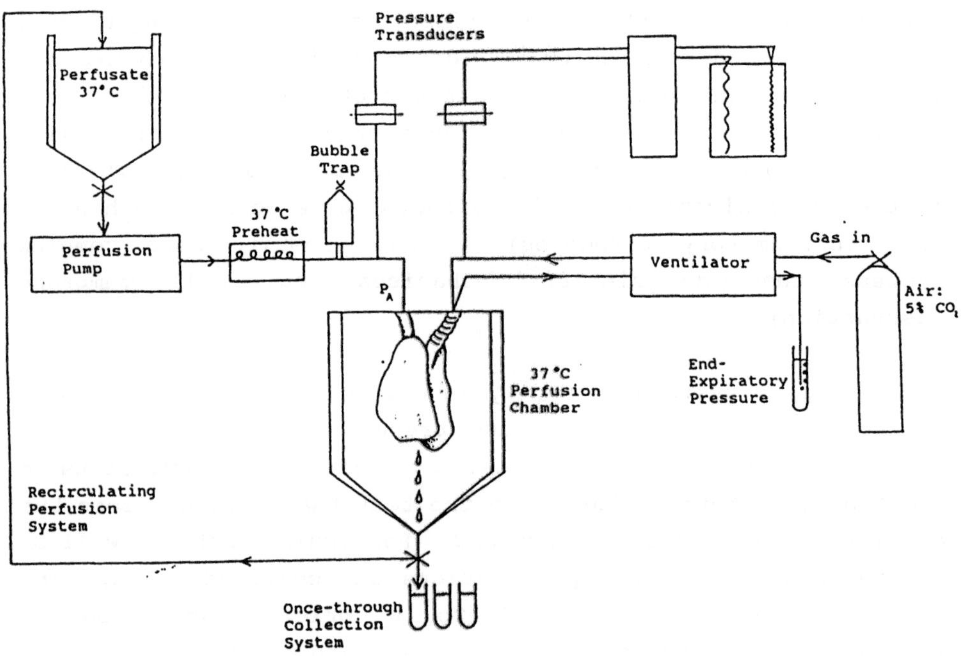

Figure 1. **Diagram of isolated lung perfusion.** Ventilation pressure and perfusion pressure are monitored by pressure transducers and a polygraph. Reprinted from Roth and Bassett, 1989.

The vasculature is perfused with suitable medium, and the lungs are ventilated with a gas mixture appropriate for the questions to be addressed. The lungs may be perfused either in a recirculating or a single-pass manner as the needs of the experimental protocol dictate. Hemodynamic (eg, vascular resistance) and ventilatory (eg, tracheal pressure) functions can be measured, and the perfusion medium can be sampled for evaluation of markers of injury (eg, enzyme release), analysis of toxicant concentration, evaluation of metabolic function, etc. Fluid accumulation (ie, edema) during perfusion may be monitored as increased tissue weight. After perfusion, the lungs may be removed for bronchoalveolar lavage or biochemical and histopathological evaluation.

Isolated lung preparations have many strengths as an experimental tool. They allow the use of relevant routes of exposure to chemicals: agents can be delivered either by the airways or via the vasculature, as appropriate. The structure of the lungs remains intact, and the relationship among the various resident cell types and their ability to communicate with each other is preserved. Multienzyme metabolic pathways within or between cells can be maintained under essentially normal control mechanisms. Since the lungs are disconnected from the neuronal and circulatory systems present in vivo, the responses of the lungs to toxicants can be studied in the absence of confounding influences from other organs. Of course, this can be desirable or undesirable, depending on the focus of investigation. The volume of perfusion medium, its composition and the concentration of toxicant can be controlled easily, large and frequent samples of perfusion medium can be taken, if necessary, and mass balance of toxicants can be determined. Finally, gas exchange and hemodynamic functioning can be evaluated, which is usually impossible in more reductionist systems in vitro.

Several limitations exist that restrict the usefulness of isolated lung preparations. The period of viability is limited; most preparations become edematous and useless within 2 to 5 hours of perfusion. The amount of toxicant needed is often less than that required in vivo but usually greater than that needed for organ or cell culture preparations. Toxic responses of specific lung cell types cannot usually be ascertained in perfused lung preparations. They are not useful for toxicologic screening studies, and their employment does not have the potential to reduce animal use as some culture methods do. Clearly, as with any in vitro preparation, these strengths and limitations should be taken into account before employing isolated lungs in toxicologic studies.

Isolated lungs have been used successfully in numerous studies of metabolism of toxicants and of toxic responses and their mechanisms. The study of acute respiratory distress provides one example. The adult respiratory distress syndrome (ARDS) in humans can be precipitated by a large number of events, including chemical exposure (Hogg, 1991). Numerous animal models have been developed for the study of this phenomenon, and some but not all of the evidence derived from these and human studies has pointed to the blood neutrophil (PMN) as playing a critical role in the lung injury. One such model involves administration of phorbol ester to animals. When phorbol myristate acetate (PMA) is administered intravenously to rabbits or sheep, respiratory distress occurs (Schraufstatter et al., 1984; Shasby et al., 1983; Dyer and Snapper, 1986). This is associated with damage to the alveolar barrier and accumulation of fluid in the alveolar spaces. Depletion of PMNs from the blood prior to administration of PMA prevent lung injury in some models in vivo but not in others (Dyer and Snapper, 1986; Johnson and Ward, 1982; Schraufstatter et al., 1984; Shasby et al., 1983), and studies in isolated lungs have also yielded disparate results (Shasby et al., 1983; Jackson et al., 1986). Thus, the role of PMNs in the pulmonary edema produced by PMA administration was unclear.

Although some of the disparities might be explained by differences in species or route of administration, it seemed that different doses of PMA could also produce contrasting results. We explored this possibility using an isolated rat lung preparation in which the composition of the medium perfusing the lungs could be easily controlled (Carpenter et al., 1987). Isolated rat lungs were perfused for thirty minutes with a recirculating buffer containing various concentrations of PMA. PMA caused dose- and time-dependent increases in perfusion pressure and lung weight, the latter signifying edema formation (Fig. 2). Transmission electron microscopy of lungs after perfusion with a damaging dose of PMA revealed blebbing of capillary endothelial cells and marked interstitial edema.

In a separate experiment, lungs were perfused similarly, but with medium to which PMNs had been added. Addition of unstimulated PMNs elevated perfusion pressure slightly but did not cause edema. However, when PMA was added to the PMN-containing medium at a dose below that which caused injury in the absence of PMNs, marked pulmonary edema and a pronounced increase in perfusion pressure occurred (Fig. 3). This study indicated that PMA is capable of injuring lungs by at least two distinct mechanisms, one of which is dependent upon PMNs.

Activated PMNs can injure tissue by a number of mechanisms. One of these involves the actions of toxic oxygen metabolites generated upon activation of PMNs by a stimulus such as PMA (Shasby et al., 1983). By adding ferricytochrome C to the perfusion medium and following its color change during perfusion, we demonstrated superoxide production during perfusion, but only when PMNs were present in the medium and activated by a small dose of PMA. Addition of catalase and superoxide dismutase (SOD) to the medium prior to perfusion prevented the increase in lung weight, indicating that the injury required toxic oxygen species. A different mechanism operates when a larger dose of PMA is given to isolated lungs perfused with buffer devoid of PMNs, since the injury was not prevented by coperfusion with catalase and SOD (Table 1).
These results indicate that the injury to lungs caused by PMA-stimulated PMNs is mediated by toxic oxygen species and that injury produced by a larger dose of PMA alone occurs by a mechanism independent of toxic oxygen species released into the perfusion medium (Carpenter et al., 1987).

Additional study revealed the importance of an arachidonic acid metabolite in the PMN-dependent injury (Carpenter and Roth, 1987). When lungs were perfused with medium containing PMNs and a low dose of PMA, a time-dependent increase in thromboxane B_2 concentration occurred in the medium (Fig. 4). This did not

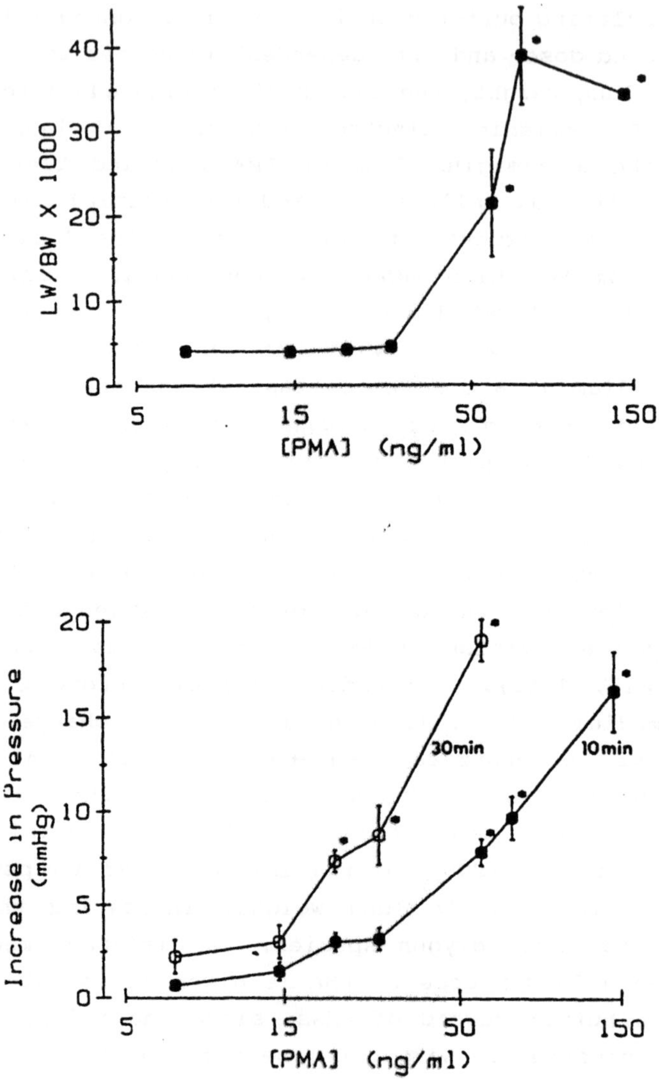

Figure 2. **Injury to the isolated rat lung preparation by phorbol myristate acetate (PMA).** Isolated lungs were perfused for 30 min with buffer containing various concentrations of PMA. PMA increased (A) lung weight/body weight ratio (LW/BW) and (B) perfusion pressure in a dose-related manner. *Significantly different from lungs perfused without PMA (P<0.05). Reprinted from Carpenter et al, 1987.

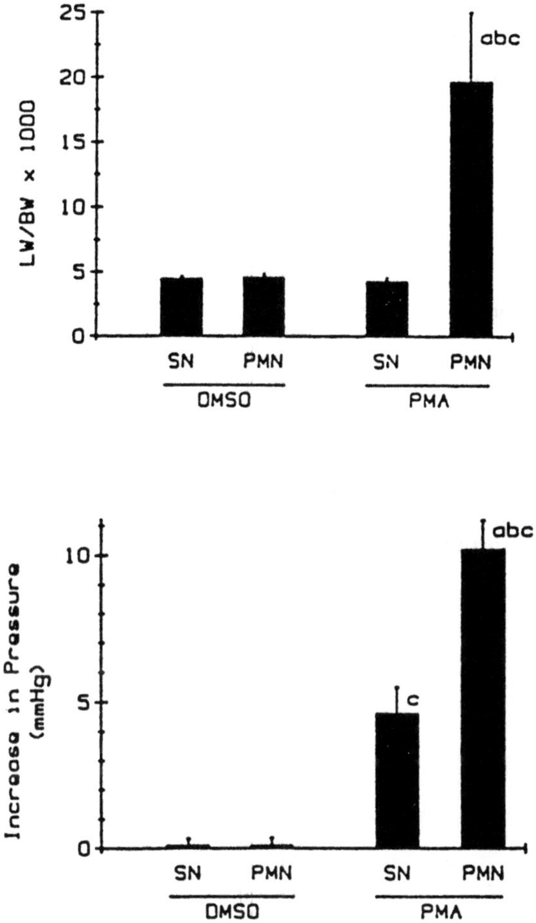

Figure 3. **Neutrophil(PMN)-dependent injury to the isolated rat lung preparation perfused with a small dose of PMA.** Isolated lungs were perfused with buffer containing PMNs (1×10^8) or supernatant vehicle (SN) and with either PMA (14 ng/ml) or dilute DMSO vehicle. PMN perfusion with the addition of PMA resulted in increased lung weight/body weight ratio (LW/BW)(top) and increased perfusion pressure (bottom). a, Significantly different from DMSO/PMN group; b, significantly different from PMA/SN group; c, significantly different from DMSO/SN group. Reprinted from Carpenter et al, 1987.

TABLE 1

Injury to Isolated Lungs Caused by Perfused PMA in the Presence or Absence of Neutrophils(PMNs)

	Injury to isolated lungs from:			
	Small dose of PMA +PMN		Large dose of PMA only	
	Edema	↑Pressure	Edema	↑Pressure
CAT/SOD protection	++	0	0	0
Protection by Tx synthesis inhibition	++	+	+	+

++, complete protection; +, partial protection; 0, no protection. PMA, phorbol myristate acetate; CAT, catalase; SOD, superoxide dismutase; Tx, thromboxane.

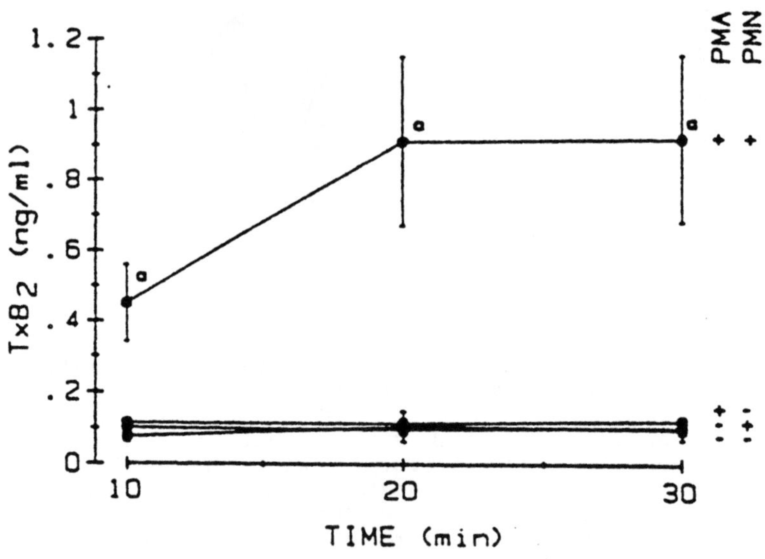

Figure 4. **Thromboxane concentration in medium of isolated lungs perfused with PMA and neutrophils (PMNs).** Perfusion of lungs with PMNs and a small dose of PMA (14 ng/ml) resulted in release of thromboxane (TxB_2) into the perfusion medium. a, Significantly different from all other treatment groups ($P<0.05$). Reprinted from Carpenter and Roth, 1987.

occur if either PMA or PMNs were omitted from the medium, indicating that their presence was needed for the generation of thromboxane. The increases in both lung weight and perfusion pressure correlated well with perfusion medium thromboxane concentration. Addition of an inhibitor of thromboxane biosynthesis to the perfusion medium abolished the increase in lung weight and attenuated the increase in perfusion pressure. Interestingly, the PMN-independent injury caused by a larger dose of PMA was also associated with increased perfusion medium thromboxane concentration, and thromboxane synthesis inhibition provided partial protection from injury.

These studies supported a role for thromboxane in injury to lungs caused by stimulated PMNs. Thromboxane is a potent vasoconstrictor, and the reduced perfusion pressure associated with the protection afforded by thromboxane synthesis inhibition is consistent with the hypothesis that vasoconstriction plays a critical role in the development of pulmonary edema in this model. Indeed, a vasodilator that prevented the increased pressure also prevented edema formation (Carpenter and Roth, 1986). However, a vasoconstrictive component alone is insufficient to cause damage, since oxygen radical scavengers attenuated the edema without affecting the increase in pressure. Thus, both thromboxane-induced vasoconstriction and the presence of toxic oxygen species appear to be needed for PMA-stimulated PMNs to injure isolated rat lungs.

Although PMA substantially increased the amount of thromboxane released from lungs perfused with PMN-containing medium, the source of thromboxane was unclear: did it originate from the PMNs or from lung tissue or both? Isolated PMNs were found to produce thromboxane tonically, but the release of this arachidonic acid metabolite was not increased by PMA. Lung tissue itself was also a probable source. By using aspirin, an irreversible inhibitor of cyclooxygenases that metabolize arachidonic acid, we were able to determine whether the thromboxane that was critical for lung injury in this model arose from lung tissue or from PMNs (Carpenter-Deyo and Roth,

1989). PMN production of thromboxane was eliminated by incubating PMNs with aspirin and washing them before using them to perfuse isolated lungs. Conversely, thromboxane production by lung tissue was inhibited by pretreating rats with aspirin before isolating their lungs for perfusion experiments. Interestingly, cyclooxygenase inhibition in _either_ lung tissue _or_ PMNs eliminated the increase in lung vascular permeability caused by PMA. Thus, _both_ sources of thromboxane were required for the lung injury to occur in this model. Thus, the isolated lung preparation allowed us to identify critical sources of an important mediator, which would have been more difficult to delineate _in vivo_.

This is but one example of how isolated lung preparations may be used to explore mechanisms of toxicity. The scientific literature is replete with others. It is important to remember, however, that even though this preparation may represent the situation _in vivo_ more closely than some other _in vitro_ models, it is still a simplified system with all the limitations thereof. Results should not be overinterpreted. Returning to the example above as a case in point, the extent to which some of the findings apply to the whole animal remains a question. The sources of thromboxane critical for lung injury may be quite different in the whole animal, the lungs from which are perfused with a much more complex medium (ie, blood) and are subject to influences not represented in isolated organ preparations. For example, activated blood platelets are prodigious producers of thromboxane A_2, but platelets were not included in the study described above. Interestingly, PMA causes injury to isolated lungs perfused with a platelet-containing buffer by a mechanism dependent on thromboxane (Wang et al., 1991). Thus, the respective roles of platelets, PMNs and other cells as producers of injurious thromboxane _in vivo_ remain open to question. This illustration serves to emphasize that the best information that such a preparation can provide is what _might_ happen _in vivo_.

Lung slices in culture. Although attempts have been made for over three decades to maintain organs in culture, success with lung parenchyma has occurred only in the past several years. Techniques to culture embryonic lung parenchyma have been developed which result in normal growth, structural organization and differentiation for several days (Hsu et al., 1992; Gross and Wilson, 1982; McAteer et al., 1983). Development of techniques using adult lungs have been relatively problematic, inasmuch as normal structure and function is difficult to maintain in culture. Recently, however, reports have appeared describing techniques that allow preparation and culture of adult lungs for extended periods (Siminski et al., 1992; Stefaniak et al., 1992; Fisher and Placke, 1988; Warren and Barton, 1992).

Methods for preparing lung slices usually involve inflating excised lungs via the trachea with an agar solution. When cooled, this solidifies and permits slices to be cut without collapse of the alveoli. Sections approximately 2-5mm in thickness are placed on a suitable substrate (eg, gelatin sponge) and incubated in an appropriate medium (Siminski et al., 1992). The slices are turned often to prevent migration of cells into the underlying matrix. Siminski and coworkers (1992) report maintenance of mouse lung slices in culture for up to 60 days in serum-free conditions with substantial retention of pulmonary architecture.

The development of lung parenchymal culture techniques has progressed to the point where they are useful in addressing toxicological questions. Advantages of lung slice systems are several. Relationships among most cells as they are _in vivo_ can now be preserved for extended periods of time. However, although pulmonary parenchymal epithelial cells tend to remain intact and alveoli appear impressively close to normal for long periods, capillaries are undiscernible and bronchial epithelium is desquamated within two weeks. Thus, certain pulmonary cell types seem to be maintained poorly in organ culture. As with perfused lungs, extrapulmonary influences that can complicate interpretation _in vivo_ are eliminated and the composition of the culture medium can be controlled. Amounts of toxicant and volumes of medium required are usually modest compared to intact lung preparations. Unlike perfused lungs, lung slices may prove useful for some toxicological screening purposes, although long-term organ culture remains difficult.

A limitation of any culture method is that it is not conducive to evaluation of gas exchange or hemodyamic functions. The nature of exposure to toxicants in lung slice systems is unlike in vivo, since all cells types are equally exposed. It is not always easy to ascribe toxicant-induced responses to a particular cell type, although this limitation might be overcome in some cases with modern imaging techniques. Finally, some cell functions and interactions seem to be compromised as the time in culture increases. Accordingly, as with any biologic system, several advantages and limitations pertain to the use of lung slices in toxicologic studies. Even though the development of lung parenchymal culture techniques has progressed to the point where they should be useful in addressing toxicological questions, to date little has been published in the peer-reviewed toxicology literature using such methods. Although not discussed here, airway explant cultures have been used in toxicologic studies and have many of the same advantages and limitations as an experimental tool (Mossman and Sesko, 1990).

Cells in culture. Various methods have been used to isolate cells of the respiratory system for studies in culture. Cells are plated on various substrata, incubated in various culture media and grown to an appropriate cell density. Numerous systems for maintaining cells and exposing them to gases or particulates have been developed, (eg, Chen et al., 1993). These methods have been the subject of numerous original articles and reviews (Christensen et al., 1993; Whitcutt et al., 1988; Leikauf and Driscoll, 1993; Massey, 1989; Beckmann et al., 1992; Dobbs, 1990; Nettesheim et al., 1990; Rennard et al., 1990; Tanswell et al., 1991; Poole and Brown, 1987; Basbaum and Finkbeiner, 1989; Voelker and Mason, 1989) and will not be discussed in detail here.

Generally, cell isolation and purification techniques employ outgrowth of cells from explants of lung tissue, surface brushing of airways, collagenase or pronase digestion, density gradient centrifugation, centrifugal elutriation and/or flow cytometry. Sometimes, agents providing selective growth advantage are added to the culture medium. Of the many cell types in the respiratory system, several have been cultured successfully. Enriched fractions of basal and mucosal cells as well as Clara cells have been prepared from airways. Methods for culture of individual airway cells such as serous and goblet cells, Clara cells and basal cells have appeared (Leikauf and Driscoll, 1993). Type I alveolar cells are terminally differentiated in vivo, are difficult to isolate and do not divide in culture. Nearly pure cultures of Type II alveolar epithelial cells have been grown in culture and used in toxicologic studies, and these will differentiate into Type I-like cells under certain conditions. Transformed cell lines (eg, L2, A549, LEC) that retain some Type II cell characteristics have been developed from human and animal lungs. Lung fibroblasts and vascular endothelial cells have been isolated from several species and cultured successfully. Pulmonary macrophages isolated from animals and humans by bronchoalveolar lavage have been used extensively in toxicological studies. The ease of isolation, extent to which cells will proliferate and can be passaged, and the length of time they maintain diffentiated function and morphology vary remarkably with cell type. These factors in some cases limit the usefulness of specific cells in studies of lung toxicology, and methods for successful isolation and culture of many types of cells of the respiratory tract are still in the developmental stage.

The advantages of cells in culture as model systems in toxicology are now well known. Human as well as animal cells can be used and comparisons can be made among them. Molecular mechanisms can be addressed, which is often difficult in less reductionist systems. Specific cells may be studied without confounding influences of other cell types and endogenous mediators present _in vivo_, and responses to toxicants can be identified with certainty with a specific cell type. Many cell culture systems are viable for prolonged periods: chronic toxic responses and repair mechanisms that require protracted times, such as cell replication or collagen synthesis, can be studied. This contrasts, for example, with isolated lung preparations for which viability is limited to periods too short to address such questions. Toxicant dosimetry and kinetics are easily controlled, and cells can be exposed to a wide range of toxicant concentrations that would be impossible _in vivo_. Some, but not all cell culture models result in a large yield of cells which can be employed efficiently to provide relatively rapid answers to toxicologic questions and reduce animal use.

Despite their clear advantages, cell culture models have several limitations. Inevitably, pulmonary architecture is destroyed and relationships with other cells and tissues are not represented. This limitation can be overcome to a degree by using coculture systems of heterologous cells, but such relationships are never entirely restored. Functions (eg, mechanical function, gas exchange) that require intact tissue cannot be evaluated as they can _in vivo_. On the other hand, biochemical or structural alterations in cells (eg, increased collagen synthesis, cell swelling) that give rise to such functional changes in intact lungs can sometimes be used to predict these changes. Secondly, the way in which cells are exposed to toxicants or their active metabolites is often different in cell culture than _in vivo_. Third, the state of morphological and functional differentiation of cells in culture may be markedly different from that _in vivo_. This may limit appropriate extrapolation of results from the culture dish to the whole animal (see below). Finally, a problem in interpretation or extrapolation may arise if an inappropriate cell type is chosen for study. This is likely to happen, for example, when there is uncertainty about which cell types are critical to a response _in vivo_. A number of studies have been performed on cell types isolated from lung but which are not really the cell of interest. Whether using an inappropriate cell just because it is derived from lung tissue will yield useful information or more relevant information than a cell derived from a completely different source (eg, another tissue) depends on the nature of the events to be studied (see below and Massey, 1989; Poole and Brown, 1987). In any case, the rationale for using such a cell just because it derives from lung if it is not the specific cell type of interest is suspect.

The limitations of cells in culture have the potential to mislead both in the exploration of mechanisms of toxcity and in toxicity screening applications. Nevertheless, cell culture methodolgies have and will continue to provide valuable information about pulmonary responses to toxic chemicals. Provided below are two examples of the usefulness and limitations of cells in culture in the exploration of toxic mechanisms:

1. Mineral dusts and alveolar macrophages. When inhaled, certain mineral dusts such as asbestos and silica, produce a pronounced inflammatory response and pulmonary fibrosis. There is evidence that pulmonary alveolar macrophages are important cellular mediators of these responses. Driscoll and coworkers (1990a) evaluated the response of pulmonary macrophages, isolated by bronchoalveolar lavage, to asbestos and silica particles. They compared responses to these materials to responses produced by aluminum oxide and titanium dioxide, two materials that are relatively innocuous after inhalation exposure (Leikauf and Driscoll, 1993).

Two of the endpoints chosen for evaluation were cytotoxicity (ie, release of lactate dehydrogenase [LDH]) and release of tumor necrosis factor-alpha (TNF-α). The latter is a differentiated cellular function of alveolar macrophages involved in inflammatory responses and possibly pulmonary fibrosis. It is noteworthy that cytotoxic responses in vitro were not particularly predictive of the fibrogenic potential of the materials after inhalation in vivo (Driscoll et al., 1990a). For example, silica caused substantial LDH release from macrophages in vitro, yet asbestos, which is highly fibrogenic in vivo, was less cytotoxic than aluminum oxide or titanium dioxide, two relatively innocuous dusts. The functional response of macrophages was more predictive: both silica and asbestos activated macrophages in vitro to release TNF, whereas aluminum oxide and titanium dioxide did not (Driscoll et al., 1990a; Fig. 5). Interestingly, stimulation of TNF release by asbestos occurred at doses that were not associated with cytotoxicity. These and related results suggest that alveolar macrophage activation in vitro may predict adverse inflammatory

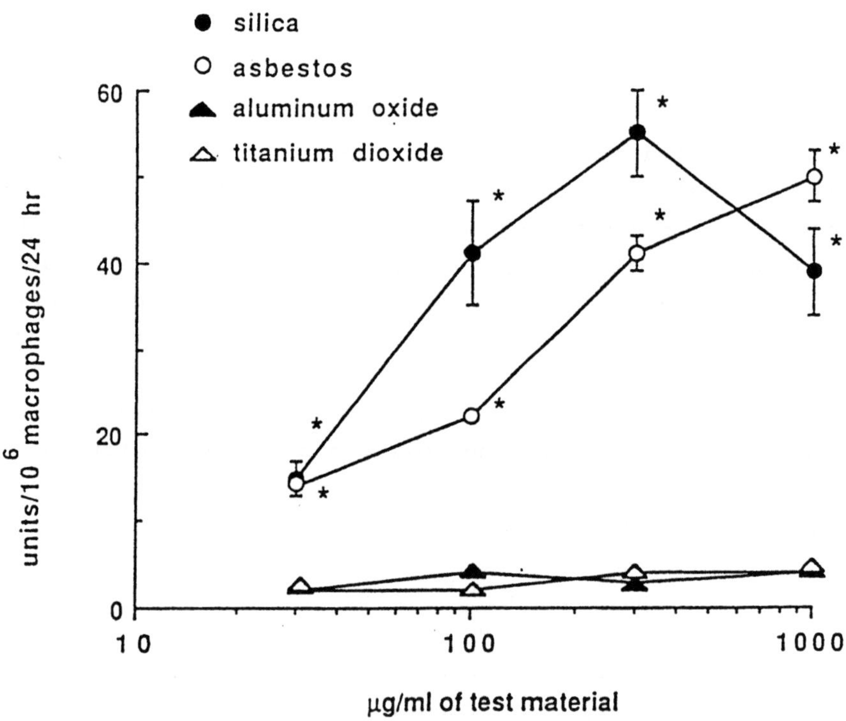

Figure 5. **Release of tumor necrosis factor (TNF) by rat alveolar macrophages exposed to mineral dusts.** Macrophages were exposed in vitro to mineral dusts for 24 hr, then TNF was measured in the culture medium. *Significantly different from unexposed control macrophages ($P<0.05$). Reprinted from Leikauf and Driscoll, 1993.

responses and/or fibrogenic potential from mineral dust exposure in vivo.

Not all alveolar macrophage functional responses in vitro predicted adequately responses to inhalation of dusts in vivo, however. For example, large lung burdens of titanium dioxide in vivo resulted in TNF release from macrophages, but in vitro even

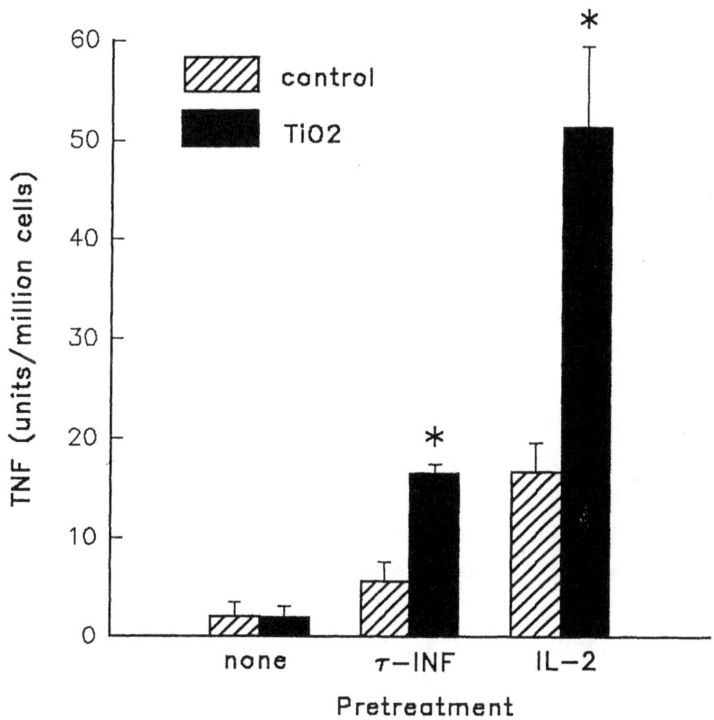

Figure 6. **Effect of pretreatment with gamma-interferon or interleukin-2 on release of TNF by rat alveolar macrophages exposed to titanium dioxide in vitro.** Macrophages were pretreated with either gamma-interferon (gamma-INF) or interleukin-2 (IL-2)(200 units/ml) for 24hr prior to exposure to a large dose of titanium dioxide (TiO_2) particles (1000 ug/ml) for an additional 24hr prior to assay of the medium for TNF. *Significantly different from unexposed control macrophages($P<0.05$). Redrawn from Leikauf and Driscoll, 1993.

large doses of this dust failed to cause TNF release from alveolar macrophages (Driscoll et al., 1990b; Leikauf and Driscoll, 1993). Thus, the system in vitro was unable to predict this cellular functional response in vivo. This shortcoming was used to gain insight into potential mechanisms. Other cytokines, such as gamma-interferon and interleukin-2, are released during inflammation and can influence macrophage responses. Such agents, released under conditions of large alveolar dust burden, might prime macrophages, or in this case convert them from nonresponders to responders. Indeed, when macrophages were exposed in vitro to gamma-interferon or to interleukin-2 they released large amounts of TNF in response to titanium dioxide challenge (Fig. 6). This result illustrates how differences in cellular environment can determine whether or not an in vitro test will adequately predict toxicant-induced changes in cellular functional responses in vivo.

2. Monocrotaline pyrrole and injury to pulmonary endothelium. We have employed cultured cells in our exploration of a model of toxin-induced pulmonary vascular injury. Monocrotaline is a plant toxin which is metabolized in the liver of rats to a toxic pyrrole (Fig. 7). This reactive pyrrole reaches the lung via the circulation, where it binds covalently to lung cells and causes pulmonary vascular injury, remodeling of pulmonary arteries, chronic pulmonary hypertension and compensatory right heart hypertrophy (Reindel et al., 1990; Roth and Reindel, 1990). These structural and hemodynamic changes do not begin immediately: after a single iv injection of chemically synthesized monocrotaline pyrrole, the onset of major lung injury does not occur for 1 to 2 days, but once it begins, the untoward effects worsen progressively even without further treatment with the toxic metabolite. Morphologic and functional studies in vivo pointed to the pulmonary vascular endothelium as an early and important target for monocrotaline pyrrole (Reindel et al., 1990). Accordingly, we have employed cultured pulmonary endothelium to gain insight into the mechanisms by which monocrotaline pyrrole produces its delayed and progressive effects.

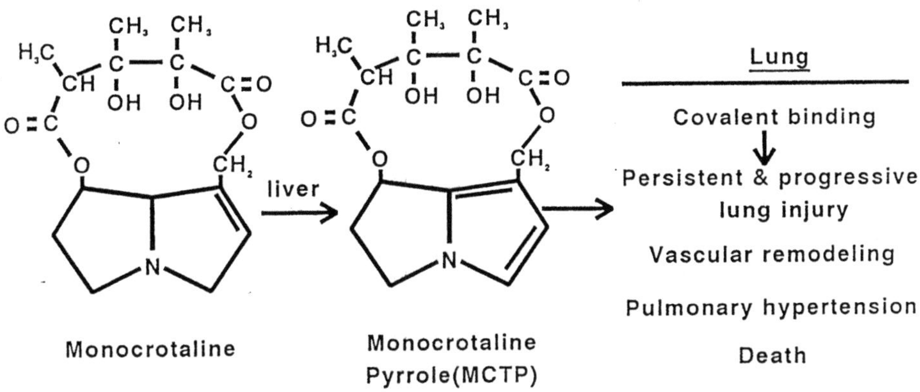

Figure 7. **Pulmonary toxicology of monocrotaline.** Monocrotaline is a plant toxin, which when ingested is metabolized to a pyrrolic metabolite that injures the pulmonary vasculature. The delayed and progressive lung injury and pulmonary hypertension can be reproduced in rats by intravenous injection of chemically synthesized monocrotaline pyrrole.

We began by exposing confluent monolayers of cultured endothelium to monocrotaline pyrrole to determine whether it produced a toxicity which, as _in vivo_, was delayed and progressive (Reindel and Roth, 1991). It did, and this result encouraged us to go forward to characterize further the cellular responses with the hope that the results might provide insight into how endothelium might be involved in the changes in the lungs that occur in this model. We found that doses similar to those to which the lung endothelium was exposed _in vivo_ resulted in some cell death after several days. Moreover, this became more pronounced with time, even without additional toxicant administration. Thus, this effect of monocrotaline pyrrole on endothelium in culture resembled that _in vivo_ in the sense that the toxicity was delayed in onset and progressively worsened with time.

One of the functions of endothelium is to participate in

repair of wounded vascular intima. When an endothelial cell monolayer is partially denuded, surrounding endothelial cells migrate into the wound, spread to cover the wounded area and divide to repopulate the area with healthy endothelium (Evans, 1982). This process prevents the leakage of plasma into the interstitium. Our results in vitro suggested that monocrotaline pyrrole compromises this repair process. As lethally affected endothelial cells died and lifted from the culture plate, remaining viable cells spread to cover the resulting gaps in the monolayer; however, they had lost their ability to undergo cell division (Reindel and Roth, 1991; Hoorn et al., 1993; Fig. 8). The loss of ability to replicate was associated with pronounced and long-lasting DNA crosslinking (Wagner et al., 1993).

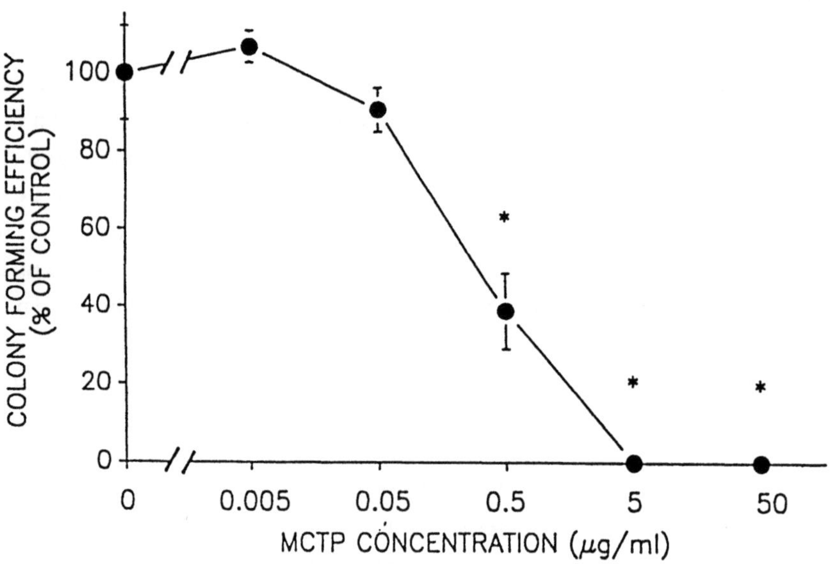

Figure 8. **Monocrotaline pyrrole inhibits proliferation of pulmonary artery endothelial cells.** Endothelial cells plated at low density were exposed to monocrotaline pyrrole (MCTP) once and the number of colonies was counted 14 days later. Even at the highest concentrations, some single cells remained attached to the culture plate that were viable according to the criterion of trypan blue exclusion. *Significantly different from control (0 ug/ml) (P<0.05). Reprinted from Reindel and Roth, 1991.

This was a potentially important finding, inasmuch as endothelial cell replication is critical to repair of denuded regions of an endothelial cell monolayer. This result in vitro led us to propose an hypothesis about how monocrotaline produces delayed and progressive vascular injury in vivo (Roth and Reindel, 1990). Viable but affected endothelial cells of the vascular intima can initially cover defects (gaps) in the monolayer of endothelium by migration and cell spreading, but ultimately their inability to complement this with replicative repair results in the progressively worsening vascular leak that characterizes monocrotaline pneumotoxicity. Thus, the fundamental change in cell biology we observed in cultured cells was consistent temporally, quantitatively and qualitatively with those occurring in vivo and allowed us to formulate a mechanistic hypothesis about early events that occur in the pathway to lung injury.

Our initial results using cultured cells caused us to wonder if alterations in differentiated functions of endothelium might participate in the pulmonary changes that occur in this model. For example, fibrin clots occur in pulmonary vessels of monocrotaline-treated rats, and it has been suggested that these might contribute to increasing pulmonary vascular resistance. The presence of such clots is controlled in part by fibrinolysis, and lung tissue from monocrotaline pyrrole-treated rats demonstrates decreased fibrinolytic activity prior to and during the onset of lung injury and pulmonary hypertension (Fig. 9; Schultze and Roth, 1993a). Fibrin is digested by the enzyme plasmin, which arises from the zymogen plasminogen via the action of plasminogen activators. The activity of plasminogen activator is regulated in turn by plasminogen activator inhibitors. Monocrotaline pyrrole treatment of rats had no effect on tissue plasminogen activator activity in the blood, but plasminogen activator inhibitor was increased, a condition favoring the development of fibrin clots.

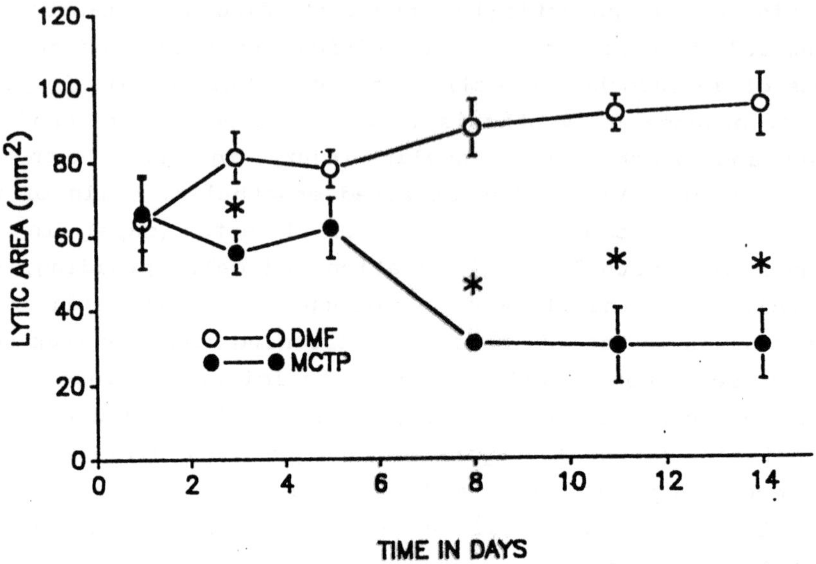

Figure 9. **Fibrinolytic activity of lung tissue from rats treated with monocrotaline pyrrole.** Rats received a single, i.v. injection of monocrotaline pyrrole (MCTP) on day 0. Fibrinolytic activity was decreased during the development and progression of lung injury and pulmonary hypertension. *Significantly different from controls treated with dimethylformamide (DMF) vehicle ($P<0.05$). Reprinted from Schultze and Roth, 1993a.

Both plasminogen activator and plasminogen activator inhibitor are produced by vascular endothelium. Accordingly, we determined whether treatment of pulmonary vascular endothelium with monocrotaline pyrrole altered the production of these in a way that might explain the increased fibrinolytic activity observed <u>in vivo</u>. To our surprise, exposure to monocrotaline pyrrole <u>increased</u> plasminogen activator activity and <u>decreased</u> plasminogen activator inhibitor activity in cultured endothelium (Fig. 10). This, of course, would favor fibrinolysis and contrasts with the reduction in fibrinolytic activity found <u>in vivo</u> (Schultze and Roth, 1993b).

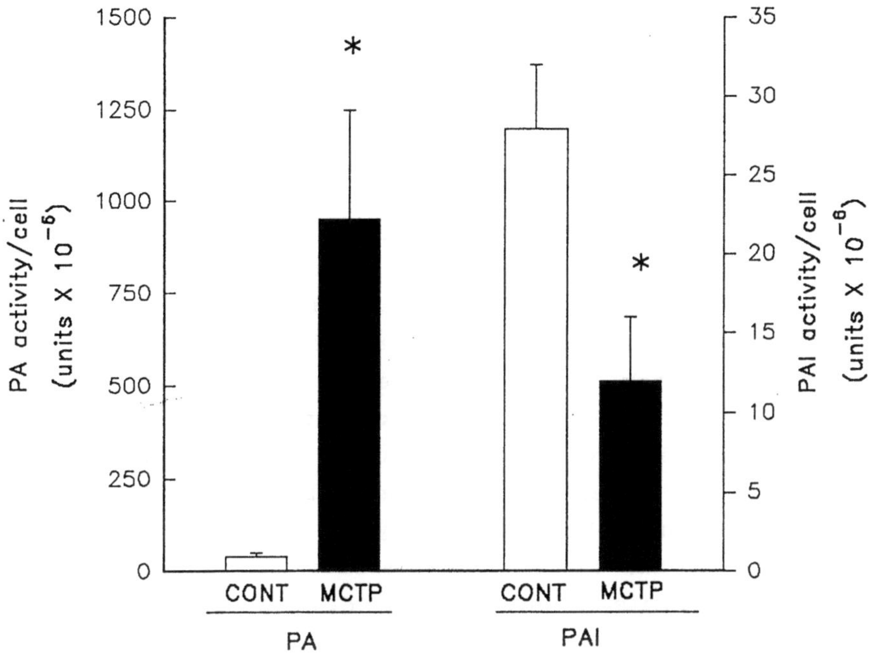

Figure 10. **Effect of monocrotaline pyrrole on plasminogen activator and plasminogen activator inhibitor activity released from pulmonary artery endothelial cells in culture.** Endothelial cells were exposed to monocrotaline pyrrole (MCTP; 5 ug/ml) or vehicle at time 0 and evaluated at five days. MCTP treatment resulted in increased plasminogen activator activity and decreased plasminogen activator inhibitor activity released into the culture medium, a change indicative of increased fibrinolytic activity. *Significantly different from vehicle control ($P<0.05$). Redrawn from Schultze and Roth, 1993b.

There are other instances in this model for which results in vivo do not match well results in cultured endothelium. For example, monocrotaline pyrrole treatment of endothelium resulted in increased release into the culture medium of 6-keto-prostaglandin F1-alpha, a stable metabolite of prostacyclin (Reindel et al., 1991). In contrast, isolated lungs taken from rats treated with monocrotaline pyrrole in vivo showed no change in production of this arachidonic acid metabolite (Ganey and Roth, 1987). Thus, one does not have to search widely to find instances in which toxicologic effects on cells in vitro differ from those in intact tissue, emphasizing the need to exercise caution in extrapolating results from systems in vitro to the situation in vivo.

The reasons underlying such differences are potentially numerous. Often, cells used for study in vitro are isolated from sources that differ from the cells responsible for an effect observed in vivo. In the example above, although endothelial cells are major producers of proteins involved in fibrinolysis, other cells produce them too and may be responsible for the decreased fibrinolytic activity observed in vivo. A species difference in the source of cells also may have played a role. Alternatively, the release of fibrinolytic proteins by endothelial cells in vivo may be influenced by other lung cell types, and these might override the direct effect of monocrotaline pyrrole on endothelium. Finally, such differences could arise if cells grown in culture for some reason respond differently than the same cells in an intact organ, perhaps due to local environmental differences.

This example illustrates that a single cell system may be useful for predicting certain responses but fail to predict other responses, even to the same toxicant. Understanding the factors that contribute to differences in predictive capacity of various in vitro systems is one of the challenges to toxicologists as such systems continue to be developed and employed.

Cell culture in predictive toxicology: a perspective. The meteoric increase in the use of in vitro methods has clearly increased our knowledge in toxicology as well as in other arenas. It is now common for new, potential mechanisms of toxicity to be discovered from studies in vitro, as opposed to arising from whole animal studies. Along with the rapid barrage of findings from in vitro studies has come a proliferation of hypotheses about how chemical exposures and other phenomena lead to injury to humans and animals. The variety of in vitro systems and the many outcomes that are now evaluated as toxic endpoints have led to theories about in vivo responses that are not only numerous but elaborate and complex. Illustrations of this can be found in schemes depicting how inflammation can lead to tissue injury: numerous cell types interacting via dozens of soluble mediators resulting in a mindboggling array of permutations and a myriad of potential mechanisms for injury.

Unfortunately, our knowledge of what occurs in vitro and the hypotheses that have derived therefrom often outpace our understanding of what actually does happen in vivo. This has the potential to mislead when the accumulation of results from in vitro studies remains unmatched by an understanding of which of them apply in vivo. It is important to remember, in this regard, that any result generated in vitro only allows us to form an hypothesis about what might apply in vivo. Whether we are considering mechanistic studies or toxicity testing, an hypothesis generated from a study in vitro should be accepted as such and nothing more. That is, to know with certainty whether it applies in vivo requires further evaluation.

In vitro toxicologic studies often use a "surrogate cell" maintained in culture to predict responses of a "principal cell," usually a cell within an organ in vivo. With this in mind, it would be useful to be able to foresee when a result generated from a surrogate cell in culture is likely to reflect what occurs in a cell or tissue in vivo. Are there conditions or outcomes of in vitro studies that are more or less likely to be predictive of toxic responses of cells in vivo? Under what conditions must a surrogate cell employed for in vitro evaluation be identical to the principal cell which it is meant to model in vivo? Conversely, when will the response of any surrogate cell grown in culture predict correctly the response to toxic exposure of one or more cell types in vivo?

To provide a conceptual framework for addressing such questions, toxicants might be viewed as causing either specific or nonspecific events that affect cells (Fig 11). Nonspecific events are ones that result in direct damage to membranes, cellular organelles or DNA and result in alterations in fundamental cellular processes such as cell replication, oxidative phosporylation, etc. Specific events affect processes that usually result in alterations in differentiated cellular function. Such events may be classified as "proximal," such as binding of toxicant to specific cellular receptors, or "distal," such as effects on signal transduction or selective gene expression. Accordingly, specific events usually affect complex

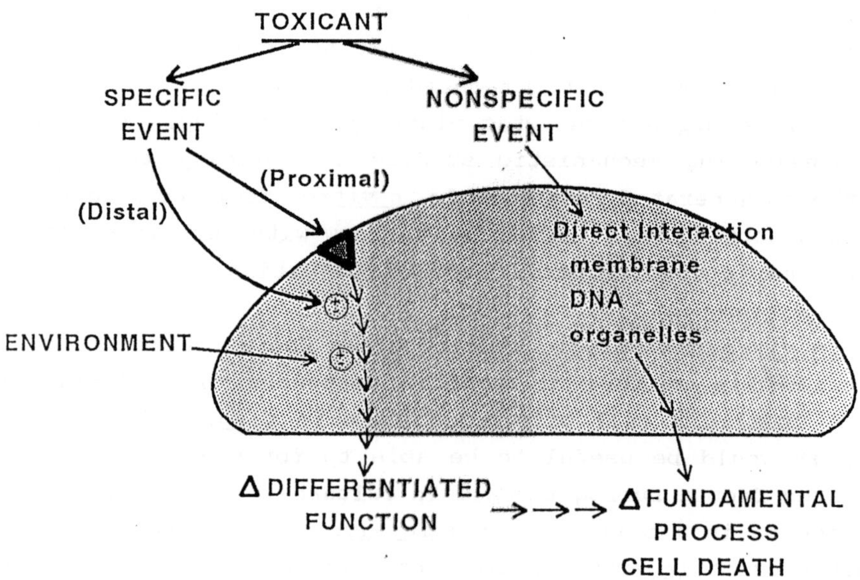

Figure 11. **Conceptual framework for toxic responses of cells in culture.** Toxicant exposure can produce nonspecific events such as damage to membranes, DNA or intracellular organelles that result in alterations in fundamental cellular processes (eg, cell division, membrane integrity) and cell death. Alternatively, toxicants can act more specifically, producing proximal events such as receptor binding or distal events such as alterations in signal transduction that alter differentiated functioning of cells. Such functional alteration may lead indirectly to cell death. Specific events are particularly influenced by cellular environment.

intracellular processes and culminate in alteration of one or more differentiated functions of a cell. From a toxicological viewpoint, such functional changes sometimes lead indirectly to changes in fundamental cellular processes and thereby to cell death. Alternatively, they may result in harmful influences on other cells, affecting their functioning and leading consequently to tissue injury and dysfunction.

Different types of cells are likely to respond similarly to nonspecific events such as direct damage to plasma membrane, DNA or critical intracellular organelles. If the damage is extensive enough to overwhelm defense mechanisms, the outcome will be predictable. For example, a direct irritant such as a strong acid damages membranes and is likely to kill any type of cell in the culture dish as well as cells of the airways in vivo. In this case, cell death in vitro is likely to predict the outcome in vivo--ie, impaired airway function due to cell death--if matching of dosimetry in the two situations can be accomplished. Qualitatively, most types of cells are likely to respond similarly, to this type of insult (eg, with cell death), although the exact doses of irritant needed to kill specific cell types may vary somewhat. Indeed, one might argue that any cell type might be used successfully in this case to predict toxicity.

In the monocrotaline pyrrole model described above, the covalent binding of toxicant to cultured endothelial cells and resultant DNA crosslinking could be classified as a nonspecific event. The persistent DNA crosslinking had a predictable outcome--inhibition of cell proliferation. One would predict further that other cell types would be affected similarly, and this is indeed the case. We compared antiproliferative responses to monocrotaline pyrrole in several different cell types, including vascular smooth muscle cells, fibroblasts, transformed cell lines (MDCK, CRFK cells) and endothelial cells from different species (Reindel et al., 1991; Wagner et al., 1993). Not only did monocrotaline pyrrole inhibit cell proliferation in each of the cells, but the dose-response curves were nearly identical across cell types. In this case, it is likely that any type of proliferating mammalian cell would have correctly predicted the response of other cell types. Importantly, inhibition of proliferation was consistent with the response to the toxicant in vivo and allowed us to form a reasonable hypothesis about the mechanism of lung injury. Thus, one characteristic of nonspecific events is that they tend to be similar across cell types and predictive of responses of cells in tissues in vivo (Table 2).

TABLE 2

Predictability of Cellular Toxic Responses *In Vivo* from Cell Culture Results

Type of Event	Example	Result	Predictive with no Mechanistic Information	Useful in Predicting with Knowledge of:
Nonspecific	Membrane or DNA damage	Cell death	Highly probable	None required
Specific (proximal)	Receptor binding	Altered differentiated function*	Only if receptors are same and coupled to same function	Receptors in SC and PC and coupled functions in PC
Specific (distal)	Altered signal transduction	Altered differentiated function*	Only if transduction paths are same and coupled to same function	Signal transduction in SC and PC and coupled function in PC

SC = Surrogate cell: the cell chosen for *in vitro* study.
PC = Principal cell: the cell *in vivo*, the toxic response of which the *in vitro* study is meant to predict.
* Often markedly influenced by cell environment.

Specific, proximal events usually require interaction with cellular receptors and culminate in one or more specific alterations in cell function. Receptor types, signal transduction mechanisms, effector mechanisms (eg, gene expression) and differentiated cellular functions differ markedly across cell types. Accordingly, it is not surprising that the ability of a toxicant to produce a response and the exact nature of the response, (ie, which cell function is affected) varies widely from one type of cell to another. In the absence of additional information, using a surrogate cell to predict what outcome there will be to a cell in vivo from exposure to a toxicant producing a specific, proximal event will be succesful only if the critical receptors on both cell types are identical and if the functional outcomes of toxicant-receptor interaction are also identical. However, predicting from studies in a surrogate cell type what the toxic outcome will be in different cell type might be accomplished with some knowledge of the receptor(s) and the response(s) involved. If two different cells have similar receptors, the end result of toxicant-receptor interaction may be quite different, but responses should be predictable if the outcome of receptor binding is known in the principal cell (ie, the cell to which results are to be extrapolated). Clearly, prediction of toxicity in other cell types and in vivo for toxicants that have specific, proximal actions is substantially more complex than for nonspecifically acting toxicants.

Toxicants that specifically influence events distal to receptor binding present similar predictive challenges. In the absence of an understanding of mechanisms in the surrogate and principal cells, responses in the surrogate cell can be predictive only if the specific, distal event affected (eg, signal transduction) as well as the outcome to which it is coupled are the same in the two cell types. However, if it is known that the specific event affected in both cells is the same and if the cellular function to which the affected event is coupled in the principal cell is known, responses in the surrogate cell might predict response in the principal cell, even if the response is different in the two cell types. In other words, if a toxicant interferes with a specific step in a signal transduction process in a surrogate cell, then such interference will have a defined outcome in that cell type in terms of its ultimate effect on cell function. However, extrapolating results to another cell type requires knowing that the cell possesses the same signal transduction mechanism as the surrogate cell and also knowing what differentiated functions are controlled by the signal transduction process in the principal cell. For example, PMA affects PMN function (eg, superoxide release) by substituting for diacylglycerol in the signal transduction process. Knowing this, we could correctly predict the effect of PMA on blood platelets by knowing (1) that platelets have the same signal transduction process and (2) that the end result of activation of that process in platelets is release of specific soluble mediators into the extracellular mileu. Clearly, for toxicants acting to influence distal events, information about how differentiated cellular functions are controlled is useful if results from cell culture studies are to predict effectively events in other cell types in vitro or in cells of organs in vivo.

Specific events (proximal and distal) are particularly susceptible to environmental influences, such as physical forces, extracellular matrix composition, endogenous soluble mediators, neighboring cells, and oxygen tension. In some cases inflammatory mediators released by cells may be responsible for untoward effects on other, nearby cells. For example, the response of an airway epithelial cell to direct interaction with asbestos fibers may be affected by release of inflammatory mediators from nearby neutrophils that are activated by asbestos exposure (Kamp et al., 1989, 1993). In the mineral dust example discussed above, the presence of cytokines in the extracellular milieu rendered an otherwise nonresponsive macrophage responsive to titanium dioxide. The composition of extracellular matrix has profound influences on differentiated functions of cells. Since the matrix to which cells in culture are attached is usually different from that occurring in vivo, it should not be surprising that highly differentiated cellular functions and how they are influenced by toxicants are likely to differ between cells in the culture dish and those in the whole animal or person. The pO_2 of culture medium is rarely identical to that of extracellular fluid that bathes cells in vivo. The pO_2 of the environment can be a critical determinant of cellular responses to toxicants and endogenous mediators. For example, leukotriene D_4 is nontoxic to hepatoctyes incubated in oxygen replete medium but is cytotoxic if the cells are made hypoxic (Keppler et al., 1985). Gap junctional intercellular communication between heterogeneous cells can influence cell function and response to toxicants. Physical forces present in vivo are seldom represented in culture systems, yet these may exert important influences. For example, intravascular shear forces influence the functioning of endothelial cells that line blood vessels and can affect their susceptiblity to injury. Differences in cellular environment may explain why monocrotaline pyrrole treatment resulted in decreased fibrinolytic activity in lungs in vivo but increased fibrinolytic activity in pulmonary endothelial cells treated in vitro (see above, Schultze and Roth, 1993b; Evans et al., 1992). The importance of extracellular environment is also apparent in the marked change in expression of drug metabolizing enzymes in cells removed from an organ and placed in a culture environment. Clearly, such environmental influences must be considered and accounted for if we expect our findings in cell culture studies to be predictive of responses in vivo.

In summary, nonspecific toxic events lead usually to cell death or to alterations in basic cell processes such as cell replication. These responses are similar across cell types, and consequently a response of this nature observed in one cell type can be expected to be the same in other cells and hence has a high likelihood of being predictive of the response of a cell in an organ in vivo. Specific events such receptor binding, effects on signal transduction or altered gene expression often result in effects on specific, differentiated cellular functions. For results in vitro to be adequately predictive in this case, critical receptor(s) of the surrogate cell and the cell to which results are to be extrapolated must be similar, and the functional outcome of receptor binding in the latter cell must be known. The environment surrounding cells in vivo is usually highly controlled and often quite different from that of cells in culture, and this can influence quantitatively and qualitatively the responses of cells to toxicants. Thus, the nature of the critical events produced by a toxicant will determine the usefulness of a particular cell system and the type of information needed in order for the results to have predictive value.

GENERAL CONCLUSIONS

It is clear that the rise of in vitro systems has contributed much to our understanding of the toxicology of the respiratory system. They will continue to do so in the future, especially as systems now under development become available and more widely employed. The complexity of the lungs in vivo present particular challenges for the in vitro toxicologist, and each in vitro system has strengths and important limitations in addressing these complexities. For any situation, the choice of a system should depend ideally on the the nature and site of injury to the respiratory system as well as its probable mechanism. The system chosen, manipulation of cells in preparation, the nature of the mechanism of toxicity and an altered environment in vitro may lead to effects of toxicants disparate from those that occur in vivo. Accordingly, it is important to keep in mind that in vitro systems are useful in toxicology only for constructing hypotheses regarding what might pertain in vivo. The testing of such hypotheses will, for the foreseeable future, continue to require studies in animals and/or people.

Acknowledgments. The author thanks Drs. John M. Frazier and Patricia E. Ganey for their helpful comments and Terri Schmidt for editorial assistance. Supported by U.S.P.H.S. Grant ES02581 and a Burroughs Wellcome Toxicology Scholar Award.

REFERENCES

Basbaum, C.B. and Finkbeiner, W.E. Mucus-producing cells of the airways. In: Lung Cell Biology, edited by Massaro, D. New York: Marcel Dekker, Inc., 1989, p. 37-79.

Beckmann, J.D., Takizawa, H., Romberger, D., Illig, M., Claassen, L., Rickard, K. and Rennard, S.I. (1992). Serum-free culture of fractionated bovine bronchial epithelial cells. In Vitro Cell. Dev. Biol. **28A**, 39-46.

Boyd, M.R. (1977). Evidence for the Clara cell as a site of cytochrome P450-dependent mixed function oxidase activity in the lung. Nature (London) **269**, 713.

Carpenter, L.J., Johnson, K.J., Kunkel, R.G. and Roth, R.A. (1987). Phorbol myristate acetate produces injury to isolated rat lungs in the presence and absence of perfused neutrophils. Toxicol. Appl. Pharmacol. **91**, 22-32.

Carpenter, L.J. and Roth, R.A. (1986). 12-tetradecanoyl-phorbol-13-acetate (PMA) produces injury to isolated rat lungs in the presence and absence of perfused neutrophils. Federation Proceedings **45**, 220. (Abstract)

Carpenter, L.J. and Roth, R.A. (1987). Involvement of thromboxane in injury to isolated rat lungs perfused with phorbol myristate acetate in the presence and absence of neutrophils. Toxicol. Appl. Pharmacol. **91**, 33-45.

Carpenter-Deyo, L. and Roth, R.A. (1989). Cyclooxygenase inhibition in lungs or in neutrophils attenuates neutrophil-dependent edema in rat lungs perfused with phorbol myristate acetate. J. Pharmacol. Exp. Ther. **251**, 983-991.

Chen, L.C., Fang, C.P., Qu, Q.S., Fine, J.M. and Schlesinger, R.B. (1993). A novel system for the in vitro exposure of pulmonary cells to acid sulfate aerosols. Fund. Appl. Toxicol. **20**, 170-176.

Christensen, T.G., Breuer, R., Haddad, C.E. and Niles, R.M. (1993). Quantitative ultrastructural analysis of the relationship between cell growth, shape change, and mucosecretory differentiation in cultured hamster tracheal epithelial cells exposed to retinoic acid. Am. J. Respir. Cell Mol. Biol. **9**, 287-294.

Devereux, T.R., Domin, B.A. and Philpot, R.M. (1989). Xenobiotic metabolism by isolated pulmonary cells. Pharmacol. Ther. **41**, 243-256.

Devereux, T.R., Jones, K.G., Bend, J.R., Fouts, J.R., Statham, C.N. and Boyd, M.R. (1981). In vitro metabolic activation of the pulmonary toxin, 4-ipomeanol, in nonciliated bronchiolar epithelial (Clara) and alveolar type II cells isolated from rabbit lung. J. Pharmacol. Exp. Ther. **220**, 223-227.

Dobbs, L.G. (1990). Isolation and culture of alveolar type II cells. Am. J. Physiol. **258(2)**, L134-L147.

Driscoll, K.E., Higgins, J.M., Leytart, M.J. and Crosby, L.L. (1990a). Differential effects of mineral dusts on the in vitro activation of alveolar macrophage eicosanoid and cytokine release. Toxicol. In Vitro **4**, 284-288.

Driscoll, K.E., Lindenschmidt, R.C., Maurer, J.K., Higgins, J.M.

and Ridder, G. (1990b). Pulmonary response to silica or titanium dioxide: inflammatory cells, alveolar macrophage-derived cytokines and histopathology. Am. J. Respir. Cell Mol. Biol. **2**, 381-390.

Dyer, E.L. and Snapper, T.R. (1986). Role of circulating granulocytes in sheep lung injury produced by phorbol myristate acetate. J. Appl. Physiol. **60**, 576-589.

Evans, M.J. Cell death and cell renewal in small airways and alveoli. In: Mechanisms in Respiratory Toxicology, Vol 1, edited by Witschi, H.P. and Netteschein, P. Boca Raton: CRC Press, 1982, p. 189-218.

Fisher, A.B. The isolated perfused lung. In: Toxicology of Inhaled Materials, edited by Witschi, H.P. and Brain, J.D. New York: Springer-Verlag, 1985, p. 149-179.

Fisher, G.L. and Placke, M.E. Cell and organ culture models of respiratory toxicity. In: Toxicology of the Lung, edited by Gardner, D.E., Crapo, J.D. and Massaro, E.J. New York: Raven Press, Ltd., 1988, p. 285-314.

Ganey, P.E. and Roth, R.A. (1987). 6-keto prostaglandin $F_{1\alpha}$ and Thromboxane B2 in isolated, buffer-perfused lungs from monocrotaline pyrrole-treated rats. Exp. Lung Res. **12**, 195-206.

Gil, J. Comparative morphology and ultrastructure of the airway. In: Mechanisms in Respiratory Toxicology, Vol 1, edited by Witschi, H. and Nettesheim, P. Boca Raton: CRC Press, 1982, p. 3-25.

Gross, I. and Wilson, C.M. (1982). Fetal lung in organ culture, IV. Supra-additive hormone interactions. J. Appl. Physiol. **52**, 1420-1425.

Hogg, J.C. Neutrophil traffic. In: The Lung. Scientific Foundations, Vol 1, edited by Crystal, R.G., West, J.B., Barnes, P.J., Cerniack, N.S. and Weibel, E.R. New York: Raven Press, 1991, p. 565-579.

Hoorn, C.M., Wagner, J.G. and Roth, R.A. (1993). Effects of monocrotaline pyrrole on cultured rat pulmonary endothelium. Toxicol. Appl. Pharmacol. **120**, 281-287.

Hsu, M-T., Dimaio, M., Reiss, O.K., Ciurea, D. and Gil, J. (1992). A novel system for the culture of human lung: lung development and the response to injury. Am. J. Physiol. **263(7)**, L308-L316.

Jackson, J.I., White, C.W., McMurtry, I.F., Berger, B.M. and Repine, J.E. (1986). Dimethylthiourea decreases acute lung edema in phorbol myristate acetate-treated rabbits. J. Appl. Physiol. **61**, 353-360.

Johnson, K.J. and Ward, P.A. (1982). Acute and progressive lung injury after contact with phorbol myristate acetate. Am. J. Pathol. **107**, 29-35.

Kamp, D.W., Dunne, M., Weitzman, S.A. and Dunn, M.M. (1989). The interaction of asbestos and neutrophils injures cultured human pulmonary epithelial cells: role of hydrogen peroxide. J. Lab. Clin. Med. **114**, 604-612.

Kamp, D.W., Dunne, M., Dykewicz, M.S., Sbalchiero, J.S., Weitzman, S.A. and Dunn, M.M. (1993). Asbestos-induced injury to cultured human pulmonary epithelial-like cells: role of neutrophil elastase. J. Leuk. Biol. **54**, 73-80.

Keppler, D., Hagmann, W., Rapp, S., Denzlinger, C. and Koch, H.K. (1985). The relation of leukotrienes to liver injury.

Hepatology **5(5)**, 883-891.
Leikauf, G. and Driscoll, K. Cellular approaches in respiratory tract toxicology. In: Toxicology of the Lung, edited by Gardner, D.E. New York: Raven Press, Ltd., 1993, p. 335-370.
Massey, T.E. Isolation and use of lung cells in toxicology. In: In Vitro Toxicology: Model Systems and Methods, edited by McQueen, C.A. Caldwell,N.J.: Telford Press, 1989, p. 35-66.
McAteer, J.A., Cavanagh, T.J. and Evan, A.P. (1983). Submersion culture of the intact fetal lung. In Vitro **19**, 210-218.
Mossman, B.T. and Sesko, A.M. (1990). In vitro assays to predict the pathogenecity of mineral fibers. Toxicology **60**, 53-61.
Murray, J.F. The Normal Lung: The Basis for Diagnosis and Treatment of Pulmonary Disease, Philadelphia:W.B. Saunders, 1976.
Netter, F.H.,The CIBA Collection of Medical Illustrations. Vol. 7. Respiratory System, Summit,N.J.:CIBA Pharmaceutical Company, 1980. Ed. 2nd pp. 3-43.
Nettesheim, P., Jetten, A.M., Inayama, Y., Brody, A.R., George, M.A., Gilmore, L.B., Gray, T. and Hook, G.E.R. (1990). Pathways of differentiation of airway epithelial cells. Environ. Hlth. Persp. **85**, 317-329.
Niemeier, R.W. (1984). The isolated perfused lung. Environ. Hlth. Persp. **56**, 35-41.
Poole, A. and Brown, R.C. In vitro methods to investigate toxic lung disease. In: In Vitro Methods in Toxicology, edited by Atterwill, C.K. and Steele, C.E. Cambridge: Cambridge University Press, 1987, p. 189-209.
Reindel, J.F., Ganey, P.E., Wagner, J.G., Slocombe, R.F. and Roth, R.A. (1990). Development of morphologic, hemodynamic, and biochemical changes in lungs of rats given monocrotaline pyrrole. Toxicol. Appl. Pharmacol. **106**, 179-200.
Reindel, J.F., Hoorn, C.M., Wagner, J.G. and Roth, R.A. (1991). Comparison of response of bovine and porcine pulmonary arterial endothelial cells to monocrotaline pyrrole. Am. J. Physiol. **261**, L406-L414.
Reindel, J.F. and Roth, R.A. (1991). The effects of monocrotaline pyrrole on cultured bovine pulmonary artery endothelial and smooth muscle cells. Am. J. Pathol. **138(3)**, 707-719.
Rennard, S.I., Daughton, D.M., Robbins, R.A., Thompson, A.B. and Von Essen, S. (1990). In vivo and in vitro methods for evaluating airways inflammaton: implications for respiratory toxicology. Toxicology **60**, 5-14.
Rhoades, R.A. (1984). Isolated perfused lung preparation for studying altered gaseous environments. Environ. Hlth. Persp. **56**, 43-50.
Roth, R.A. and Bassett, D.J.P. The isolated perfused lung in toxicological studies. In: In Vitro Models In Toxicology, edited by McQueen, C.A. Caldwell,N.J.: Telford Press, 1989, p. 3-33.
Roth, R.A. and Reindel, J.F Lung vascular injury from monocrotaline pyrrole, a putative hepatic metabolite. In: Biological Reactive Intermediates IV, edited by Witmer, C.M. New York: Plenum Press, 1990, p. 477-487.
Roth, R.A. and Vinegar, A. (1990). Action by the lungs on

circulating xenobiotic agents, with a case study of physiologically based pharmacokinetic modeling of benzo(a)pyrene disposition. Pharmacol. Ther. **48**, 143-155.

Ryan, U.S. and Li, A.P. Metabolism of endogenous and xenobiotic substances by pulmonary vascular endothelial cells. In: Metabolic Activation and Toxicity of Chemical Agents to Lung Tissues and Cells, edited by Gram, T.E. New York: Pergamon Press, 1993, p. 89-105.

Schraufstatter, I.V., Revak, S.D. and Cochrane, C.G. (1984). Proteases and oxidants in experimental pulmonary inflammatory injury. J. Clin. Invest. **73**, 1175-1184.

Schultze, A.E. and Roth, R.A. (1993a). Fibrinolytic activity in blood and lungs of rats treated with monocrotaline pyrrole. Toxicol. Appl. Pharmacol. **121**, 129-137.

Schultze, A.E. and Roth, R.A. (1993b). Procoagulant and fibrinolytic properties of bovine endothelial cells treated with monocrotaline pyrrole. Toxicol. Appl. Pharmacol. **122**, 7-15.

Shasby, D.M., Shasby, S.S. and Peach, M.J. (1983). Granulocytes and phorbol myristate acetate increase permeability to albumin of cultured endothelial monolayers and isolated perfused lungs: role of oxygen radicals and granulocyte adherence. Am. Rev. Respir. Dis. **127**, 72-76.

Siminski, J.T., Kavanagh, T.J., Chi, E. and Raghu, G. (1992). Long-term maintenance of mature pulmonary parenchyma cultured in serum-free conditions. Am. J. Physiol. **262(6)**, L105-L110.

Smith, B.R. and Bend, J.R. (1981). Lung perfusion techniques for xenobiotic metabolism and toxicity studies. Methods In Enzymology **77**, 105-120.

Spencer, H. The anatomy of the lung. In: Pathology of The Lung, Oxford: Pergamon Press, 1985, p. 17-77.

Stefaniak, M.S., Krumdiek, C.L., Spall, R.D., Gandolfi, A.J. and Brendel, K. (1992). Biochemical and histological characterization of agar-filled precision cut rat lung slices in dynamic organ culture as an in vitro tool. In Vitro Toxicology **5(1)**, 7-19.

Tanswell, A.K., Byrne, P.J., Han, R.N.N., Edelson, J.D. and Han, V.K.M. (1991). Limited division of low-density adult rat type II pneumocytes in serum-free culture. Am. J. Physiol. **260(4)**, L395-L402.

Voelker, D.R. and Mason, R.J. Alveolar type II epithelial cells. In: Lung Cell Biology, edited by Massaro, D. New York: Marcel Dekker,Inc., 1989, p. 487-538.

Waddell, W.J. and Marlowe, C. (1980). Tissue and cellular disposition of paraquat in mice. Toxicol. Appl. Pharmacol. **56**, 127.

Wagner, J.G., Petry, T.W. and Roth, R.A. (1993). Characterization of monocrotaline-induced DNA cross-linking in pulmonary artery endothelium. Am. J. Physiol. **264(8)**, L517-L522.

Wang, D., Chou, C-L., Hsu, K. and Chen, H.I. (1991). Cyclooxygenase pathway mediates lung injury induced by phorbol and platelets. J. Appl. Physiol. **70(6)**, 2417-2421.

Warren, J.S. and Barton, P.A. (1992). In vitro analysis of pulmonary inflammation using rat lung organ cultures. Exp. Lung Res. **18**, 55-67.

Whitcutt, M.J., Adler, K.B. and Wu, R. (1988). A biphasic chamber system for maintaining polarity of differentiation of cultured respiratory tract epithelial cells. In Vitro Cell. Dev. Biol. **24(5)**, 420-428.

NATO ASI Series H

Vol. 21: **Cellular and Molecular Basis of Synaptic Transmission.** Edited by H. Zimmermann. 547 pages. 1988.

Vol. 22: **Neural Development and Regeneration. Cellular and Molecular Aspects.** Edited by A. Gorio, J. R. Perez-Polo, J. de Vellis, and B. Haber. 711 pages. 1988.

Vol. 23: **The Semiotics of Cellular Communication in the Immune System.** Edited by E.E. Sercarz, F. Celada, N.A. Mitchison, and T. Tada. 326 pages. 1988.

Vol. 24: **Bacteria, Complement and the Phagocytic Cell.** Edited by F. C. Cabello und C. Pruzzo. 372 pages. 1988.

Vol. 25: **Nicotinic Acetylcholine Receptors in the Nervous System.** Edited by F. Clementi, C. Gotti, and E. Sher. 424 pages. 1988.

Vol. 26: **Cell to Cell Signals in Mammalian Development.** Edited by S.W. de Laat, J.G. Bluemink, and C.L. Mummery. 322 pages. 1989.

Vol. 27: **Phytotoxins and Plant Pathogenesis.** Edited by A. Graniti, R. D. Durbin, and A. Ballio. 508 pages. 1989.

Vol. 28: **Vascular Wilt Diseases of Plants. Basic Studies and Control.** Edited by E. C. Tjamos and C. H. Beckman. 590 pages. 1989.

Vol. 29: **Receptors, Membrane Transport and Signal Transduction.** Edited by A. E. Evangelopoulos, J. P. Changeux, L. Packer, T. G. Sotiroudis, and K.W.A. Wirtz. 387 pages. 1989.

Vol. 30: **Effects of Mineral Dusts on Cells.** Edited by B.T. Mossman and R.O. Begin. 470 pages. 1989.

Vol. 31: **Neurobiology of the Inner Retina.** Edited by R. Weiler and N.N. Osborne. 529 pages. 1989.

Vol. 32: **Molecular Biology of Neuroreceptors and Ion Channels.** Edited by A. Maelicke. 675 pages. 1989.

Vol. 33: **Regulatory Mechanisms of Neuron to Vessel Communication in Brain.** Edited by F. Battaini, S. Govoni, M.S. Magnoni, and M. Trabucchi. 416 pages. 1989.

Vol. 34: **Vectors asTools for the Study of Normal and Abnormal Growth and Differentiation.** Edited by H. Lother, R. Dernick, and W. Ostertag. 477 pages. 1989.

Vol. 35: **Cell Separation in Plants: Physiology, Biochemistry and Molecular Biology.** Edited by D. J. Osborne and M. B. Jackson. 449 pages. 1989.

Vol. 36: **Signal Molecules in Plants and Plant-Microbe Interactions.** Edited by B.J.J. Lugtenberg. 425 pages. 1989.

Vol. 37: **Tin-Based Antitumour Drugs.** Edited by M. Gielen. 226 pages. 1990.

Vol. 38: **The Molecular Biology of Autoimmune Disease.** Edited by A.G. Demaine, J-P. Banga, and A.M. McGregor. 404 pages. 1990.

NATO ASI Series H

Vol. 39: Chemosensory Information Processing.
Edited by D. Schild. 403 pages. 1990.

Vol. 40: Dynamics and Biogenesis of Membranes.
Edited by J. A. F. Op den Kamp. 367 pages. 1990.

Vol. 41: Recognition and Response in Plant-Virus Interactions.
Edited by R. S. S. Fraser. 467 pages. 1990.

Vol. 42: Biomechanics of Active Movement and Deformation of Cells.
Edited by N. Akkas. 524 pages. 1990.

Vol. 43: Cellular and Molecular Biology of Myelination.
Edited by G. Jeserich, H. H. Althaus, and T. V. Waehneldt. 565 pages. 1990.

Vol. 44: Activation and Desensitization of Transducing Pathways.
Edited by T. M. Konijn, M. D. Houslay, and P. J. M. Van Haastert. 336 pages. 1990.

Vol. 45: Mechanism of Fertilization: Plants to Humans.
Edited by B. Dale. 710 pages. 1990.

Vol .46: Parallels in Cell to Cell Junctions in Plants and Animals.
Edited by A. W Robards, W. J . Lucas, J . D. Pitts, H . J . Jongsma, and D. C. Spray. 296 pages. 1990.

Vol. 47: Signal Perception and Transduction in Higher Plants.
Edited by R. Ranjeva and A. M. Boudet. 357 pages. 1990.

Vol. 48: Calcium Transport and Intracellular Calcium Homeostasis.
Edited by D. Pansu and F. Bronner. 456 pages. 1990.

Vol. 49: Post-Transcriptional Control of Gene Expression.
Edited by J. E. G. McCarthy and M. F. Tuite. 671 pages. 1990.

Vol. 50: Phytochrome Properties and Biological Action.
Edited by B. Thomas and C. B. Johnson. 337 pages. 1991.

Vol. 51: Cell to Cell Signals in Plants and Animals.
Edited by V. Neuhoff and J. Friend. 404 pages. 1991.

Vol. 52: Biological Signal Transduction.
Edited by E. M . Ross and K . W. A. Wirtz. 560 pages. 1991.

Vol. 53: Fungal Cell Wall and Immune Response.
Edited by J. P. Latge and D. Boucias. 472 pages. 1991.

Vol. 54: The Early Effects of Radiation on DNA.
Edited by E. M. Fielden and P. O'Neill. 448 pages. 1991.

Vol. 55: The Translational Apparatus of Photosynthetic Organelles.
Edited by R. Mache, E. Stutz, and A. R. Subramanian. 260 pages. 1991.

Vol. 56: Cellular Regulation by Protein Phosphorylation.
Edited by L. M. G. Heilmeyer, Jr. 520 pages. 1991.

NATO ASI Series H

Vol. 57: **Molecular Techniques in Taxonomy.**
Edited by G . M . Hewitt, A. W. B. Johnston, and J. P. W. Young . 420 pages. 1991.

Vol. 58: **Neurocytochemical Methods.**
Edited by A. Calas and D. Eugene. 352 pages. 1991.

Vol. 59: **Molecular Evolution of the Major Histocompatibility Complex.**
Edited by J. Klein and D. Klein. 522 pages. 1991.

Vol. 60: **Intracellular Regulation of Ion Channels.**
Edited by M. Morad and Z. Agus. 261 pages. 1992.

Vol. 61: **Prader-Willi Syndrome and Other Chromosome 15q Deletion Disorders.**
Edited by S. B. Cassidy. 277 pages. 1992.

Vol. 62: **Endocytosis. From Cell Biology to Health, Disease and Therapie.**
Edited by P. J. Courtoy. 547 pages. 1992.

Vol. 63: **Dynamics of Membrane Assembly.**
Edited by J. A. F. Op den Kamp. 402 pages. 1992.

Vol. 64: **Mechanics of Swelling. From Clays to Living Cells and Tissues.**
Edited by T. K. Karalis. 802 pages. 1992.

Vol. 65: **Bacteriocins, Microcins and Lantibiotics.**
Edited by R. James, C. Lazdunski, and F. Pattus. 530 pages. 1992.

Vol. 66: **Theoretical and Experimental Insights into Immunology.**
Edited by A. S. Perelson and G. Weisbuch. 497 pages. 1992.

Vol. 67: **Flow Cytometry. New Developments.**
Edited by A. Jacquemin-Sablon. 1993.

Vol. 68: **Biomarkers. Research and Application in the Assessment of Environmental Health.** Edited by D. B. Peakall and L. R. Shugart. 138 pages. 1993.

Vol. 69: **Molecular Biology and its Application to Medical Mycology.**
Edited by B. Maresca, G. S. Kobayashi, and H. Yamaguchi. 271 pages. 1993.

Vol. 70: **Phospholipids and Signal Transmission.**
Edited by R. Massarelli, L. A. Horrocks, J. N. Kanfer, and K. Löffelholz. 448 pages. 1993.

Vol. 71: **Protein Synthesis and Targeting in Yeast.**
Edited by A. J. P. Brown, M. F. Tuite, and J. E. G. McCarthy. 425 pages. 1993.

Vol. 72: **Chromosome Segregation and Aneuploidy.**
Edited by B. K. Vig. 425 pages. 1993.

Vol. 73: **Human Apolipoprotein Mutants III. In Diagnosis and Treatment.**
Edited by C. R. Sirtori, G. Franceschini, B. H. Brewer Jr. 302 pages. 1993.

NATO ASI Series H

Vol. 74: **Molecular Mechanisms of Membrane Traffic.**
Edited by D. J. Morré, K. E. Howell, and J. J. M. Bergeron. 429 pages. 1993.

Vol. 75: **Cancer Therapy. Differentiation, Immunomodulation and Angiogenesis.**
Edited by N. D'Alessandro, E. Mihich, L. Rausa, H. Tapiero, and T. R. Tritton. 299 pages. 1993.

Vol. 76: **Tyrosine Phosphorylation/Dephosphorylation and Downstream Signalling.**
Edited by L. M. G. Heilmeyer Jr. 388 pages. 1993.

Vol. 77: **Ataxia-Telangiectasia.** Edited by R. A. Gatti, R. B. Painter. 306 pages. 1993.

Vol. 78: **Toxoplasmosis.** Edited by J. E. Smith. 272 pages. 1993.

Vol. 79: **Cellular Mechanisms of Sensory Processing. The Somatosensory System.**
Edited by L. Urban. 514 pages. 1994.

Vol. 80: **Autoimmunity: Experimental Aspects.**
Edited by M. Zouali. 318 pages. 1994.

Vol. 81: **Plant Molecular Biology. Molecular Genetic Analysis of Plant Development and Metabolism.**
Edited by G. Coruzzi, P. Puigdomènech. 579 pages. 1994.

Vol. 82: **Biological Membranes: Structure, Biogenesis and Dynamics.**
Edited by Jos A. F. Op den Kamp. 367 pages. 1994.

Vol. 83: **Molecular Biology of Mitochondrial Transport Systems.**
Edited by M. Forte, M. Colombini. 420 pages. 1994.

Vol. 84: **Biomechanics of Active Movement and Division of Cells.**
Edited by N. Akkaş. 587 pages. 1994.

Vol. 85: **Cellular and Molecular Effects of Mineral and Synthetic Dusts and Fibres.**
Edited by J. M. G. Davis, M.-C. Jaurand. 448 pages. 1994.

Vol. 86: **Biochemical and Cellular Mechanisms of Stress Tolerance in Plants.**
Edited by J. H. Cherry. 616 pages. 1994.

Vol. 87: **NMR of Biological Macromolecules.**
Edited by C. I. Stassinopoulou. 616 pages. 1994.

Vol. 88: **Liver Carcinogenesis. The Molecular Pathways.**
Edited by G. G. Skouteris. 502 pages. 1994.

Vol. 89: **Molecular and Cellular Mechanisms of H^+ Transport.**
Edited by B. H. Hirst. 504 pages. 1994.

Vol. 90: **Molecular Aspects of Oxidative Drug Metabolizing Enzymes: Their Significance in Environmental Toxicology, Chemical Carcinogenesis and Health.**
Edited by E. Arınç, J. B. Schenkman, E. Hodgson. 623 pages. 1994.

NATO ASI Series H

Vol. 91: **Trafficking of Intracellular Membranes: From Molecular Sorting to Membrane Fusion.** Edited by M. C. Pedroso de Lima, N. Düzgüneş, D. Hoekstra. 371 pages. 1995.

Vol. 92: **Signalling Mechanisms – from Transcription Factors to Oxidative Stress.** Edited by L. Packer, K. Wirtz. 466 pages. 1995.

Vol. 93: **Modulation of Cellular Responses in Toxicity.** Edited by C. L. Galli, A. M. Goldberg, M. Marinovich. 379 pages. 1995.

MIX
Papier aus verantwortungsvollen Quellen
Paper from responsible sources
FSC® C105338

If you have any concerns about our products,
you can contact us on
ProductSafety@springernature.com

In case Publisher is established outside the EU,
the EU authorized representative is:
Springer Nature Customer Service Center GmbH
Europaplatz 3, 69115 Heidelberg, Germany

Printed by Libri Plureos GmbH
in Hamburg, Germany